Disaster Risk Reduction

Methods, Approaches and Practices

Disaster risk reduction is a process that leads to the safety of communities and nations. After the 2005 World Conference on Disaster Reduction, held in Kobe, Japan, the Hyogo Framework for Action (HFA) was adopted as a framework for risk reduction. The academic research and higher education in disaster risk reduction has made, and continues to make, a gradual shift from pure basic research to applied, implementation-oriented research. More emphasis is being given to multi-stakeholder collaboration and multi-disciplinary research. Emerging university networks in Asia, Europe, Africa, and the Americas have urged process-oriented research in the disaster risk reduction field. With this in mind, this new series will promote the output of action research on disaster risk reduction, which will be useful for a wide range of stakeholders including academicians, professionals, practitioners, and students and researchers in related fields. The series will focus on emerging needs in the risk reduction field, starting from climate change adaptation, urban ecosystem, coastal risk reduction, education for sustainable development, community-based practices, risk communication, and human security, among other areas. Through academic review, this series will encourage young researchers and practitioners to analyze field practices and link them to theory and policies with logic, data, and evidence. In this way, the series will emphasize evidence-based risk reduction methods, approaches, and practices.

All submitted proposals are carefully reviewed by the series editor.

Publication Ethics and Publication Malpractice Statement: https://www.springernature.com/gp/policies/book-publishing-policies

Ethical, Conflict-of-Interest, and Editorial Policy Statement: https://www.springer.com/us/authors-editors/journal-author/journal-author-helpdesk/before-you-start/before-you-start/1330#c14214

Chiho Ochiai · Sandra Carrasco · Sung Lun Tsai
Editors

Disaster and Displacement

Exploring the Physical and Social Dimensions of Resettlement and Reconstruction

Springer

Editors
Chiho Ochiai
Graduate School of Global Environmental Studies
Kyoto University
Kyoto, Japan

Sandra Carrasco
School of Humanities and Social Sciences
Deakin University
Burwood, Melbourne, VIC, Australia

Sung Lun Tsai
Shimizu Corporation
Tokyo, Japan

ISSN 2196-4106 ISSN 2196-4114 (electronic)
Disaster Risk Reduction
ISBN 978-981-96-2582-6 ISBN 978-981-96-2583-3 (eBook)
https://doi.org/10.1007/978-981-96-2583-3

This Springer imprint is published by the registered company Springer Nature Singapore Pte Ltd.
The registered company address is: 152 Beach Road, #21-01/04 Gateway East, Singapore 189721, Singapore

Preface

The global displacement of people has steadily increased in the past few years. According to the 2024 Displacement Report (IDMC, 2024), nearly 76 million people have been displaced across 151 countries. While armed conflicts and disasters are among the primary driving factors for displacement, climate change exacerbates the situation as the unprecedented impacts of weather-related disasters in the past years are magnified by the unusually high intensity and frequency of climate events in the context of climate change. Moreover, most people forcibly displaced by conflict reside in countries that are highly vulnerable to climate change and have limited capacities for climate adaptation (UNHCR, 2024).

In addition to conflicts, disasters, and climate change, development projects also cause displacement, referred to as development-induced displacement, as these are viewed as means to alleviate poverty and promote economic growth in countries of the global south (Tan, 2020). Some estimations state that over the past two decades, around 300 million people have been displaced worldwide due to development and major infrastructure projects which ironically left people in impoverished and marginalized conditions (IDMC, 2017).

Displaced people require a place to live and recover from crises and trauma including the physical reconstruction or recovery of houses, neighborhoods, and cities. The failure, delay, or inappropriateness of these would enable or hinder people's recovery as individuals and communities. For instance, the impact of disasters on housing resulting in displacement has cascading effects on affected people at individual and collective levels. Displaced people, whether temporarily or permanently, often find support for their socio-economic and cultural recovery in the recovery of their physical environment. In post-disaster contexts, resettlement implies the relocation of people to housing and necessary infrastructure built in safer locations where the vulnerabilities to disaster can be minimized. In the context of conflict-driven displacement, resettlement refers to transferring displaced people from the country in which they sought protection to another state that has agreed to admit them with permanent residence status (IOM, 2019).

Resettlement involves creating conditions for the re-establishing of housing or appropriate living environments, which are considered the backbone of recovery impacting different dimensions of individual, household, and community recovery (Peacock et al., 2018; Kroll-Smith et al., 2015). However, resettlement, also referred to as relocation, retreat, or realignment, is often seen as the least desirable approach for reconstruction and recovery as it has long-term impacts on the displaced communities, recreation of similar pre-resettlement vulnerabilities, exposure to new hazards, challenging their capacities to rebuild their livelihoods, risks of impoverishment and negative health impacts and impacting people's opportunities for further development. The complexity of the implications of resettlement on the lives of affected people after being placed in "safer locations" is often overlooked, as resettlement evaluations rarely consider a long-term perspective. Additionally, discussions mainly focus on policy analysis and socio-economic conditions without considering the linkage to the built environment.

This book provides insight into multiple case studies covering various drivers for displacement emphasizing their physical environments. Titled *Disaster and Displacement: Exploring the Physical and Social Dimensions of Resettlement and Reconstruction*, the book aims to connect home environments and recovery. The case studies in this book emphasize people's agency to improve their built environments which impacts their individual and collective recovery. The authors of this book employ diverse research methods and areas of expertise, including architecture, planning, sociology, engineering, anthropology, and disaster research. Contributors range from seasoned researchers to graduate students, bringing a variety of perspectives that enrich the discussion about displacement and resettlement and the content of this book.

Our aim, in this book, is to enhance the discussion about displacement and resettlement from diverse perspectives such as human rights, comfort, social ties, traditions, religions, new technologies, and the adaptation of traditional technologies to new conditions in achieving the future built environment. This can be accomplished by understanding the historical and social backgrounds represented by ethnic minorities and traditional architecture, as well as the social networks and bonds between people living in different environments. This book targets a wide variety of readers, including those seeking to deepen their understanding of societal issues such as discrimination and poverty through the lens of displacement and resettlement.

Kyoto, Japan — Chiho Ochiai
Melbourne, Australia — Sandra Carrasco
Tokyo, Japan — Sung Lun Tsai

References

IDMC (2017) Dams and internal displacement. Geneva, Switzerland: Internal Displacement Monitoring Centre—IDMC

IDMC (2024) 2024 Global report on internal displacement, Geneva, Switzerland, Internal Displacement Monitoring Centre

IOM (2019) Glossary on migration. International migration law Geneva, Switzerland: International Organization for Migration—IOM
Kroll-Smith S, Baxter V & Jenkins P (2015) Left to chance: Hurricane Katrina and the story of two New Orleans neighborhoods, University of Texas Press
Peacock W G, Dash N, Zhang Y & Zandt S V (2018) Post-disaster sheltering, temporary housing and permanent housing recovery. Handbook of disaster research, 569–594
Tan Y (2020) Development-induced displacement and resettlement: An overview of issues and interventions. In: Bastia, T. & Skeldon, R. (eds) Routledge Handbook of Migration and Development. London, UK: Routledge
UNHCR (2024) Climate change and displacement [Online]. Geneva, Switzerland: United Nations High Commissioner for Refugees—UNHCR. Available: https://www.unhcr.org/what-we-do/build-better-futures/climate-change-and-displacement#:~:text=Displaced%20populations%20frequently%20have%20no,%2C%20drought%2C%20storms%20and%20heatwaves. [Accessed 14 June 2024]

Contents

Editors and Contributors

About the Editors

Chiho Ochiai associate professor in the Graduate School of Global Environmental Studies, Kyoto University, Japan. Her research interests are community-based disaster management, post-disaster rebuilding, and traditional and local knowledge. She has worked closely with the disaster-affected local communities in rural sites and traditional communities at world heritage sites like Shirakawa-go and Kumano Hongu Taisha Shrine in Japan, Indian Ocean Tsunami in Aceh, Indonesia, and Wildfire in California, USA.

Sandra Carrasco is a Peruvian Australian academic at the University of Newcastle. She holds a Ph.D. and master's in environmental management, Kyoto University, Japan, a master's in education, Victoria University, Australia, and a bachelor's in architecture, Saint Augustine National University, Peru. Her research focuses on disaster risk reduction, community resilience, incremental housing, sustainable development, diversity and inclusion, and migrant and refugee studies.

Sung Lun Tsai he completed a Doctoral Program at the Graduate School of Global Environmental Studies, Kyoto University, Japan, and is currently working in the Ocean Programs, Emerging Frontiers Division of Shimizu Corporation in Tokyo, Japan. He is in charge of the floating structure design and business development. His research interests are architectural and social issues of post-disaster relocation and resettlement of Taiwanese indigenous people and the community after the Morakot Typhoon.

Contributors

Miwa Abe Undergraduate School of Humanities and Science Department of Community and Social Studies, Tokai University, Shibuya, Japan

Nil Akdede Department of Architecture, Atilim University, Ankara, Turkey

Penny Allan School of Architecture, University of Technology Sydney, Ultimo, Australia

Bekir Özer Ay Department of Architecture, Middle East Technical University, Ankara, Turkey

Martin Bryant School of Architecture, University of Technology Sydney, Ultimo, Australia

Cinthya Butrón-Revilla Faculty of Architecture and Urbanism, National University of San Agustín, Arequipa, Peru

Sandra Carrasco School of Humanities and Social Sciences, Deakin University, Melbourne, Australia

Guido Cimadomo Department of Art and Architecture, University of Malaga, Málaga, Spain

Neeraj Dangol Melbourne School of Design, The University of Melbourne, Melbourne, Australia

Ipek Gürsel Dino Department of Architecture, Middle East Technical University, Ankara, Turkey

Majdi Faleh College of Engineering, Architectural Engineering, University of Sharjah, Sharjah, United Arab Emirates

Francis Hubbard School of Built Environment, University of New Zealand, Wellington, New Zealand

Shinji Kajita Graduate School of Engineering, Kyoto University, Kyoto, Japan

Ricelli Laplace Graduate School of Global Environmental Studies, Kyoto University, Kyoto, Japan

Tsai Sung Lun Deloitte Tohmatsu Venture Support Co., Ltd., Tokyo, Japan; Shimizu Corporation, Tokyo, Japan

Gabriela Manchego-Huaquipaco Faculty of Architecture and Urbanism, National University of San Agustín, Arequipa, Peru

Alan March Melbourne School of Design, The University of Melbourne, Melbourne, Australia

Nathalia González Meixueiro Department of Art and Architecture, University of Malaga, Málaga, Spain

Edward Ng School of Architecture, Chinese University of Hong Kong, Ma Liu Shui, Hong Kong SAR

Chiho Ochiai Graduate School of Global Environmental Studies, Kyoto University, Kyoto, Japan

Augusto Cesar Oyama Disaster Prevention Research Institute, Kyoto University, Kyoto, Japan

Maria Isabel Perez Rodriguez Graduate School of Global Environmental Studies, Kyoto University, Kyoto, Japan

Regan Potangaroa School of Built Environment, University of New Zealand, Wellington, New Zealand

Yasodhara Ranasinghe Melbourne School of Design, The University of Melbourne, Melbourne, Australia

Hannah Robertson Melbourne School of Design, The University of Melbourne, Melbourne, Australia

Kei Saito Graduate School of Engineering, Kyoto University, Kyoto, Japan

Andrew Toland School of Architecture, University of Technology Sydney, Ultimo, Australia

Tadashi Uchiyama Undergraduate School of Humanities and Science Department of Community and Social Studies, Tokai University, Shibuya, Japan

Jingying Wang School of Architecture, Chinese University of Hong Kong, Ma Liu Shui, Hong Kong SAR

Suzanne Wilkinson School of Built Environment, University of New Zealand, Wellington, New Zealand

Carlos Zeballos-Velarde Faculty of Architecture and Urbanism, National University of San Agustín, Arequipa, Peru

Yuqi Zhang Graduate School of Global Environmental Studies, Kyoto University, Kyoto, Japan

Chapter 1
Introduction: Displacement and Resettlement

Sandra Carrasco

Abstract The intensifying impacts of climate change, armed conflicts, and economic instability are driving increasing global displacement, necessitating a deeper understanding of resettlement as a multifaceted process. This chapter introduces the diverse contexts and studies included in this book, which explores the complexities of displacement and the subsequent challenges of resettlement, emphasizing the interplay between the built environment, political structures, and social recovery. The chapter identifies distinctions in relevant terminology related to relocation and resettlement and the implications for displaced communities. Additionally, the chapter outlines the current state of knowledge on resettlement, identifying research and policy gaps that impact community recovery. The chapter outlines the book's structure and the diverse studies which highlight the need for interdisciplinary methodologies and context-specific solutions to address the diverse realities of displaced communities worldwide.

1.1 Introduction

The impacts of disasters on cities and housing globally are anticipated to intensify due to climate change, posing a threat of temporary or permanent displacement for individuals, which in turn triggers cascading effects at both individual and communal levels. Despite the efforts in disaster research in the last decades, numerous aspects remain unexamined, including the validation of survivors' ability to actively contribute to their own recovery efforts. Disaster-induced displacement is becoming more complex as the effects of climate change are threatening people and forcing them to leave their homes. Also, the multiple armed conflicts, political and economic instability, and breaches of human rights are forcing people to seek safe places to live beyond their countries' borders creating more complex conditions for an eventual return to their homeland in the future.

S. Carrasco (✉)
School of Humanities and Social Sciences, Deakin University, Melbourne, Australia
e-mail: s.carrasco@deakin.edu.au

C. Ochiai et al. (eds.), *Disaster and Displacement*, Disaster Risk Reduction,
https://doi.org/10.1007/978-981-96-2583-3_1

Recognizing the multifaceted nature of displacement and resettlement impacts and the diverse paths for community recovery, this book aims to explore and elucidate the physician environment (i.e., built environment, resettlement sites, neighborhoods, housing), political, and social dimensions of displacement and reconstruction induced by disasters, conflicts, or development projects. Embracing a long-term perspective and a focus on self-recovery acknowledging the crucial role of community agency in navigating such challenges. In the analysis of recovery, this book focuses on the analysis of empirical evidence from case studies in various geographical locations. Furthermore, the analysis is enriched with case studies addressing displacement in the context of increasing frequency of disasters, armed conflict, and development projects that are highly relevant in the context of increasing global forced displacement.

This chapter centers on discussing the terminology that is key to understanding the dimensions, implications, and challenges of resettlement. The following section focuses on the state of the current knowledge and the gaps that are necessary to better understand resettlement which provide a roadmap for future research and policy development. The last section of this chapter provides insight into the approaches taken to understand the multi-faceted nature of displacement aiming to articulate built environment, resettlement planning and management, and social and community recovery.

1.2 From Conceptualization to Research, Planning, and Action

Academic and non-academic literature does not find a consensus on the definition of the phenomena of placing or housing displaced people in a considered "safe place" or resettlement or relocation site. Relocation and resettlement have been frequently used as equivalent terms evidencing the lack of clarity and the implications of the use of certain terminology (Ferris and Bower 2023; Ferris and Weerasinghe 2020; McAdam and Ferris 2015). Furthermore, the complexities emerging from the causes or drivers of displacement add more confusion to the definition and scope of the terms. For instance, World Bank publications refer to relocation in the context of disaster as "a process in which a community's housing, assets and public infrastructure are rebuilt in another location" (Jha et al. 2010, p 77). Similarly, Correa (2011, p ix) points out "planned resettlement" as a term associated with development projects involving mandatory displacement of people and productive activities. In the context of conflict-driven displacement, the International Organization for Migration (IOM 2019) links "relocation" with internal humanitarian evacuation, within the same country, as a large-scale movement of people whose safety is threatened. Conversely, "resettlement" is associated with the transfer of displaced people to another country where safety should be guaranteed to them (IOM 2019).

Jain et al. (2017) make some distinctions between relocation and resettlement. "Relocation" is related to a non-systematic movement of people from hazard-prone locations, which, however, involves less upheaval in terms of access to livelihood and social networks. Conversely, "resettlement" is an integrated and comprehensive movement of people with significant distance from the original to a new location. Resettlement, unlike relocation, is not limited to housing and services but includes the creation of new social and economic relations, and opportunities to deal with new challenges such as access to work and social cohesion (Jain et al. 2017). Ferris and Bower (2023) claim that the terminology refers to the combination of two components: an intention term such as planned, strategic, or managed; and a movement term such as relocation, resettlement, retreat, or alignment. Therefore, considering that the phenomena should not be limited to the mere movement of people but imply the need for a more holistic process, the most accurate terminology should be resettlements or planned relocations. Additionally, resettlement and planned relocation, unlike evacuation are considered permanent (McAdam and Ferris 2015).

The process of resettlement or planned relocations in the recovery of displaced people needs to acknowledge the complexities inherent to the *"reconstruction of economic lives, leadership, and political structure, environmental relations and reinterpretation of social and cultural values and priorities"* (Oliver-Smith 2023). However, the lack of understanding among responsible agencies, governments, and organizations of the complexities and implications makes it crucial to highlight the process nature of resettlement. Resettlement projects imply much more than infrastructural projects that are produced by planners and construction companies (Oliver-Smith 2023). In many cases, people are mere observers of the process, and in some resettlement projects that are labeled as "participatory," people's inputs are often limited to marginal participation through consulting, informing, and in worst scenarios can lead to manipulation of people to accept the proposed relocation or resettlement plans (Carrasco and Dangol 2019; Carrasco et al. 2016a).

1.3 What We Know, Don't Know, and Need to Know About Resettlement

The rising trends in global displacement, combined with the potential impacts of climate change, are creating environmental and social processes that will significantly increase the number of permanently displaced people (Oliver-Smith 2016). This situation calls for a reevaluation of our understanding and strategies for addressing these global challenges. In particular, it is important to identify knowledge gaps and critical areas that require further exploration. Ferris and Bower (2023) propose to synthesize the current state of knowledge and gaps in resettlement considering three components: What we know, what we don't know, and what we need to learn. Figure 1.1 presents the main concerns in these three components about resettlements

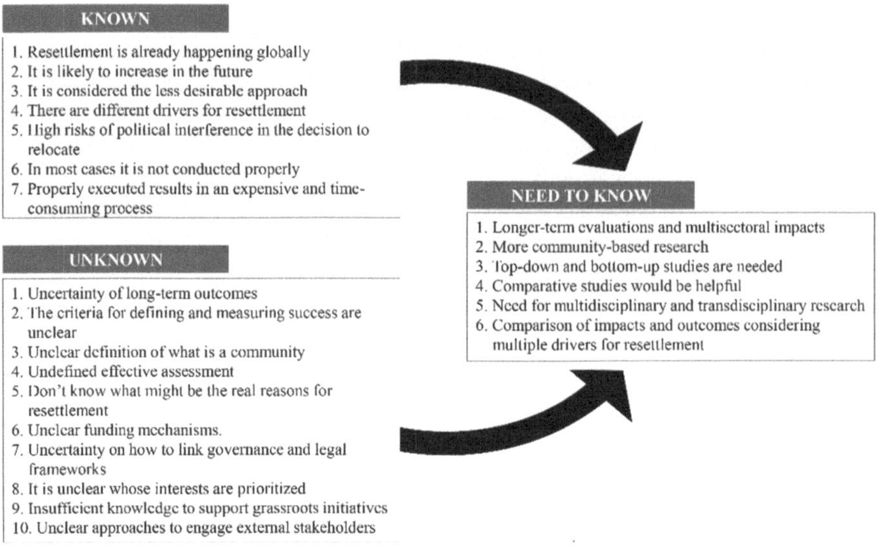

Fig. 1.1 Status of knowledge, gaps, and needed research for resettlement or planned relocations. *Source* Author

or planned relocations based on Ferris and Bower (2023), McAdam and Ferris (2015), (Ferris 2014), (Oliver-Smith 1991), (Oliver-Smith 2016), and (Oliver-Smith 2023).

1.3.1 Known

Studies documenting resettlement globally evidence that there are at least 591 cases of identified "planned relocations," "managed retreats," or "disaster-induced relocations" in the world since 1970 (Ajibade et al. 2022; Balachandran et al. 2022; Bower and Weerasinghe 2021; IOM 2022). This global mapping reveals that planned relocation is more geographically widespread than the few cases typically highlighted in the media. The identified cases were conducted in different regions of the world; however, certain areas, like the Pacific, stand out as focal points (Ferris and Bower 2023). Furthermore, the data collected in these studies only include publicly available sources which might present a portion of the actual resettlement projects conducted, underway, or planned to happen in the future.

Considering the increasing frequency and intensity of extreme climate events, the number of displaced people globally is expected to rise. Furthermore, the changed environmental conditions in the affected areas might have inevitably lost their pre-disaster attractiveness, their vulnerabilities might remain, or new risks might have emerged. Thus, displaced people would make returning to their original places extremely hard or impossible. For instance, in 2022, almost 33 million people were displaced due to disasters; of these, 21 million were displaced by flooding and drought which might make the land unusable due to desertification or submergence, making return impossible (Oliver-Smith 2023).

Resettlement is considered to be the last resort due to the immense impact on displaced people that might undermine their capacity to recover (Ferris 2014), and it is important to highlight that despite the efforts "*resettlement can't replace what is lost*" (Oliver-Smith 2023). In fact, despite the site-related risks, these might not necessarily linked to a specific location, and it is important to consider the inherent risks of carrying out relocations (Jain et al. 2017). Oliver-Smith (2023) refers to two kinds of losses when conducting resettlement: The quantified losses are those that can be expressed in terms of economic value or tangible, including loss of land, houses, jobs, access to resources, and livelihoods. Conversely, non-economic losses such as uprooting conditions, loss of social networks, culturally significant symbols, and direct and indirect impacts or displacement and resettlement such as diseases, and trauma, among others. Even though some of the non-economic losses can be quantified for certain purposes, such as insurance or compensations, those metrics don't come close to capturing the nature or gravity of the losses they cause and are often dismissed in resettlement theory, policy, and practice (Oliver-Smith 2023).

Ferris and Bower (2023) discuss the differences inherent in the process of resettlement. Differences in the drivers for resettlement arise from environmental risks, spatial patterns, and demographic differences. Differences are also related to decision-making approaches which consider resettlement as a reactive measure prompted by disastrous events such as landslides or it is the result of proactive planning or pre-emptive protective measures. Furthermore, differences are also referred to as who takes the leadership and the decisions to relocate. Some people might decide to relocate on their own, while others might be forced to relocate by governments or other actors. These differences impact policy and practice, as there's no universal approach due to the diverse nature of planned relocations.

The decision to relocate or not a group of people results in complex and controversial cases where the targeted communities to be moved are uncomfortable for the authorities such as informal settlers. For instance, in the city of Cagayan de Oro in the Philippines, the local government was highly criticized for relocating residents of informal settlements along the main city river after a major typhoon hit the area. Unlike informal settlers, formal residents were not part of this scheme. Similarly, during the resettlement process, local authorities removed inconvenient not-for-profit organizations from the conversations and the projects of housing construction in resettlement sites whose land was owned by the local government (Carrasco et al. 2016a). Thus, many people are reluctant to move and often question decisions by authorities to relocate a community and the decision to resettle may be seen with suspicion, especially when there is a lack of trust in the government (Ferris and Bower 2023). Furthermore, local government authorities, in particular, may be concerned about losing their tax base or jeopardizing their positions if a community relocates (Ferris and Bower 2023).

The planned relocation of 35 families in Louisiana in the USA which cost \$48 million, or at least \$25 million per community in Alaska requested for relocation gives an example of how expensive planned relocation projects can be (Ferris and

Bower 2023). Furthermore, resettlement projects are often designed and constructed with a focus on economic efficiency and the perspectives of the elite, rather than addressing the actual needs of displaced communities. This approach is ultimately more costly, as the settlements are often abandoned or destroyed, leading to social disintegration that hinders productivity and self-sufficiency (Oliver-Smith 2016).

1.3.2 Unknown

There is a lack of long-term and longitudinal approaches to understanding and evaluating the impacts of resettlement projects. Evaluations and studies on resettlement often focus on short-term outcomes, emphasizing the successes or failures at the end of relocation projects rather than the recovery and development process (Schilderman and Parker 2014). This reflects a misunderstanding of donors' and humanitarian agencies' roles as these are often seen merely as housing providers. However, these agencies should also empower communities and help local governments address obstacles to securing decent, durable, and affordable housing that supports daily activities with dignity and reduces vulnerability to future disasters (Schilderman and Parker 2014). Ferris and Bower (2023) also claim that most follow-up monitoring is typically short-term, often lasting less than a year, and focuses on the construction of physical infrastructure and homes rather than on human well-being and livelihoods.

Oliver-Smith (1991) claims that the successes in post-disaster resettlement are more than planning and constructing houses and towns and focus on the problem towards recovery from a holistic perspective including social, cultural, and economic factors. However, a single dimension is still prevalent in the evaluation and measuring success of resettlement projects. For instance, Oliver-Smith and Arenas (2015) found that focusing on the economic rebirth of a community as a way to measure success or failure could also create conflicts between government or agencies' expectations and communities' real economic preferences and plans. Therefore, Ferris and Bower (2023) question the focus of "success" and how to measure it. Thus, talking about "successful resettlement" should consider asking successful for whom? Successful in comparison with previous sites or others? Successful immediately after resettlement or after years or decades? Does it imply the changes over time, or not? What is the goal, to survive or to thrive? Who is the most entitled to define and evaluate "success"? Isn't it the people themselves?

A key element that is frequently referred to is related to the impacts of resettlement in a community or communities. However, Ferris and Bower (2023) question the definition of community as a forced collective unit rather than individuals. The dynamics within a group of people, especially those facing crises, are not necessarily of unity and common interests to deal with challenging conditions. It is more common to see heterogeneous groups with non-consensus views whose interactions in the resettlement sites might be highly influenced by their pre-displacement conditions as was seen among refugees from the Horn of Africa in Australia (Carrasco et al. 2023). Interestingly, even in cases when people have a strong pre-relocation sense of community and when they decide on their neighbors in the new site, there is always the risk of disintegration of community ties (Carrasco and O'Brien 2021).

When deciding to resettle people in the post-resettlement stages, it is still unclear whether the assessments of the new sites and conditions of resettled communities are conducted appropriately and the implications of post-resettlement. The sites need to be assessed considering their suitability, security, and implications for the future of resettled people, the reduction of existing risks, or the creation of new ones. Furthermore, it is still unclear how to encompass the assessment of tangible and non-tangible impacts and opportunities. For instance, how the reconstruction of physical housing and neighborhoods, generations, or livelihoods could enable or hinder non-tangible dimensions such as histories of displacement and cultural heritage loss (Ferris and Bower 2023). Also, in terms of assessments, it is crucial to question whose knowledge matters? What happens when community members don't agree with the assessment or don't trust the decision-maker or the person who is conducting the assessment? (Ferris and Bower 2023).

Arguably, the reasons behind the resettlement are not always clear, and in many cases, the suspicions and lack of trust between communities and trust might reflect unfair targeting of people to be resettled and what would happen with the original land once people leave (Carrasco and O'Brien 2019). Shumaker and Conti (1985, p 240) claim that the poor and powerless are often displaced when their homes or lands become valuable to others. Another example of hidden reasons for resettlement is illustrated by Kothari (2014) as she suggests that the Maldives government's proposal to move people from 200 islands to 10–15 central ones is driven not solely by sea-level rise but also by concerns about the inefficiency of delivering services and infrastructure to small, scattered populations. Thus, planned relocations are not limited to reducing natural hazard risks, instead, they may be shaped by existing plans and power dynamics, often benefiting some while impacting many (Ferris and Bower 2023).

Financing resettlement remains a significant challenge where it is characterized by different approaches and funding mechanisms. International funding through the World Bank, the Inter-American Development Bank, and international cooperation agencies, differs from NGOs or other humanitarian agencies including faith-based organizations and even volunteer groups or even the private sector. Also, there are instances of bilateral aid or calls for international or national financing schemes such as Fiji's Trust Fund or the United Nations Framework Convention on Climate Change–UNFCCC's Green Climate Fund and Adaptation Fund. Together with evaluating whether there are hidden expectations and agendas and appropriateness in the outcomes, there is also the need to think of mechanisms to scale up, avoid funding misuse, and ensure that human rights standards are implemented (Ferris and Bower 2023).

There is a lack of specific guidance for policymakers regarding the management of planned relocations. Existing guidelines and tools are often broad and global in scope, making localization a significant challenge, despite some countries having national guidance in place. Considering the vast differences in relocation cases, it is more appropriate to develop tailored guidance for specific situations. For example, guidance could be developed for community-initiated Indigenous relocations or government-initiated urban relocations of informal settlements in large cities with multiple origin and destination sites (Ferris and Bower 2023). However, there is always the risk that local guidelines and practices do not respect minimum standards or focus on national standards that might be detrimental to people's well-being. For instance, national standards in the Philippines for emergency housing specify a minimum area of 21 m^2 which for large households of up to twelve members resulted in overcrowding and substandard considering international standards (Carrasco et al. 2016b).

Ferris and Bower (2023) claim that while top-down governance is crucial, it is equally important to understand grassroots, community-driven mobilization related to planned relocation. However, more evidence is required to identify successful strategies and pitfalls in obtaining funding and international support for these grassroots initiatives. Additionally, technical support for governments and communities on planned relocation is currently ad hoc and unsystematic. However, the international community has no clear focal point, with different organizations leading based on existing relationships. There is an increasing need for technical assessments, decision support tools, human rights-based policies, standard operating procedures, and funding applications (Ferris and Bower 2023).

1.3.3 Need to Know

Most research has focused on the decision-making about whether and how to relocate or on the immediate circumstances following planned relocations (Ferris and Bower 2023). Although some advances have been made by focusing on the impacts of resettlement rather than solely considering the process itself there is a significant gap in the research about monitoring and evaluating outcomes over time. Filling the knowledge gaps is the responsibility of multiple actors such as academics, donors, and implementing agencies such as humanitarian organizations, governments, and also communities, and others from civil society.

Community engagement is essential for effective governance of resettlement and for addressing knowledge and data gaps. Research in this field presents an opportunity for affected communities to participate in research. Participatory Action Research (PAR) methodologies and other approaches to co-producing knowledge are critical opportunities but also present challenges for both communities and funders (Ferris and Bower 2023). Similarly, global agendas such as the 2030 Development Agenda, the Sendai Framework for Disaster Risk Reduction, the New Urban Agenda, and the Agenda for Humanity call for active community engagement and focus on empowering them and building capacities to build resilience.

The studies about resettlement mainly focus on isolated case studies. Although these provide unique narratives of local challenges, it is important to consider mechanisms to compare studies and identify lessons and good practices. Analyzing multiple cases allows for the identification of similarities and differences in the decisions and implementation of resettlement schemes (Ferris and Bower 2023). For instance, understanding the various financing models used to support planned relocations is crucial. Also, a cross-country analysis can provide valuable lessons to refine and update existing guidelines such as humanitarian standards among others.

Planned relocations or resettlements are multidimensional, involving environmental, displacement, development, livelihood, planning, governance, human rights, human security, law, and policy considerations (Ferris and Bower 2023). Research should use diverse methodological and theoretical tools, combining natural and social scientific disciplines like climate science and protection issues. Transdisciplinary approaches, integrating policymakers', community members', and academic disciplines' perspectives, are valuable.

The diverse drivers for resettlement generate different conditions for conducting resettlement. Despite these differences, it is valuable to identify commonalities that can enhance dialogue, offering insights and opportunities to reassess past failures and shortcomings. For instance, disaster-induced resettlement has been widely studied and might provide valuable insights into climate change-induced displacement and resettlement. The predominantly negative perceptions surrounding post-disaster resettlement resulted from widespread evidence suggesting that affected communities often view displacement as a "second disaster" (Oliver-Smith 1991). In contrast, approaches like voluntary relocation in response to climate change or planned resettlement are viewed more favorably, potentially serving as tools for community protection. Moreover, insights from refugee studies offer valuable long-term perspectives on how permanently displaced individuals and groups adapt and integrate into new socioeconomic and cultural environments.

1.4 How This Book Addresses the Multidimensional Nature of Displacement and Resettlement

The book features chapters covering international case studies using qualitative, quantitative, and mixed-methods research to illustrate diverse recovery pathways addressing displacement primarily driven by disasters but also includes case studies from conflict-driven and development-induced displacement and resettlement. The book contains 20 chapters that discuss displacement and resettlement theory and case studies, in addition to introduction and conclusion chapters. The relevant theory is presented in Chapter 2 which discusses the multiple dimensions of displacement and resettlement by reviewing the past and current literature to give us a basic understanding of this book.

The subsequent case study chapters in this book are divided into three parts:

Part I addresses the architectural and technical aspects of relocated or rebuilt houses, detailing the modifications made by people to suit their needs. It focuses on the design, construction, and technical adaptations necessary for effective resettlement.

Part II discusses the planning, management, and political dimensions of displacement and housing or facilities in post-disaster situations, covering both top-down and bottom-up approaches, as well as university or institution-led programs. The chapters address how different planning and management strategies can be implemented to facilitate recovery and rebuilding efforts.

Part III focuses on the community's efforts to adapt to new living environments, addressing the challenges of relocation and resettlement following various disasters. This part emphasizes the role of community agency and collective action in overcoming the challenges of displacement.

1.4.1 Part 1: Architectural and Technical Components of the Displacement and Reconstruction

Chapter 3 illustrates how resettled Indigenous communities in Taiwan adapted their permanent housing after the 2009 Typhoon Morakot. The indigenous people have their unique lifestyles and cultural practices that are very much reflected in their houses and space design of the open areas. However, standardized post-disaster housing dismisses communities' needs, prompting the construction of extensions to adapt modern houses to fulfill their needs. The chapter provides extensive graphic documentation of various modifications and emphasizes the importance of pre-planning coordination and consultation among the government, architects, and Indigenous community to realize the housing design considering the special needs and rights of the people.

Chapter 4 explains the long-term impacts of resettlement and reconstruction projects hit by the 2011 Typhoon Washi in the Philippines. The study addresses the importance of housing beneficiaries to include at the planning and implementation stage of the reconstruction of their houses, by understanding their needs and future livelihood possibilities. The resident-driven progressive adaptation and modification including enlargement of the spaces are explained from the architectural point of view and people's satisfaction levels.

Chapter 5 introduces the transmigration project implemented in Indonesia to relocate people from a high population density area to rural sites by providing farmland, a single house, and sometimes cattle. One of these project sites of origin was a large dam project site to resettle in the site where living conditions were different from their original place, but people had a hope to have a better life and become homeowners. The study explains the differences in housing and infrastructure of newly resettled sites that had been impacted by the flood. Also, the social relationship and its structural mechanism in the new sites have an impact on people's adaptation strategy and current living conditions.

Chapter 6 explores the evolution of traditional Sumu-Mayangna housing in Nicaragua, detailing the history of hardships caused by outside interventions and armed conflict. It focuses on the thermal comfort of housing and illustrates three different types: Traditional Sumu-Mayangna housing, refugee camp housing, and current-style housing. The study highlights the Sumu-Mayangna people's traditional understanding and relationship with nature in housing design. It stresses the importance of conserving the ancestral cultures and knowledge of the Sumu-Mayangna for future generations.

1.4.2 Part 2: Resettlement Planning and Management

Chapter 7 looks at the Japanese vernacular architecture and the opportunities for post-disaster housing considering their low environmental impacts and cultural adaptability. This research examines the evolution of the Itakura construction system, derived from traditional storehouses, that were used for temporary housing in the aftermath of the 2011 Great East Japan Earthquake and Tsunami. The study focuses on three challenges: legal, institutional, and technical. Overcoming these challenges required decades of effort to establish legitimacy and a stable supply chain. The use of prefabricated components for efficiency in disaster contexts and traditional carpentry skills during non-disaster periods ensures that Itakura construction addresses temporary housing issues while preserving traditional craftsmanship.

Chapter 8 introduces a three-step methodology for creating livable neighborhoods in temporary housing settlements during disaster response. It involves reviewing design codes, identifying layout elements, and formulating requirements with spatial design considerations. This approach aims to improve shelter response efforts and support humanitarian workers in designing post-disaster and post-conflict settlements.

Chapter 9 examines government-led resettlement in Sri Lanka post-2016 landslide in Aranayaka. The study compares two projects: one heavily influenced by the government, the other more shaped by end-users. Findings reveal limitations in government strategies, stressing the need for community involvement. Early engagement with communities in government proposals is recommended to improve outcomes, considering flexible regulatory approaches.

Chapter 10 explores post-disaster housing reconstruction in vulnerable communities after the 2017 earthquakes in southeastern Mexico. It focuses on a case study in Puebla, Mexico, and a university-led Housing Reconstruction Project (UHRP). Through data analysis, literature review, observations, interviews, and a housing survey, the study highlights the challenges and dynamics of community-university cooperation. Findings underscore the UHRP's strengths in disaster recovery, including robust social networks and interdisciplinary knowledge, demonstrating the potential for effective collaboration between universities and communities to address vulnerabilities.

Chapter 11 examines the role of Residents' Associations (RAs) in post-disaster resettlement sites in Japan. It highlights the impact of involving residents and RAs in decision-making during housing reconstruction planning. The study highlights that community development council-led projects facilitate community integration, contrasting with challenges in city-led initiatives, especially in public housing-only settlements. Engaging prospective residents in discussions enhances RA vitality and improves local dynamics and living conditions.

Chapter 12 analyzes the relocation of the Kumano Hongu Taisha Shrine in Wakayama Prefecture, Japan, after a devastating 1889 flood. It explores the decision-making process and involvement of various parties. The shrine moved from its originally frequently flooded location to a hilltop site in 1891 due to flood damage. Carpenters from Wakayama City, overseen by government offices, led reconstruction, aided by local residents/parishioners affected by the disaster. This collaborative effort enabled relocation and rebuilding, emphasizing the shrine's community significance and resilience.

Chapter 13 examines the challenges and importance of selecting reliable data and indicators for managing disaster-induced displacements. The study discusses the need for accurate information to guide emergency response and recovery efforts, in line with the "Build Back Better" principle. The chapter suggests using a hierarchical decomposition algorithm and thematic analysis of disaster intervention reports to improve the selection of evidence-based indicators, considering the complexity of displacement scenarios and the necessity for a coordinated resilience approach.

1.4.3 Part 3: Reconstructing the Sense of Home and Community Agency in Resettlement Recovery

Chapter 14 analyzes two historical case studies of long-distance collective relocations in Japan: the 1889 flood in Totsukawa Village and the 1914 eruption of Sakurajima. It examines the feasibility of such relocations and their impact on communities. The study results indicate that while government efforts supported infrastructure and livelihoods, many struggled to adapt to new environments. However, over time, migrants integrated, fostering cultural tolerance. Even after a century, interactions persist between original and relocated communities, suggesting that well-implemented migration policies can sustain human relationships over long distances.

Chapter 15 examines how local knowledge shapes decisions on displacement after disasters, contrasting community-driven responses with government-led strategies. The chapter focuses on floods in the Northern Rivers Region, New South Wales, Australia. The study highlights the importance of community insights, critiques oversimplified views of disasters, and advocates for tailored approaches to enhance resilience.

Chapter 16 examines post-disaster resettlement in Maca, a Peruvian town in the Colca Valley which is prone to volcanic and seismic risks. Despite deep distrust of authorities due to unmet promises, residents display resilience tied to their cultural heritage, fostering community cohesion. The study stresses the need for community involvement in disaster management, promoting collaborative approaches that respect local knowledge and values for sustainable development.

Chapter 17 examines post-disaster livelihood development in Indigenous communities through two case studies following Typhoon Morakot in 2009: one relocated and one non-relocated settlement. The study analyzes challenges and compares livelihood strategies, identifying common issues like road connectivity and government interactions. The relocated settlement faces land-ownership problems, while non-relocated communities fully utilize their resources. Livelihood strategies vary based on resources, organization, and location, emphasizing the need for tailored approaches and collaboration with external stakeholders.

Chapter 18 discusses the reconstruction of the sense of home, and community, and the integration of refugees resettled in Australia. The study focuses on public housing in Melbourne and serves as a focal point in housing discussions, particularly for vulnerable groups like migrants and refugees from the Horn of Africa. The study investigates how living conditions affect community cohesion, highlighting the need for interventions to promote integration and rebuild social support. Additionally, the discussion contrasts the need for refugees' integration and the challenges they face in preserving their cultural identity and combating discrimination.

Chapter 19 discusses how mega-events, like the Olympic Games, lead to displacement and resettlement and their impacts on urban development. Using the Rio 2016 Games as a case study, which displaced over 77,000 favela residents, the study argues that while these events promise urban development, they often overlook their negative social impacts. By examining experiences in Brazil and Japan, the chapter suggests that mega-events prioritize agendas like gentrification, posing as public interest and sustainability initiatives. This highlights the need to rethink planning practices to mitigate such consequences.

Chapter 20 focuses on communities relocated due to the decline of industries such as coal mining, within the broader context of continuous urban shrinkage. This study examines two representative communities in Fuxin, a typical shrinking coal city in China, to evaluate the current community livability more than a decade after government-led resettlement from shantytowns near coal mines. It investigates residents' perceptions of urban and community decline, their satisfaction with recent government-led renovations in the resettled areas, and how these perceptions and satisfaction levels influence livability evaluations. By emphasizing the residents' perspective, this study uses community livability as a metric to assess the effectiveness of resettlement and renovation strategies, providing insights for improving community livability in shrinking cities.

Finally, chapter 21 summarizes the most relevant outcomes from the case studies presented and proposes opportunities for future studies to better understand the challenges of displacement and resettlement and its implications in building resilient communities and strengthening the relationships of various stakeholders involved.

1.5 Conclusions

This book on displacement and resettlement offers a comprehensive exploration of the complexities following disasters, conflicts, and development projects worldwide. By analyzing diverse international case studies and research methodologies, the book covers architectural, planning and management, and social dimensions of reconstruction, emphasizing culturally sensitive housing designs, community involvement in planning, and the importance of fostering a sense of home. From Indigenous communities in Taiwan adapting post-disaster housing to government-led resettlement projects in Indonesia and Sri Lanka, each case points out the necessity of collaboration among stakeholders for sustainable outcomes. Long-term impacts are examined, highlighting the need for inclusive, community-centered approaches to resilience and development.

References

Ajibade I, Sullivan M, Lower C, Yarina L, Reilly A (2022) Are managed retreat programs successful and just? A global mapping of success typologies, justice dimensions, and trade-offs. Glob Environ Chang 76(102576). https://doi.org/10.1016/j.gloenvcha.2022.102576

Balachandran B, Olshansky RB, Johnson LA (2022) Planning for disaster-induced relocation of communities. J Am Plann Assoc 88(3):288–304. https://doi.org/10.1080/01944363.2021.1978855

Bower E, Weerasinghe S (2021) Leaving place, restoring home: enhancing the evidence base on planned relocation cases in the context of hazards, disasters, and climate change. Platform on disaster displacement (PDD) and Andrew and Renata Kaldor Centre for International Refugee Law. https://disasterdisplacement.org/portfolio-item/leaving-place-restoring-home

Carrasco S, Dangol N (2019) Citizen-government negotiation: cases of in riverside informal settlements at flood risk. Int J Disaster Risk Reduct 38:101195. https://doi.org/10.1016/j.ijdrr.2019.101195

Carrasco S, Dangol N, Faleh M (2023) Rethinking social networks in responding to COVID-19: the case of African migrants in Melbourne's public housing. Int J Disaster Risk Reduct 98:104073. https://doi.org/10.1016/j.ijdrr.2023.104073

Carrasco S, O'Brien D (2021) Beyond the freedom to build: long-term outcomes of elemental's incremental housing in Quinta Monroy. Urbe. Revista Brasileira de Gestão Urbana 13. https://doi.org/10.1590/2175-3369.013.e20200001

Carrasco S, O'Brien D (2019) Urbanism of emergency: use and adaptation of public open spaces in disaster-induced resettlement sites. In: Resettlement Challenges for Displaced Populations and Refugees, pp 163–174. Springer

Carrasco S, Ochiai C, Okazaki K (2016a) Disaster induced resettlement: multi-stakeholder interactions and decision making following tropical storm Washi in Cagayan de Oro, Philippines. Procedia–Soc Behav Sci 218:35–49. https://doi.org/10.1016/j.sbspro.2016.04.008

Carrasco S, Ochiai C, Okazaki K (2016b) A Study on Housing Modifications in Resettlement Sites in Cagayan de Oro, Philippines. J Asian Arch Build Eng 15(1):25–32. https://doi.org/10.3130/jaabe.15.25

Correa E (2011) Populations at risk of disaster: a resettlement guide. The World Bank, The international bank for reconstruction and development. http://documents1.worldbank.org/curated/en/612501468045040748/pdf/653550WP0Box360settlement0guide0150.pdf

Ferris E (2014) Planned relocations, disasters and climate change: consolidating good prac-tices and preparing for the future. In: Background document, Sanremo consultation, pp 12–14 march 2014. Geneva, Switzerland: UNHCR Division of International Protection, Brookings and Georgetown University

Ferris E, Bower E (2023) Planned relocations: what we know, don't know, and need to learn. Internal displacement research programme (IDRP), London, UK

Ferris E, Weerasinghe S (2020) Promoting human security: planned relocation as a protection tool in a time of climate change. J Migr Hum Secur 8(2):134–149. https://doi.org/10.1177/2331502420909305

IOM (2019) Glossary on migration. International migration law. International organization for migration–IOM, Geneva, Switzerland

IOM (2022) Leaving place, restoring home II: a review of French, Spanish and Portuguese literature on planned relocation in the context of hazards, disasters, and climate change. International organization for migration (IOM). https://environmentalmigration.iom.int/sites/g/files/tmzbdl1411/files/documents/pub2021_183_r_2022_final-version-march-2022.pdf

Jain G, Johnson C, Lavell A, Lwasa M, Oliver-Smith A, Wilkinson E (2017) Risk-related Resettlement and Relocation in Urban Areas. In. Climate and Development Knowledge Network–CDKN Global, Cape Town, South Africa

Jha AK, Barenstein JD, Phelps P, Pittet D, Sena S (2010) Safer homes, stronger communities: a handbook for reconstructing after natural disasters. World Bank Publications

Kothari U (2014) Political discourses of climate change and migration: resettlement policies in the Maldives. Geogr J 180(2):130–140. https://doi.org/10.1111/geoj.12032

McAdam J, Ferris E (2015) Planned relocations in the context of climate change: unpacking the legal and conceptual issues. Camb Int Law J 4(1):137–166. https://doi.org/10.7574/cjicl.04.01.137

Oliver-Smith A (1991) Successes and failures in post-disaster resettlement. Disasters 15(1):12–23. https://doi.org/10.1111/j.1467-7717.1991.tb00423.x

Oliver-Smith A (2016) Resettlement Principles. In. United Nations Framework Convention on Climate Change–UNFCCC, Bonn, Germany

Oliver-Smith A (2023) Why speak about planned relocation? United Nations University's institute for environment and human security (UNU-EHS) and Munich Re Foundation (MRF). https://www.youtube.com/watch?v=5SzcMuDGZ8U&list=PLmql2xc4rFB37ATKqLJMP-KDiOq7cL4MG&index=2

Oliver-Smith A, Arenas C (2015) Post-disaster resettlement: the transition to the new community in Gramalote, Colombia. Natural hazards center quick response research report series, report 266. https://hazards.colorado.edu/quick-response-report/post-disaster-resettlement

Schilderman T, Parker E (2014) Still Standing? : Looking Back at Reconstruction and Disaster Risk Reduction in Housing. Practical Action Publishing. https://practicalactionpublishing.com/book/2078/still-standing

Chapter 2
Exploring the Multiple Dimensions of Displacement and Resettlement

Sandra Carrasco

Abstract Displacement and resettlement are increasingly prevalent globally, driven by factors such as disasters, conflicts, climate change, and development projects. This chapter explores the complex relationship between displacement and resettlement, analyzes the risks and opportunities arising from understanding the various displacement drivers, and the challenges for displaced people, and emphasizes the need for holistic approaches to address the multifaceted nature of displacement and resettlement. The analysis focuses on the discussion of the complexities of resettlement planning and implementation, highlighting the importance of participatory approaches, integration of human rights and gender considerations, investment in livelihood restoration, enhanced monitoring and evaluation, fostering international cooperation, and addressing environmental sustainability. The discussion emerging from the analysis acknowledges the multidimensional nature of displacement and resettlement and observes that by adopting inclusive, rights-based, and environmentally sustainable strategies, policymakers and practitioners can mitigate risks, promote resilience, and enable sustainable integration and development of displaced populations in their new locations.

2.1 Introduction

Displacement and resettlement are multifaceted phenomena impacting millions worldwide. The drivers for displacement include disasters, armed conflicts and violence, climate change, and development projects. By the end of 2023, almost 76 million people worldwide were living in conditions of displacement due to disasters and conflicts (IOM 2024).

Disasters result from existing risks, hazard exposure, vulnerabilities, and capacities that cause loss of life, injury, or destroyed or damaged assets in a society or a community in a specific period of time (UNDRR n.d.). Governments often view resettlement as the best strategy to mitigate risks when site-specific vulnerabilities

S. Carrasco (✉)
School of Humanities and Social Sciences, Deakin University, Melbourne, Australia
e-mail: s.carrasco@deakin.edu.au

C. Ochiai et al. (eds.), *Disaster and Displacement*, Disaster Risk Reduction,
https://doi.org/10.1007/978-981-96-2583-3_2

are unmanageable (Jha et al. 2010). For instance, the 2004 Indian Ocean tsunami displaced over 1.7 million people across 14 countries (Inderfurth et al. 2005). Also, the 2010 Haiti earthquake displaced an estimated 1.5 million people, with nearly 33,000 still in evacuation camps a decade later, facing health and security risks (Charles and Iglesias 2020). Forced displacement from resettlement exacerbates vulnerabilities, stripping communities of homes, livelihoods, and social networks, often seen as a second disaster (Oliver-Smith 1991). Resettlement is the least desirable disaster recovery approach due to its lasting impact on communities (Gaillard 2015), highlighting the need for effective resettlement strategies (Thomas and Frankenberg 2014).

Displacement trends show a steady rise, exacerbated by climate change, with weather-related disasters tripling in the last forty years (Bellizzi et al. 2023). The increasing intensity and frequency of climate-change-related disasters are becoming a major cause of global displacement (Bellizzi et al. 2023; UNOCHA 2023). Gradual environmental changes, including rising sea levels, desertification, extreme weather, and wildfires, result in 'people displaced by climate change,' 'climate refugees,' or 'environmental migrants' (Offner and Marlowe 2021). Climate-driven mobility could displace 25 million to 1 billion people by 2050 (IOM 2008), challenging regions to accommodate migrants (Vince 2022). Climate-driven displacement is often slow, making it less visible but equally devastating. Pacific Island nations, including Fiji, Tuvalu, and Kiribati, face gradual land loss due to rising sea levels, forcing communities to consider relocation (Alam and Alam 2022). Fiji, Tuvalu, and Kiribati are experiencing gradual but irreversible land loss due to rising sea levels, forcing communities to consider relocation options (Alam and Alam 2022). Thus, disasters, migration, and climate change are complex and interlinked environmental and societal issues (Alam and Alam 2022) that require proactive planning and international cooperation to ensure sustainable resettlement solutions.

Armed conflicts and political instability are also drivers for mass displacement, with refugees and internally displaced persons (IDPs) seeking safety. Conflict-induced displacement leads to prolonged resettlement issues and heightened vulnerability. The protracted nature of conflicts means that displaced populations often spend years or even decades in temporary shelters, facing ongoing insecurity and limited access to basic services. Furthermore, Bohnet et al. (2021) and (Neef et al. 2023) explored the relationships between conflict and disaster-driven resettlement and its implications for security, vulnerabilities in hazard-prone areas, and challenges for recovery and integration due to the complexities in the interactions with host communities.

Displacement creates conditions of vulnerability within the uprooted communities which is acknowledged by various studies covering diverse drivers for displacement and resettlement as presented earlier. However, the conditions and impacts of human mobility are seen differently by the nature of displacement. Studies focusing on displacement resulting from disasters often assert that it is the worst approach that challenges the recovery of affected people. In contrast, climate-related and conflict-driven displacement may offer opportunities for resilience through migration and relocation, as communities may have more time to prepare and can potentially benefit from new environments or opportunities in the long term.

Therefore, this study aims to analyze the interconnections between different drivers of displacement and resettlement to identify resulting vulnerabilities and potential opportunities for effective resettlement planning. By understanding the different impacts of various forms of displacement, policymakers and practitioners can better address the specific needs and challenges faced by uprooted communities, ultimately leading to more sustainable and resilient outcomes.

2.2 Methodology

This chapter provides a comprehensive review of the literature on displacement and resettlement, exploring areas that overlap the diverse drivers of displacement, including disasters, climate change, conflicts, and development-driven displacement. This study further analyses the impacts on affected populations, including risks caused by displacement, challenges in recovery, and alternatives for improvements in the planning, relocation, and resilience building. By analyzing these factors, the chapter explores the extent of conceptual alignment and argues for the value of re-thinking strategies and policies, considering the nexus between the discourses in favor of and against resettlement.

The sources selected were analyzed through a traditional literature review process, employing narrative synthesis to interpret and elucidate similarities, differences, and complementary aspects among the studies (Barnett-Page and Thomas 2009; Cardoso Ermel et al. 2021; Ferrari 2015). By synthesizing existing scholarship, this chapter contributes to a more comprehensive discourse on effective responses to displacement challenges, emphasizing the need for holistic approaches that account for the diverse needs and circumstances of affected populations, considering the importance of addressing these issues in a holistic and informed manner.

2.3 Risks Associated with Displacement and Resettlement

The decision to resettle people results in temporary or permanent displacement, often forced as it results from government-led decisions, which disrupts the complex ties built among people and their environment through long cultural, social, economic, and political histories, which are impossible to transfer to new settlements (Gaillard 2015). Cernea (1997) conceptualized the impacts of "unmitigated forced displacement economic and social uprooting" as "impoverishment risks." The social hazards existing before displacement can potentially become impoverishment disasters if mitigation measures are not practiced (Cernea 1997). In one of the seminal studies on the impacts of displacement, Cernea (1997) defines eight impoverishment hazards resulting from the process of social and economic deprivation in the aftermath of resettlement:

2.3.1 Landlessness

The loss of land is seen to have tremendous impacts on people's productive systems, commercial activities, and livelihoods, as well as their sense of security and ownership, which is forcibly removed from them. Studies such as (Oliver-Smith 2010) highlight the people's resistance to moving out as they fear losing the foundation of their lives, especially marginalized people such as informal dwellers (Carrasco and Dangol 2019). Also, the loss of land is considered the principal form of decapitalization for displaced rural people (Cernea 1997). Oliver-Smith (1991) observes the poor choice of sites for resettlement which are often driven by the urgency to acquire land or limited to government-owned land and lacking the appropriateness of the areas to withstand potential hazards. For instance, resettlement sites in flat areas might provide little shelter from snow and wind (Oliver-Smith 1991), or houses built in hilly areas might be prone to landslides, which trigger a second displacement (Carrasco et al. 2016a).

2.3.2 Joblessness

The loss of jobs occurs in urban and rural displaced communities. However, in urban areas where competition for jobs is fierce, low-income and marginalized people often work in unskilled jobs such as laborers, service workers, or small businessmen (Cernea 1997). People removed from their settlements also leave behind their livelihood. For instance, a study in the Philippines showed that people living near urban commercial areas were resettled from 5 to almost 20 km from their original homes and struggled to financially recover from threatening conditions of long-term poverty (Carrasco et al. 2016a, 2016c) as it is unlikely that people re-establish the same livelihood activities as before the move (Vanclay 2017). In some cases, governments or other stakeholders provide compensations, subsidies, or other kinds of financial support aiming to revitalize local economies of resettled communities. However, these financial incentives alone are insufficient to ensure that people can re-establish their lives and livelihoods in their new location (Vanclay 2017). For instance, microcredits provided to survivors of the 2004 Indian Ocean tsunami were not enough to restore people's livelihoods (Ismail et al. 2018). Conversely, Ismail et al. (2018) also found out that if financial assistance is accompanied by training and upskilling programs, it can have a larger impact on people's livelihood restoration in the long term.

2.3.3 *Homelessness*

The loss of housing could be temporary for some displaced people. However, for many, homelessness can become a chronic condition (Cernea 1997). Formal ownership of land and housing is not always guaranteed to resettled communities, limiting their sense of home and security and hindering future investments in improving their living conditions (Carrasco et al. 2016b). Furthermore, in case resettled people are not granted formal ownership, they might be considered informal settlers, subject to potential evictions and removing the government's responsibilities towards these communities, and no compensation needs to be paid (Vanclay 2017). The lack of formal ownership might also prevent non-government agencies from supporting displaced communities hindering their development opportunities (Carrasco et al. 2016b).

2.3.4 *Marginalization*

Marginalization occurs when families lose their economic stability and experience downward mobility: middle-income farming households become small landowners, and small shopkeepers and craftsmen fall into poverty (Cernea 1997). Many are unable to utilize their skills in new locations, rendering their human capital inactive. Displacement also harms self-esteem, social status, and confidence, creating a sense of injustice. Economic marginalization often begins before displacement due to disinvestment or insufficient investment in the affected areas (Cernea 1997). Resettlement sites can also be clusters of stigma and marginalization to their residents even in contexts where resettlement is a common practice for development projects, such as in China (Zhang et al. 2021). People displaced due to conflicts are also considered socio-economic vulnerable and are housed in social housing in host countries, which already have a stigma associated with these areas (Raynor et al. 2020). This further highlights the stigma and marginalization that are unfairly associated with refugees (Carrasco et al. 2023).

2.3.5 *Increased Morbidity and Mortality*

Displacement causes significant health issues due to social stress, insecurity, psychological trauma, and diseases such as malaria and schistosomiasis (Cernea 1997). Poor water quality and inadequate sewage systems heighten the risk of epidemics, chronic diarrhea, and dysentery, with infants, children, and the elderly being the most vulnerable. Charles and Iglesias (2020) reported that despite the considerable support from the international community, displaced people due to the 2010 Haiti earthquake were living in evacuation camps with precarious conditions exposed to torrential rain, heavy winds, and a deadly cholera epidemic that killed at least 10,000 people.

2.3.6 Food Insecurity

Forced displacement increases the risk of chronic undernourishment, characterized by insufficient calorie-protein intake for normal growth and work, leading to food insecurity (Cernea 1997). The lack of coping methods results in cutting food and household expenses, borrowing, and asset sales (Alam et al. 2020). The loss of livelihoods has a direct impact on communities' expenditure on food and education for children (Ismail et al. 2018). Those facing agricultural challenges adapt by trying new crops, adjusting planting times, afforestation, home gardening, and diversifying livelihoods (Alam et al. 2020).

2.3.7 Loss of Access to Common Property

The loss of access to common property assets, such as forests, water bodies, and grazing lands, particularly impacts the income and livelihoods of poor and landless individuals (Cernea 1997). Government resettlement schemes rarely compensate for this loss. Additionally, the lack of access to basic public services is often associated with these risks (Mathur 1997). Access to common areas, which are crucial for community interaction and essential for recovery and building resilience, is also impacted by resettlement (Carrasco and O'Brien 2019). Remarkably, even resettled entire communities in sites with common spaces, such as courtyards, are subject to social conflicts and social disarticulation as there is a competition between individual and collective interests at the moment to invest in improving their living environments (Carrasco and O'Brien 2021).

2.3.8 Social Fragmentation

Displacement fractures the fabric of society, dispersing and fragmenting communities while dismantling social structures and interpersonal bonds (Cernea 1997). The affected people's interactions and relationships, which happen at individual and group levels, create networks that support the recovery of people affected by disasters and other crises (Carrasco et al. 2023). These supporting networks are developed over longer periods of time, sometimes during a lifetime, and some groups struggle more than others to restore them. Additionally, social isolation is linked with a higher risk of poor mental health, including depression (Cacioppo et al. 2010). Some groups within society are more at risk of social isolation, such as older persons with tragic consequences. For instance, many older persons resettled in temporary housing after the 1995 Kobe Earthquake in Japan were unable to reestablish a social network, resulting in them dying alone in what is called "isolated" or "unaccompanied" deaths, and their corpses were discovered only several days after (Dahl 2020).

2.4 Resettlement Planning and Implementation

Despite the tremendous impacts of displacement and resettlement, there is consistent evidence that planned resettlement is geographically widespread (Ferris and Bower 2023). The differences in terminology, such as planned relocations, retreat, realignment, strategic or managed, evidence that these processes are underway and are expected to happen more in the future, especially considering the increasing frequency and intensity of climate-related events (Ferris and Bower 2023).

Therefore, together with the discussions about the risks of displacement and resettlement in the affected people, it is crucial to focus on how the resettlement is done. Oliver-Smith (1991) argues that the difference between successes and failures in resettlement relies on how the process is done and what or how much is done. In other words, the impacts of resettlements are not only limited to providing shelter. Ignoring the local capacities of displaced communities, inadequate attention to social and cultural issues, and the lack of experienced resettlement practitioners and organizations are some of the barriers to achieving long-term sustainable outcomes (Vanclay 2017). When people distrust the government, decisions to relocate a community can be met with suspicion. Municipal authorities may be especially concerned about losing their tax base or jeopardizing their own positions if a community relocates (Ferris and Bower 2023). Interference in the process of resettlement, such as unfair land expropriation, selection of beneficiaries, allocation of resources and land, and the establishment of partnerships, is perceived as a lack of transparency in the process or corruption (Carrasco and O'Brien 2018; Vanclay 2017).

In the literature and in the decision-makers' discussions about resettlement, the impacted people are often referred to as communities, considering that there is a collective mindset and that the challenges and benefits are shared by a group of people. However, Ferris and Bower (2023) claim the lack of clarity of what a community is, as they might be dispersed and heterogeneous. For instance, Indigenous groups could be part of the displaced population, but have specific cultural values and requirements that might challenge further the process of resettlement and their recovery and development.

Gaillard (2015) observed particular ethnic factors that drove the resettlement of Bacolor people, which is an ethnic minority in the Philippines, which were related to the very strong attachment to their ancestral land and challenged their integration into their new settlements. Similarly, Carrasco et al. (2023) observed the challenges among refugees from the Horn of Africa who resettled in Australia to create social support networks within their communities as the differences between ethnic backgrounds in their homeland remained after resettlement. Furthermore, these differences also prevented them from integrating with the people in their host country as they kept small social networks and mainly relied on them to deal with their daily challenges. The livelihood recovery of affected people can also be impacted by the lack of consideration of their local culture and values that are closely linked to their ancestral land and the resources available to them (Tsai et al. 2022a, b).

Displacement can also be the result of exclusion, insecurity, and persecution of people whose gender identity do not conform to the dominant social and cultural norms (Mkhize and Mambondiani 2023; Türk 2013). The strong psychological impact of forcibly displaced LGBTI + individuals and the stigma attached to them in conservative societies, deprive them of economic opportunities and religious practices and deny assistance from their communities and government services (Mkhize and Mambondiani 2023).

Climate change is becoming a crucial factor for human mobility, and it is globally acknowledged its increased impact as a driver for displacement in the twenty-first century (IPCC 2014). Unlike the negative concerns about resettlement, Ferris and Weerasinghe (2020) claim that planned relocation or resettlement will become a tool for human security in the context of the emerging risks of climate change and become imperative as people living in high-risk areas continue to grow. These concerns have led to calls for improved governance of community displacement and resettlement in the context of climate change (Goodwin-Gill and McAdam 2017).

The conventional top-down approaches have rightly emphasized the benefits of community participation in resettlement decision-making (UNHCR 2015). However, these approaches have also been criticized for being implemented in a "tokenistic" manner (Satiroglu and Choi 2015). Therefore, Arnall et al. (2019) propose a "claims making" model "from below," which encourages communities to take the initiative to achieve specific goals regarding climate displacement and resettlement. For instance, some communities may prefer to relocate entire groups of people to a new area and receive financial support to achieve this. Others may focus on avoiding forced resettlement. Alternatively, if resettling the entire community is impractical, they may seek other forms of compensation, such as individual relocation payments (Arnall et al. 2019).

Resettling communities is a complex process that requires meticulous planning to achieve positive outcomes (Ferris and Weerasinghe 2020). Although, most studies highlight the failure of resettlement projects. Vanclay (2017) claims that resettlement can become a development opportunity if project developers need to shift their focus away from simply minimizing the immediate cost of resettlement for the project. Instead, they should embrace a mindset that prioritizes the development of awareness and commitment to creating value for both the project and the community by minimizing the negative impacts on the affected community and ultimately enhancing outcomes for all stakeholders involved. However, Ferris and Bower (2023) argue that there is no clear understanding of what success looks like or how to measure its outcomes. Thus, (Ferris and Bower 2023) propose a comprehensive approach considering four dimensions that can provide a clearer understanding of successful outcomes: (1) understanding outcomes from affected populations' perceptions and host-communities and supporting actors, (2) comparison of pre and post-resettlement conditions, (3) short and long-term assessments, and (4) domains related to risk reduction, funding, and livelihood recovery and evaluate whether people are merely surviving or thriving.

2.5 Discussion

The complex process of resettlement profoundly impacts affected communities and challenges the capacities of the stakeholders involved in planning, implementing, and providing aid to displaced people. To minimize risks and enhance the effectiveness of displacement and resettlement programs, it is crucial to consider the following aspects:

2.5.1 Promote Participatory Approaches

Encourage the use of participatory approaches in the planning and implementation of resettlement programs. Engaging displaced populations in active decision-making processes can ensure that their needs and preferences are considered and that resettlement outcomes are sustainable and culturally appropriate. These approaches are highly valuable for the people, the governments, and other stakeholders involved. Although they might require additional resources and extended timeframes, the impacts tend to be positive in the long term (Tsai et al. 2022a, b).

2.5.2 Integrate Human Rights, Equity, and Gender Considerations

Ensure that resettlement programs are grounded in human rights principles and address gender-specific vulnerabilities, also considering minority groups such as Indigenous or Aboriginal people and non-binary people. Integrating human rights and gender considerations into resettlement planning can help protect the rights of displaced populations and promote equitable and inclusive solutions. Considering diversity and inclusion are not limited to considering the particular cultural or identity factors of specific social groups but also considering their capabilities. For instance, Aboriginal Australians are particularly sensitive to climate change, as their living patterns in remote areas prone to draught challenge their livelihoods and wellbeing (Askland et al. 2022). Their vulnerabilities are magnified as Aboriginal people do not consider resettlement as a viable adaptation strategy, as it contradicts their attachment to the land (Nash et al. 2018). Thus, legitimating their voices might untapped potential in the cultural knowledge of the Aboriginal people to deal with climate change planning (Askland et al. 2022).

2.5.3 Invest in Livelihood Restoration

Prioritize livelihood restoration as a key component of resettlement programs. Providing displaced populations with access to training and upskilling, accompanied by microfinance opportunities and support for small enterprises, can help them rebuild their livelihoods and achieve economic independence (Ismail et al. 2018).

2.5.4 Enhance Monitoring and Evaluation

Develop robust monitoring and evaluation frameworks to assess the progress and impact of resettlement programs. Continuous monitoring and evaluation can help identify challenges, measure outcomes, and inform future planning, ensuring that resettlement programs are responsive to the needs of displaced populations. It is important to rethink how the success and failure of resettlement are measured by considering longer timeframes and diverse perspectives from affected people and stakeholders, evaluating if people at least achieved pre-resettlement conditions or improved them, and evaluating risk reduction and avoiding the creation of new risks.

2.5.5 Foster International Cooperation

Enhance national and international cooperation to boost countries' abilities to address displacement crises and execute efficient resettlement programs. This involves leveraging multilateral agreements, funding mechanisms, and joint initiatives to support host nations and adequately resource resettlement efforts (Ferris and Bower 2023). Additionally, reinforce international frameworks to offer clear guidelines and strong protections for displaced populations. Strengthening the implementation of existing frameworks, like the Guiding Principles on Internal Displacement (UNHCR 2004), The Peninsula Principles on Climate Displacement Within States (Displacement Solutions 2013), the Brookings-IOM-UNHCR Toolbox (UNHCR 2016), and the IFRC guide (IFRC 2021), can guarantee the protection of displaced persons' rights and bolster support for resettlement programs.

2.5.6 Address Environmental Sustainability

Ensure that resettlement programs incorporate environmental sustainability considerations (Vanclay 2017). Conducting environmental impact assessments and implementing mitigation measures can help ensure that resettlement sites are environmentally viable and that displaced populations have access to natural resources.

2.6 Conclusion

Displacement and resettlement are increasing in the world due to diverse factors such as disasters, conflicts, climate change, and development projects. Even though the approaches to resettlement and their impacts are differentiated, it is important to explore lessons across various contexts and displacement drivers, also considering multidisciplinary research and integrating the perspectives of policymakers and community members.

Displacement and resettlement represent complex and multifaceted challenges that require coordinated efforts across multiple sectors and governance scales. By understanding the risks linked to displacement and resettlement and implementing strategies combining top-down and bottom-up approaches fostering the engagement of displaced communities in the decision-making. Thus, the risks of failure in resettlement programs can be minimized and promote resilient and equitable resettlement solutions that also focus on long-term impacts.

Multi-stakeholder engagement, including policymakers, practitioners, and communities, is aimed at fostering collaborative efforts to plan context-specific strategies that address the diverse needs of displaced populations and promote sustainable integration in their new settlements and with host communities. Through incorporating principles of human rights, participatory planning, and international cooperation, together with considering resettlement or relocation guidelines the effectiveness of resettlement programs can be enhanced focusing on the long-term well-being of displaced populations.

References

Alam GMM, Alam K, Mushtaq S, Sarker MNI, Hossain M (2020) Hazards, food insecurity and human displacement in rural riverine Bangladesh: implications for policy. Int J Disaster Risk Reduct 43:101364. https://doi.org/10.1016/j.ijdrr.2019.101364

Alam MA, Alam M (2022) Disasters, climate change, and migration concerns in Fiji and the pacific: an overview. In: Siddiqui AR, Sahay A (eds) Climate Change, Disaster and Adaptations: Contextualising Human Responses to Ecological Change. Springer International Publishing, pp 35–46. https://doi.org/10.1007/978-3-030-91010-5_4

Arnall A, Hilson C, McKinnon C (2019) Climate displacement and resettlement: the importance of claims-making from below. Clim Policy 19(6):665–671. https://doi.org/10.1080/14693062.2019.1570065

Askland HH, Shannon B, Chiong R, Lockart N, Maguire A, Rich J, Groizard J (2022) Beyond migration: a critical review of climate change induced displacement. Environ Sociol 8(3):267–278. https://doi.org/10.1080/23251042.2022.2042888

Barnett-Page E, Thomas J (2009) Methods for the synthesis of qualitative research: a critical review. BMC Med Res Methodol 9(1):1–11

Bellizzi S, Popescu C, Panu Napodano CM, Fiamma M, Cegolon L (2023) Global health, climate change and migration: the need for recognition of climate refugees. J Glob Health 13:03011. https://doi.org/10.7189/jogh.13.03011

Bohnet H, Cottier F, Hug S (2021) Conflict versus disaster-induced displacement: similar or distinct implications for security? Civil Wars 23(4):493–519. https://doi.org/10.1080/13698249.2021.1963586

Cacioppo JT, Hawkley LC, Thisted RA (2010) Perceived social isolation makes me sad: 5-year cross-lagged analyses of loneliness and depressive symptomatology in the Chicago health, aging, and social relations study. Psychol Aging 25(2):453–463. https://doi.org/10.1037/a0017216

Cardoso Ermel AP, Lacerda DP, Morandi MIW, Gauss L (2021) Literature reviews: modern methods for investigating scientific and technological knowledge. Springer Nature. https://doi.org/10.1007/978-3-030-75722-9

Carrasco S, Dangol N (2019) Citizen-government negotiation: cases of in riverside informal settlements at flood risk. Int J Disaster Risk Reduct 38:101195. https://doi.org/10.1016/j.ijdrr.2019.101195

Carrasco S, Dangol N, Faleh M (2023) Rethinking social networks in responding to COVID-19: the case of African migrants in Melbourne's public housing. Int J Disaster Risk Reduct 98:104073. https://doi.org/10.1016/j.ijdrr.2023.104073

Carrasco S, O'Brien D (2021) Beyond the freedom to build: long-term outcomes of elemental's incremental housing in Quinta Monroy. Urbe. Revista Brasileira de Gestão Urbana 13. https://doi.org/10.1590/2175-3369.013.e20200001

Carrasco S, O'Brien D (2018) The role of humanitarian agencies in reconstruction and development of disaster affected communities in Japan and the Philippines. Procedia Eng

Carrasco S, O'Brien D (2019) Urbanism of emergency: use and adaptation of public open spaces in disaster-induced resettlement sites. In: Resettlement Challenges for Displaced Populations and Refugees. Springer, pp 163–174

Carrasco S, Ochiai C, Okazaki K (2016a) Disaster induced resettlement: multi-stakeholder interactions and decision making following tropical storm Washi in Cagayan de Oro, Philippines. Procedia - Soc Behav Sci 218:35–49. https://doi.org/10.1016/j.sbspro.2016.04.008

Carrasco S, Ochiai C, Okazaki K (2016b) Impacts of resident-initiated housing modifications in resettlement sites in Cagayan de Oro, Philippines. Int J Disaster Risk Reduct 17:100–113. https://doi.org/10.1016/j.ijdrr.2016.05.001

Carrasco S, Ochiai C, Okazaki K (2016c) A study on housing modifications in resettlement sites in Cagayan de Oro, Philippines. J Asian Arch Build Eng 15(1):25–32. https://doi.org/10.3130/jaabe.15.25

Cernea M (1997) The risks and reconstruction model for resettling displaced populations. World Dev 25(10):1569–1587. https://doi.org/10.1016/S0305-750X(97)00054-5

Charles J, Iglesias JA (2020) Ten Years after Haiti's Earthquake: A Decade of Aftershocks and Unkept Promises. The Miami Herald. https://www.miamiherald.com/news/nation-world/world/americas/haiti/article238836103.html

Dahl N (2020) Governing through kodokushi. Japan's lonely deaths and their impact on community self-government. Contemp Jpn 32(1):83–102. https://doi.org/10.1080/18692729.2019.1680512

Displacement solutions (2013) The Peninsula Principles on Climate Displacement within States. Displacement Solutions, Geneva, Switzerland

Ferrari R (2015) Writing narrative style literature reviews. Med Writ 24(4):230–235. https://doi.org/10.1179/2047480615Z.000000000329
Ferris E, Bower E (2023) Planned Relocations: What We Know, Don't Know, and Need to Learn. Internal Displacement Research Programme (IDRP), London, UK
Ferris E, Weerasinghe S (2020) Promoting human security: planned relocation as a protection tool in a time of climate change. J Migr Hum Secur 8(2):134–149. https://doi.org/10.1177/2331502420909305
Gaillard J C (2015) Post-disaster resettlement. In: People's Response to Disasters in the Philippines: Vulnerability, Capacities, and Resilience. Palgrave Macmillan US, pp 143–155. https://doi.org/10.1057/9781137484291_10
Goodwin-Gill GS, McAdam J (2017) UNHCR and climate change, disasters and displacement. The United Nations Refugee Agency (UNHCR), Geneva, Switzerland
IFRC (2021) Planned Relocation in the Context of Disasters and Climate Change: A guide for Asia Pacific National Societies. International Federation of Red Cross and Red Crescent Societies, Geneva, Switzerland
Inderfurth KF, Fabrycky D, Cohen S (2005) The 2004 Indian Ocean tsunami: one year report. Washington, D.C. https://www2.gwu.edu/~sigur/assets/docs/scap/SCAP25-Tsunami2.pdf
IOM (2008) Migration and Climate Change. International Organization for Migration, Geneva, Switzerland
IOM (2024) Forced migration or displacement. Global Migration Data Portal, International Organization for Migration. https://www.migrationdataportal.org/themes/forced-migration-or-displacement#:~:text=In%202023%2C%20there%20were%2046.9,and%20violence%20(ibid). Accessed 21 May 2024
IPCC (2014) Summary for policymakers. In: Field CB, Barros VR, Dokken DJ, Mach KJ, Mastrandrea MD, Bilir TE, Chatterjee M, Ebi KL, Estrada YO, Genova RC, Girma B, Kissel ES, Levy AN, MacCracken S, Mastrandrea PR, White LL (eds) Climate Change 2014: Impacts, Adaptation, and Vulnerability. Part A: Global and Sectoral Aspects. Contribution of Working Group II to the Fifth Assessment Report of the Intergovernmental Panel on Climate Change. Cambridge University Press, Cambridge, pp 1–32. https://www.ipcc.ch/site/assets/uploads/2018/02/ar5_wgII_spm_en.pdf
Ismail N, Okazaki K, Ochiai C, Fernandez G (2018) Livelihood changes in Banda Aceh, Indonesia after the 2004 Indian Ocean Tsunami. Int J Disaster Risk Reduct 28:439–449. https://doi.org/10.1016/j.ijdrr.2017.09.003
Jha AK, Barenstein JD, Phelps P, Pittet D, Sena S (2010) Safer homes, stronger communities: a handbook for reconstructing after natural disasters. World Bank Publications
Mathur H (1997) Loss of access to basic public services. Impoverishment risks in resettlement. Sage, New Delhi
Mkhize G, Mambondiani L (2023) LGBTI+ persons concealed displacement in Zimbabwe. Agenda 37(4):87–104. https://doi.org/10.1080/10130950.2023.2336341
Nash D, Memmott P, Reser J, Suliman S (2018) We're the same as the Inuit!: exploring Australian Aboriginal perceptions of climate change in a multidisciplinary mixed methods study. Energy Res Soc Sci 45:107–119. https://doi.org/10.1016/j.erss.2018.06.027
Neef K, Jones E, Marlowe J (2023) The conflict, climate change, and displacement nexus revisited: the protracted Rohingya refugee crisis in Bangladesh. J Peacebuilding Dev 18(3):231–247. https://doi.org/10.1177/15423166231190040
Offner S, Marlowe J (2021) Reconceptualising climate-induced displacement in the context of terminological uncertainty. Environ Hazards 20(5):477–492. https://doi.org/10.1080/17477891.2020.1867492
Oliver-Smith A (1991) Successes and failures in post-disaster resettlement. Disasters 15(1):12–23. https://doi.org/10.1111/j.1467-7717.1991.tb00423.x
Oliver-Smith A (2010) Defying displacement: grassroots resistance and the critique of development. University of Texas Press. https://doi.org/10.7560/717633

Raynor K, Panza L, Ordóñez C, Adamovic M, Wheeler MA (2020) Does social mix reduce stigma in public housing? A comparative analysis of two housing estates in Melbourne. Cities 96:102458. https://doi.org/10.1016/j.cities.2019.102458
Satiroglu I, Choi N (2015) Development-induced displacement and resettlement. Routledge, Taylor and Francis Group
Thomas D, Frankenberg E (2014) Boxing day tsunami: the resilience and recovery that followed. The Conversation. https://theconversation.com/boxing-day-tsunami-the-resilience-and-recovery-that-followed-35803
Tsai SL, Ochiai C, Deng CZ, Tseng MH (2022a) A sustainable post-disaster housing development framework for an indigenous Hao-Cha community in Taiwan: considering culture and livelihood in housing extensions. Int J Disaster Resil Built Environ 13(5):583–600. https://doi.org/10.1108/IJDRBE-02-2021-0019
Tsai SL, Ochiai C, Tseng MH, Deng CZ (2022) A framework for NGO-led post-disaster reconstruction programs in the indigenous community: a case study of the 2009 Typhoon Morakot in Taiwan. Int J Disaster Risk Reduct 73:102879. https://doi.org/10.1016/j.ijdrr.2022.102879
Türk V (2013) Ensuring protection to LGBTI persons of concern. Int J Refug Law 25(1):120–129. https://doi.org/10.1093/ijrl/eet005
UNDRR (n.d.) Disaster risk. https://www.undrr.org/terminology/disaster-risk
UNHCR (2004) Guiding Principles on Internal Displacement. United Nations High Commissioner for Refugees, Geneva, Switzerland
UNHCR (2015) Guidance on Protecting People from Disasters and Environmental Change through Planned Relocation. United Nations High Commissioner for Refugees—UNHCR, Geneva
UNHCR (2016) A toolbox: Planning Relocations to Protect People from Disasters and Environmental Change. United Nations High Commissioner for Refugees, International Organization for Migration, Geneva Switzerland
UNOCHA (2023) Global Humanitarian Overview 2024. United Nations Office for the Coordination of Humanitarian Affairs, New York, USA
Vanclay F (2017) Project-induced displacement and resettlement: from impoverishment risks to an opportunity for development? Impact Assess Proj Apprais 35(1):3–21. https://doi.org/10.1080/14615517.2017.1278671
Vince G (2022) The century of climate migration: why we need to plan for the great upheaval. The Guardian. https://www.theguardian.com/news/2022/aug/18/century-climate-crisis-migration-why-we-need-plan-great-upheaval
Zhang M, Qiao S, Yeh AG-O (2021) Blemish of place: Territorial stigmatization and the depreciation of displaced villagers resettlement houses in Chengdu, China. Cities 117:103330. https://doi.org/10.1016/j.cities.2021.103330

Part I
Architectural and Technical Components of the Displacement and Reconstruction

Chapter 3
Post-Disaster Reconstruction in the Indigenous Context: Investigation of Housing Modification in Taiwan

Tsai Sung Lun

Abstract Given the lack of permanent housing extension research on the post-disaster indigenous context, by using the case study of the Haocha community—a relocated community after the 2009 Typhoon Morakot in Taiwan, the research aims to 1. Understand the permanent housing extension in the indigenous communities from the macro and micro scale. 2. Understand the housing extension from the perspective of social and built environment. First, the chapter confirms that indigenous groups tend to extend their permanent house more than the Chinese groups due to their unique livelihood and cultural requirements, while the Rinari settlement—one of the reconstructed settlements—had most of the extension area. Second, by studying the 28 Haocha households—which relocated to the Rinari settlement after Typhoon Morakot—five different extension patterns are confirmed with detailed architectural drawings. Third, three case studies are shown to illustrate that housing extension behavior can be motivated by the built-environmental factor, pre-disaster livelihood factor, post-disaster livelihood factor, and cultural factor. Lastly, some suggestions from the preparedness stage, design and planning, and post-construction are given.

3.1 Introduction

Most studies on post-disaster reconstruction (PDR) have focused on short-term disaster recovery, given that immediate rehabilitation and reconstruction receive immediate attention from authorities. However, without long-term consideration regarding the disaster risk reduction policy, PDR countermeasures are doomed to fail. Therefore, this chapter aims to understand the change in the post-disaster living environment (permanent housing modification) of affected people from a long-term perspective. Moreover, as mentioned in previous chapters, built environmental and socioeconomic factors play an essential role in post-disaster reconstruction and recovery. Hence, this chapter also explores the modification of housing motivation

---

T. S. Lun (✉)
Deloitte Tohmatsu Venture Support Co., Ltd., Tokyo, Japan
e-mail: tsai.sung.lun@fuji.waseda.jp

C. Ochiai et al. (eds.), *Disaster and Displacement*, Disaster Risk Reduction,
https://doi.org/10.1007/978-981-96-2583-3_3

from the perspective of these factors, which have not been understood thus far (Tsai et al. 2021). Eventually, this chapter provides future direction and suggestions for NGOs, communities, and the government, given that the modification of permanent housing by affected communities has become a ubiquitous phenomenon reported in different PDR settlements.

3.1.1 From Short-Term to Long-Term Recovery

The history of PDR can be traced back to the 1970s. For instance, the Peru and Turkey governments implemented large-scale PDR projects in the aftermath of earthquakes (Schilderman 2010). As concurrence issues occurred between the government and affected people during this PDR implementation, the discussion and research of PDR gradually shifted toward sustainable development under the long-term recovery perspective. For instance, Dias et al. (2015) discussed the indicators affecting the long-term satisfaction of relocated communities after the 2004 Indian Ocean earthquake and tsunami. The study stated that from a long-term perspective, a sustainable resettlement program does not merely include the reconstruction of housing but also entails re-establishing people's socioeconomic and cultural lives. Based on this study, Ali et al. (2020) studied the 2010 floods in Sindh, Pakistan; from a retrospective viewpoint, they identified that issues regarding community-level involvement, local administration, community capacity, disaster governance, different stakeholders and coordination, and information and knowledge management could hamper the transition from short-term to long-term PDR.

3.1.2 Built Environmental Factors in PDR Projects

According to the discussion on PDR's history, it can be deduced that one of the primary objectives of PDR is to cope with housing development and reconstruction—related to the built environmental issues. Moreover, under the built environment category, PDR can be further discussed in terms of settlement planning and housing levels. First, regarding the settlement planning level, it is inevitable to discuss the concept of relocation, given that disaster-affected communities may have to relocate after the disaster. As Jha et al. (2010) stated, relocation is "a process whereby a community's housing, assets, and public infrastructure are rebuilt in another location." Regarding the housing level, Quarantelli (1982) first proposed that the PDR process should include: 1. Emergency shelters, 2. Temporary shelters, 3. Temporary housing, 4. Permanent housing. Shao et al. (2018) explained that these four housing stages had different roles and characteristics in post-disaster reconstruction and recovery. Many studies have also provided post-disaster housing guidelines. For instance, Tucker et al. (2014) developed design guidelines for constructing sustainable permanent housing by analyzing the PDR project implemented after the 2004 Indian Ocean tsunami in Sri Lanka.

3.1.3 Socioeconomic Factors in PDR Projects

As stated by Dias et al. (2015), socioeconomic factors are also essential factors that significantly influence the outcomes of PDR projects. From the socioeconomic perspective, the factor of vulnerability has been mentioned frequently by the related studies (Masterson et al. 2014). On the other hand, many studies have focused on the impact on post-disaster livelihood. Ahmed (2011) advocated that "livelihood deserves priority in recovery, as it enables disaster victims to become self-sufficient rather than relying on continued relief aid." Therefore, Speranza et al. (2014) suggested that establishing livelihood stability requires incorporating resilience concepts into basic livelihood approaches. Recently, the concept of resilient livelihood has been widely discussed (Speranza et al. 2014).

Some researchers also found that different ethnic groups have different post-disaster recovery trajectories and vulnerabilities (Airriess et al. 2008; Monteil et al. 2020). Although the definition of indigeneity remains ambiguous—it is certain that indigeneity retains a distinct culture and identity apart from the globalized culture (United Nations General Assembly 2007), it is widely believed that indigenous peoples especially value their human land, human nature, and human relationships (Lin 2019).

3.1.4 Indigenous Groups in Taiwan

Currently, 16 ethnic groups are recognized by the Taiwanese government, including Amis, Atayal, Paiwan, Bunun, Puyuma, Rukai, Tsou, Saisiyat, Yami, Kavalan, Truku, Sakizaya, Sediq, Hla'alua, and Kanakanavu. They are concentrated in the mountainous area. Every group has its distinct culture and language. Similar to the global context, these groups had experienced several natural disasters and were subject to reconstruction and relocation. For instance, Hsu et al. (2015) argued that the top-down institutional PDR process could enhance the "procedural vulnerability" of the indigenous groups in Taiwan. Lin and Lin (2016) highlighted the misunderstanding and miscommunication between the indigenous people and the government during the 2009 Typhoon Morakot PDR project. Taiban et al. (2020) underscored the post-disaster livelihood development of indigenous groups, in which indigenous households were unable to develop their post-disaster livelihood due to the lack of arable land and coordination and support from the government.

3.1.5 Housing Extension

The relationship between the built environmental and socioeconomic factors has been discussed in the housing development process. For instance, Rapoport (1999) stated that the housing environment should reflect human wants and needs. The housing can embody the culture of its inhabitants, reflected in its size, construction material, location, and type. A mismatch between the housing design, layout, and size with the household's culture and livelihood can lead to housing alteration and modifications.

Recently, the housing extension issue under the PDR project context has been widely discussed. For instance, after the 2004 Indian Ocean tsunami in Hambantota New Town, Sri Lanka, permanent housing was extended after the disaster (Ahmed and McEvoy 2014). Carrasco et al. (2016) focused on the PDR project implemented after the 2011 Typhoon Washi, revealing that resident-initiated housing extension also occurred in this case. However, in this instance, the housing extension spurred a dispute between the government and residents due to safety concerns regarding the extended housing.

3.2 Research Question and Methodology

Thus, based on the previously mentioned literature, it can be considered that studying the permanent housing extension can provide a lens for understanding the intertwined relationship between the built environmental and the socioeconomic perspective in the post-disaster context. This chapter focuses on the PDR of indigenous groups, given their unique characteristics and vulnerabilities. Therefore, this chapter aims to accomplish the following:

1. To understand the permanent housing extension undertaken by the indigenous communities at both macro and micro scales;
2. To understand the housing extension from the perspective of socioeconomic and built environmental factors.

Several research methods were adopted to explore the housing extension issues. The methods included drone survey in 29 permanent housing settlements and semi-structured interviews with 28 households in the Rinari settlement. Moreover, the detailed housing measurement was applied to those interviewed households' housing units.

3.3 The Aftermath of Typhoon Morakot

Typhoon Morakot hit Taiwan from August 6–August 8, 2009. During that time, the precipitation level broke the previous record in Taiwan and approached the highest rainfall record in the world (Chen 2009). Due to the topography and structure of the typhoon, the stark and continuous rainfall caused severe flooding and landslides in the south-central Taiwan mountainous regions, causing 699 deaths and many missing people; a total of 1,766 households suffered from housing damage. Given that most of the rainfall was concentrated in mountainous areas, more than 73% of the disaster victims were indigenous peoples (Hsieh et al. 2012). After the disaster, the central government swiftly launched the contingency relief countermeasure, in which the local government and NGOs provided emergency shelter for the affected people. Subsequently, due to budgeting concerns and the construction period, the construction of permanent housing was handed over to the three major NGOs in Taiwan—Tzu Chi, Red Cross, and World Vision. Consequently, the PDR project implemented after Typhoon Morakot became an NGO-led project.

3.4 Overall Housing Extension: Settlement Level (Macro Perspective)

The author conducted a drone survey in 29 permanent housing settlements and recorded the housing extension area based on indigenous permanent housing settlements, indigenous Chinese mixed permanent housing settlements, and Chinese permanent housing settlements. According to Tables 3.1, 3.2 and 3.3[1], it is evident that ten years after the disaster, indigenous settlements had the most significant average extension area per household in the indigenous permanent housing settlements, of which the Rinari settlement—a permanent housing settlement that belongs to this category—had the most average extension area per household (47.6 m^2). In the indigenous-Chinese-mixed settlements, the extension area per household ranged from 2.7–15.2 m^2 (7.9 m^2 on average). The average extension area per household in the Chinese settlements was the smallest among the three groups (9.58 m^2 on average). Thus, it is clear that indigenous groups tended to modify their permanent house the most among the observed groups.

3.5 Housing Extension Implemented by the Haocha People (Micro-Perspective)

As mentioned before, given that the most housing extension area per household among the surveyed settlements was in the Rinari settlement, this section focuses on the relationship of extension behavior with the built environmental and socioeconomic issues. Therefore, this section first focuses on the pre-disaster livelihood needs. The Rinari settlement accommodated three disaster-affected communities, including 177 households from Haocha village in Wutai Township (Rukai), 174 households from Dashe village in Sandiman Township (Paiwan), and 132 households from Majia

[1] Data were not collected for some settlements due to the local regulation of drone usage.

Table 3.1 The extension area in the indigenous permanent housing settlements

Jurisdiction		Pingtung				
Name of the settlement		Rinari	Ulalijuc	Changzhi Baihe	Xinlaiyi	Zhongjuanlu
Extension area (m^2)	Front yard	16055.7	1587.6	1897.8	2104.9	346.8
	Backyard	6954.1	2643.2	3331.1	4902.1	128.1
	2nd to 3rd floors	0.0	0.0	0.0	0.0	0.0
	Total	23009.8	4230.8	5229.0	7007.0	474.9
	Average per household	47.6	26.8	19.4	24.3	10.6
Jurisdiction		Kaohsiung		Chiayi		
Name of the settlement		Baoshan	Lele	Shanmei	Zhulu	Laiji
Extension area (m^2)	Front yard	4.5	194.8	69.0	970.1	91.7
	Backyard	0.0	5.4	306.6	513.4	23.4
	2nd to 3rd floors	0.0	0.0	0.0	0.0	0.0
	Total	4.5	200.2	375.6	1483.6	115.1
	Average per household	0.3	10.0	13.4	9.9	2.7
Jurisdiction		Chiayi	Taitung			
Name of the settlement		Leye	Daniao	Dazhu	Dawu	Jialan
Extension area (m^2)	Front yard	270.3	153.0	375.2	33.3	664.3
	Backyard	239.0	33.5	451.8	23.1	1107.7
	2nd to 3rd floors	0.0	0.0	0.0	0.0	0.0
	Total	509.2	186.5	827.0	56.5	1772.1
	Average per household	11.1	13.3	27.6	1.8	36.9

Table 3.2 The extension area in the indigenous-Chinese-mixed permanent housing settlement

Jurisdiction		Kaohsiung			Nantou	
Name of the settlement		Daai	Wulipu	Riguangxiaolin	Shenmu	Mingjuanxian
Extension area (m^2)	Front yard	2046.2	477.9	876.4	0.0	19.1
	Backyard	5305.1	397.4	120.1	316.1	111.6
	2nd to 3rd floors	0.0	497.0	150.2	0.0	0.0
	Total	7351.3	1372.2	1146.7	316.1	130.7
	Average per household	7.3	15.2	9.6	2.7	4.7

Table 3.3 The extension area in the Chinese permanent housing settlements

Jurisdiction		Pingtung	Chiayi			Nantou
Name of the settlement		Xinfeng	Rian	Rihao	Riman	Changliyuan
Extension area (m^2)	Front yard	0.0	197.8	28.3	19.7	0.0
	Backyard	134.3	700.7	126.7	165.3	0.0
	2nd to 3rd floors	0.0	0.0	0.0	23.5	99.1
	Total	134.3	898.5	155.0	208.4	99.1
	Average per household	16.8	5.0	6.0	13.4	5.5

Jurisdiction		Nantou	Yunlin	Tainan	
Name of the settlement		Honglixincun	Dongxing	Yantianli	Yujing
Extension area (m^2)	Front yard	25.6	0.0	153.5	125.2
	Backyard	171.6	61.1	243.2	233.4
	2nd to 3rd floors	0.0	0.0	0.0	0.0
	Total	197.2	61.1	396.7	358.6
	Average per household	9.9	2.2	13.7	13.8

village in Majia Township (Paiwan; Chuan (2018)). This chapter's research focus is the Haocha people, given that they account for the most significant number of households in the settlement as well as have a more extensive pre-disaster livelihood record than other communities. The pre-disaster livelihood information is helpful for understanding the reason and the motivation for housing modification.

3.5.1 Pre-Disaster Livelihood of the Haocha People

In this section, the pre-disaster livelihood of the Haocha people will be introduced in two phases—the Old Haocha era and the New Haocha era.

3.5.1.1 Old Haocha Era

In the Old Haocha settlement, people typically lived in slate houses, given that slate was easily obtainable from the nearby mountains and is characterized by robustness against natural disasters such as typhoons and earthquakes. The slate housing was usually built under the supervision of the elderly in the community. Typically, the width of the slate wall was 25 cm, with a room layout of 8 m by 8 m.

The Haocha people's slate houses can be divided into the main space, subsidiary space, and front yard (Fig. 1a, b and c). In the Old Haocha era, the family's main activities, including dining and social activities, took place in the main space. The subsidiary space was located behind the main space, being used as agricultural storage, pigpen, toilet, and sleeping areas. People in this era practiced slash-and-burn agriculture. Since they were isolated from the market system, they subsisted on the reaped agricultural products. They practiced alternate or mixed planting of crops such as millet, sweet potatoes, taro, and peanuts. The traditional Haocha community was a hierarchical society. The chieftain led the Old Haocha settlement; this title was hereditary, passing to the son. Following the chieftain in the social hierarchy was the noble class—predominantly composed of the chieftain's relatives. The rest of the people belonged to the civilian class.

Handcrafting and weaving were also part of the people's livelihood. Such activities were conducted in the front yard space. For the people in the Old Haocha era, the front yard was not only a working place but a place for residents to socialize with each other and other community members. As shown in Fig. 1b, the front yard space was significantly huge compared to the size of the interior area of the slate house, with a proportion of 1:1. Aligned with the housing design, the pavement of the front yard was decorated with large panels of slate, including slate tables and chairs.

3.5.1.2 New Haocha Era

Given the isolated geographical conditions and the difficulty of resource distribution in mountainous regions, many indigenous tribes were relocated from 1950 onward, aligning with the "modernization policy." Accordingly, the Haocha people were also relocated. The Haocha people were compelled to relocate to the New Haocha settlement in 1977—a place designated by the government and accessible by road. According to the government policy, universally designed housing that was considered "fashionable and civilized" at the time was applied in the settlement. The universally designed housing featured a brick structure with a sloped tiled roof and had a layout of approximately 4 m to 4.5 m by 10 m, which was much smaller than the traditional slate housing (Fig. 2a). After four decades, Lee (2012) found that most households had extended this universally designed housing due to a lack of cultural concern in the original housing and inadequate future space. Figure 2b shows a backyard extension added to the original housing to expand the living area. An annex to the main housing unit served as storage for agricultural products, which was evidence of the agricultural livelihood and an echo of the subsidiary space in the previous slate house. Moreover, the front yard space with the slate table remained in the front yard of the New Haocha housing—an inheritance from the Old Haocha era.

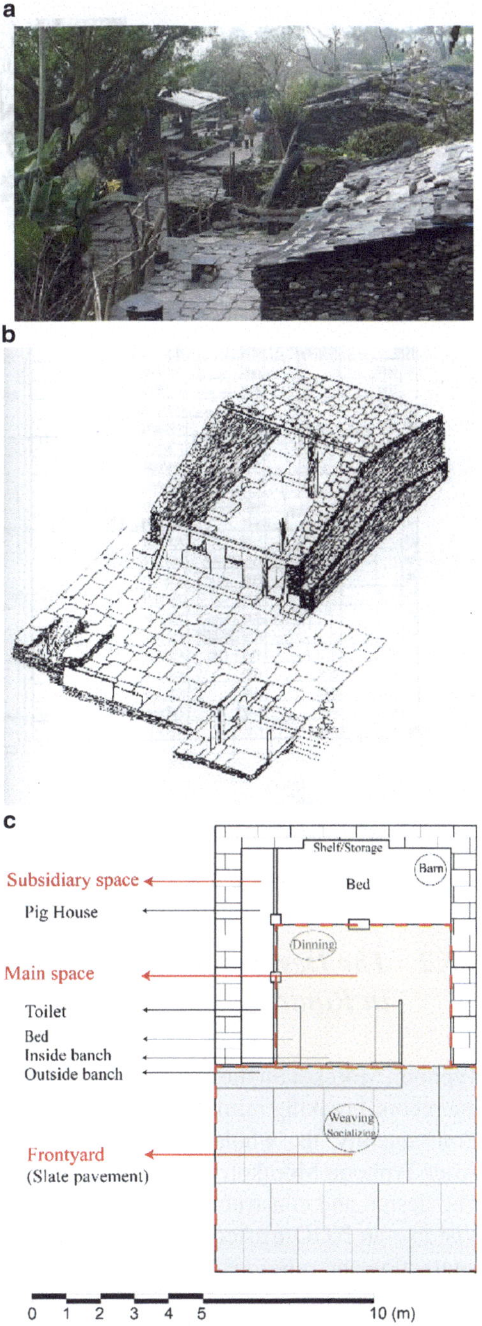

Fig. 3.1 **a** The well-preserved housing in the Old Hao-Cha settlement. *Source* Huang SM. **b** An axonometric illustration of the Old Haocha slate house (Taiban 2016). **c** Traditional house plan in the Old Haocha settlement (Taiban 2016)

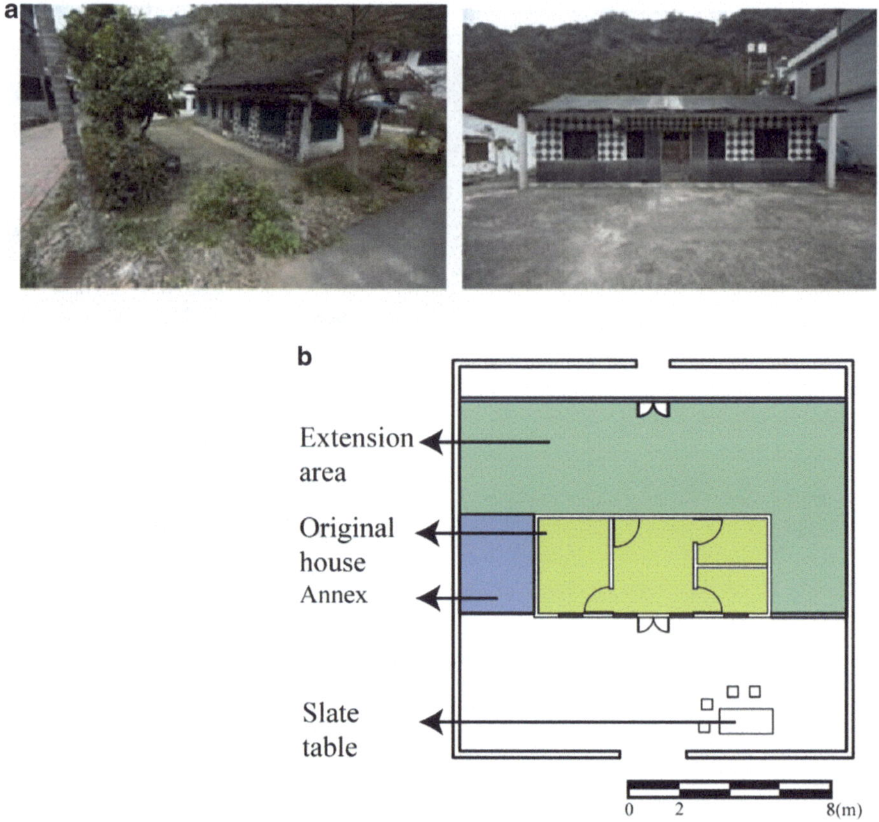

Fig. 3.2 **a** Original housing provided in the New Haocha settlement. **b** Original housing layout and housing extension patterns in the New Haocha settlement. *Source* Taiban S

3.5.2 The Design and Construction of Permanent Housing in Rinari

Typhoon Morakot hit the New Haocha settlement severely on August 8, 2009. Due to the record-breaking rainfall, the river flooded, and a landslide struck the settlement. Consequently, the whole village was buried under mud, rendering it inhabitable. After Typhoon Morakot, Haocha households were relocated to the Rinari settlement. The design and construction of new housing were conducted by the team of World Vision—an NGO trusted by the indigenous communities due to their pre-disaster interaction and religious background. To boost the engagement of local residents in the design and planning of permanent housing, many workshops were held. Initially, the World Vision team and residents attempted to apply slate material to the housing façade to reflect the community's cultural identities; however, it was difficult to obtain slate in this area. Eventually, the residents accepted using light steel for the structure and wooden planks for the housing façades.

The permanent houses were designed in two styles: single and duplex. The number of single and duplex units was determined via discussion between the community and the World Vision team. A unified layout of 105.6 m^2 or 7.32 m x 7.32 m with two floors was allocated to each eligible household, regardless of the number of members in the household. Each housing unit included the following: a kitchen, living room, bathroom, and bedroom on the first floor. The second floor had a bathroom and three bedrooms. The layout of the single and duplex types of permanent housing provided by the World Vision team is shown in Fig. 3a.

Though the residents had been consulted about the permanent housing design, the design was still distinct from the Old Haocha slate housing. Given the design mismatch, after the completion of the permanent housing construction, the government dispensed a 3,594.5 USD[2] subsidy, titled "indigenous housing façade renovation budget," to indigenous households to renovate their housing façades further (Fig. 3.3b). However, the budget was limited to renovating and redesigning the housing façades, which could not sustain the large-scale layout modification and extension intended by the residents.

3.5.3 *Attributes of Basic Housing Extension Implemented by the Haocha Community*

This section analyzes the fundamental housing extension implemented by Haocha households based on the author's observation from 2017 to 2021—approximately ten years after the disaster. A total of 28 households were selected by considering their occupation and family structure. These 28 households were further categorized into five extension types: 1. front yard extension; 2. front and back yard extension; 3. front yard, left side, right side, and back yard extension; 4. front yard, left side, right side, back yard extension, and inside renovation; 5. multiple floor extension (Fig. 3.4).

Figure 3.5 depicts the plots of the 28 measured households, aligning with the five extension patterns defined above. Additionally, the detailed housing layout, extended area, and extended functions are shown (the extended function is shown in different designated colors). The household information is presented in Table 3.4. The extension patterns are discussed below.

Pattern 1 (Front yard extension)

All of the 28 surveyed households had expanded their front yard space. Though they had moved to the Rinari settlement, the Haocha people continued chatting and socializing with community members in the front yard. Given that the front yard in the original layout was too narrow for daily gatherings, the households decided to

[2] According to the exchange rate on January 28, 2021.

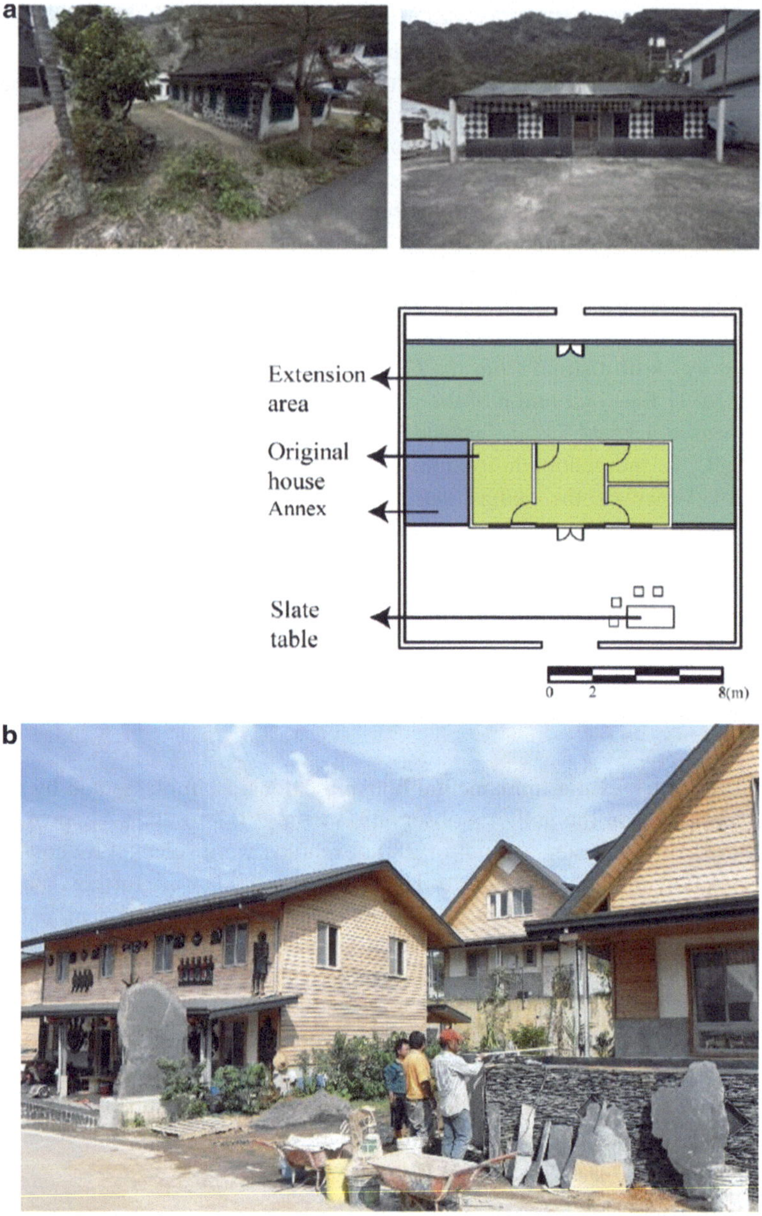

Fig. 3.3 **a** The layouts of single- and duplex-type permanent houses in the Rinari settlement. *Source* Author. **b** Implementation of the indigenous housing façade renovation project (Taiban 2018)

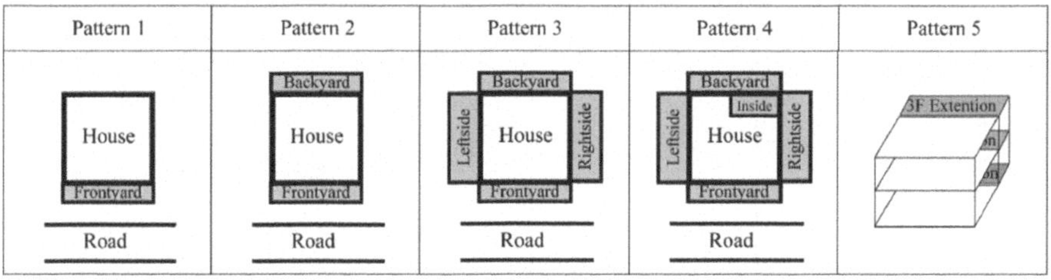

Fig. 3.4 Five patterns of housing extensions. *Source* Author

extend the front yard. Moreover, as mentioned before, since the "indigenous housing façade renovation budget" was allocated to the households for the housing façade renovation, many households also utilized the allocated budget to expand their front yard space.

Pattern 2 (Front and back yard extension)

Households that needed more living space spent the budget allocated to them to renovate their façade to expand the backyard space. According to Fig. 3.11, the backyard extension was usually used as the kitchen and dining area. The residents informed the author that, given the original layout placed the kitchen at the back of the housing unit, it was reasonable to use the backyard extension to expand the inadequate kitchen or dining space. Nine households had extended both their front and back yards.

Pattern 3 (Front yard, left side, right side, and back yard extension)

For households needing more living space, left- and right-side extensions are possible options. Upon surveying the usage of the extensions, it was observed that most households extended the left and right sides of the house, using the extended space as a workspace, studio, garage, and storage.

Pattern 4 (Front yard, left side, right side, back yard extension, and inside renovation)

It is interesting to note that except for the living space expansion, two households further renovated their original kitchen space into a working space, as the new kitchen had already been placed in the backyard extension. Since a new kitchen was added, the original kitchen was renovated to be used as a studio or storage.

Pattern 5 (Multiple floor extension)

Finally, three households with a restaurant, shop, and hostel business extended their housing units with additional floor(s). Two households added a second floor, and one extended to a third floor—the tallest in the community.

It is observed that these five patterns of housing extensions were made gradually and continuously. Since the housing extension in the Rinari settlement is a household-initiated process, the housing expansion was consistent with the households' desire

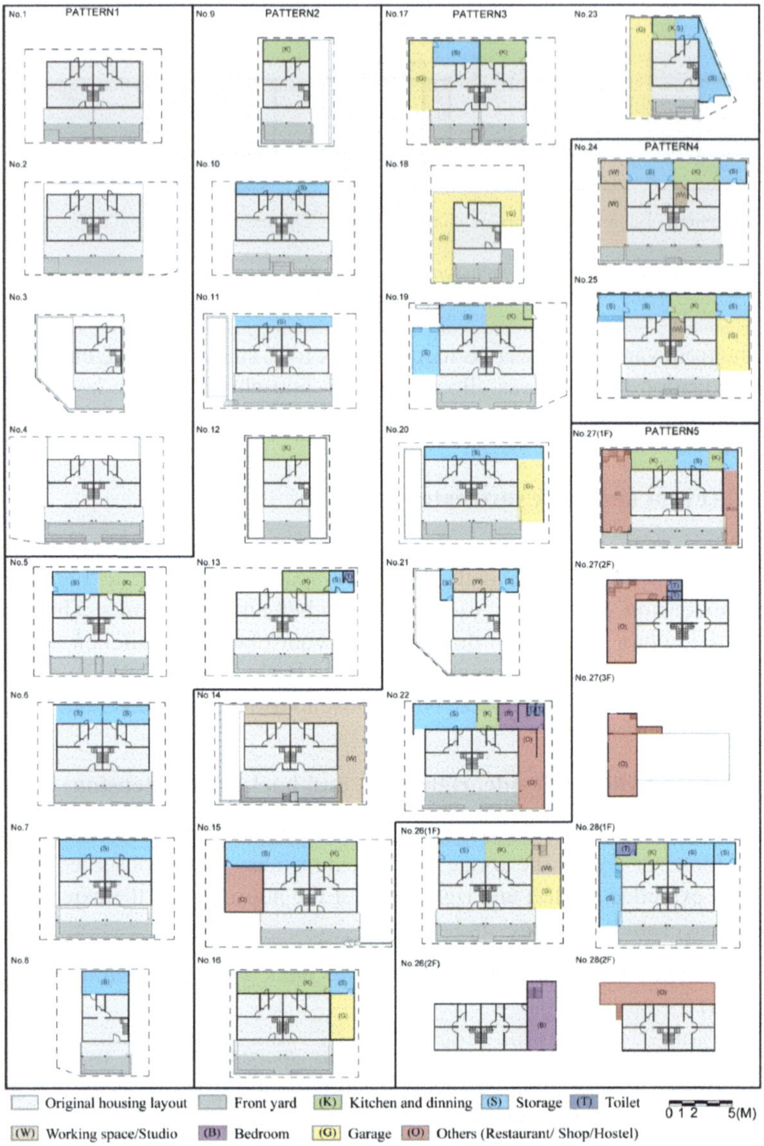

Fig. 3.5 The floor plans of the surveyed 28 Haocha households. *Source* Author

for living space and aligned with their economic capability. Therefore, it is fair to say that the housing extension discrepancy correlates to the household livelihood recovery.

Table 3.4 The housing extension profile of the surveyed 28 Haocha households

Housing code	Gender	Occupation	Extended functions	Number of extended functions	Extended floor area	Extension pattern	Number of family members	Extension area (m^2)
(D)	F	Housekeeper	None	0	1F	1	1	43.73
2(D)	M	Factory worker	None	0	1F	1	4	43.88
3(S)	F	Housekeeper	None	0	1F	1	2	24.52
4(D)	M	Office worker	None	0	1F	1	0	48.92
5(D)	M	Teacher (retired)	KS	2	1F	2	2	98.4
6(D)	M	Priest	S	1	1F	2	N/A	90.84
7(D)	M	Office worker	S	1	1F	2	4	88.82
8(S)	M	Office worker	S	1	1F	2	0	49.76
9(S)	F	Coffee retailer	K	1	1F	2	2	51.35
10(D)	M	Writer	S	1	1F	2	3	70.33
11(D)	M	Military (retired)	S	1	1F	2	3	70.48
12(S)	M	Community worker	K	1	1F	2	2	46.67
13(D)	M	Priest	KST	3	1F	2	2	85.32
14(D)	M	No occupation	W	1	1F	3	2	150.68
15(D)	M	University professor	KSO	3	1F	3	1(Rent)	164.28
16(D)	F	Office worker	KSG	3	1F	3	3	138.63
17(D)	F	Restaurant owner	KSG	3	1F	3	3	137.5
18(S)	M	Restaurant owner	G	1	1F	3	6	107.5
19(D)	M	Government related (retired)	KS	2	1F	3	9	143.12
20(D)	M	Former village leader	SG	2	1F	3	2	122.91

(continued)

Table 3.4 (continued)

Housing code	Gender	Occupation	Extended functions	Number of extended functions	Extended floor area	Extension pattern	Number of family members	Extension area (m^2)
21(S)	M	Restaurant owner	SW	2	1F	3	4	73.92
22(D)	F	Shop owner	KSTBO	5	1F	3	4	174.47
23(S)	M	Retired	KSG	3	1F	3	3	130.48
24(D)	M	Government related	KSW	3	1F	4	2	193.88
25(D)	M	Carpenter	KSWG	4	1F	4	8	183.57
26(D)	F	Teacher (retired)	KSWGB	5	1F, 2F	5	3	198.71
27(D)	M	Association leader	KSTO	4	1F, 2F, 3F	5	4	319.27
28(D)	M	Shop owner	KSTO	4	1F, 2F	5	9	216.29
Note								
Interviewee gender	F: female; M: male.							
Housing type	Housing code (S): single-type housing; (D): duplex-type housing.							
Extended parts	K: kitchen extension and dining room; S: storage; T: toilet; W: workspace; B: bedroom; G: garage; and O: other (business-related).							
Rent	The house was rented to people who were not eligible to live in the permanent house.							
Number of family members	Only members who resided permanently in the housing were tallied as family members; The value of 0 indicates that the households did not permanently reside in the housing.							

Regarding the distribution and number of extended functions, 20 (15) households chose to extend the storage (kitchen). This indicated that storage, kitchen, and dining functions were indispensable requirements for most Haocha households.

3.5.4 The Relationship between Housing Extension and the Residents' Culture and Livelihood

The previous section introduced the overall housing extension of the surveyed 28 Haocha households. Subsequently, in this section, three representative households are extracted for a detailed explanation. In Fig. 6a, the alphabet in black denotes the extension's function, while the alphabet in red corresponds to the numbered photos presented after the housing plan.

Case Study 1: C.Z.'s housing

The house owner had extended his front and back yard with additional structures on the right and left sides of the house. As shown in Fig. 6b, given the lack of space, the house owner had extended the storage area at the back of the house. Additionally, as shown in Fig. 6c, due to the house owner's occupation, a larger private office was needed to facilitate working from home, leading to the extension of a workspace. Furthermore, because many guests visited the house owner frequently, he decided to have a reception area on the right side of the house (Fig. 6d). Although not shown in the photo, it is clear that kitchen space has been added at the back of the house. According to the house owner, traditional foods such as the "chinavu" and "abai"—the traditional Rukai food made from millet bound by shell ginger leaves—are staples during important ceremonies or activities (e.g., weddings, harvest festivals, and formal community gatherings). However, because the chinavu and abai require steaming in steam cases, and as the original kitchen was completely modern without sufficient space for steam cases, the house owner thought it necessary to construct a kitchen in the backyard.

Case Study 2: L. J.'s housing

The house owner had extended his front yard, back yard, and right and left sides of the house; the house owner had also extended the house vertically (Fig. 7a, b and c). Among the surveyed households, this house had the most extended area. First, since the house owner's father owned arable land in the Rinari settlement for growing millet, the backyard was extended as a barn and agricultural equipment storage—a reflection of the ancestors' livelihood (Fig. 7d). Conversely, to boost the post-disaster economy, the house owner had decided to start a restaurant to sell indigenous food. Therefore, a modern kitchen for business purposes was extended (Fig. 7e). In the front yard, to attract tourists as well as demonstrate the community and family identity, the parapet was decorated with slate material. The slate table and chairs can be seen in the front yard—a legacy from the Old Haocha period (Fig. 7f). Moreover, given

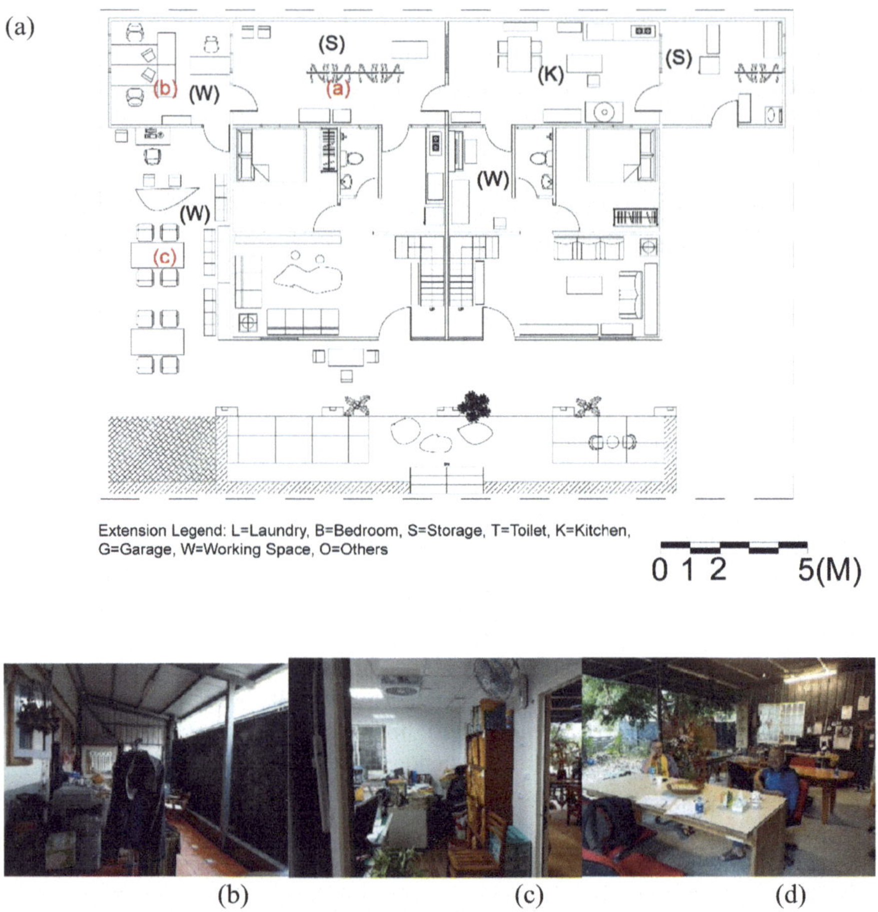

Fig. 3.6 **a** C.Z.'s housing plan. *Source* Author. **b, c, d** Photos of C.Z.'s house. *Source* Author

the limited space of the original layout, the household extended the left side of the house into a three-floor structure. The three-floor structure provided sufficient space for tourists (Fig. 7g, h and i). The house owner explained that the income from the restaurant is vital for the household to sustain its post-disaster livelihood.

Case Study 3: L.M.'s housing

The house owner had extended the front yard, back yard, and left side of the house; the house owner had also extended the house vertically (Fig. 8a and b). Similar to Case Study 1, due to the traditional food culture and lack of cooking space, the backyard had been extended to include a large-sized kitchen, with sufficient shelves and space for the steam case (Fig. 8c and d). On the other hand, it is evident that the kitchen in the original layout is used in a limited manner by the household (Fig. 8e). The house owner stated that the original kitchen was used only for the storage of

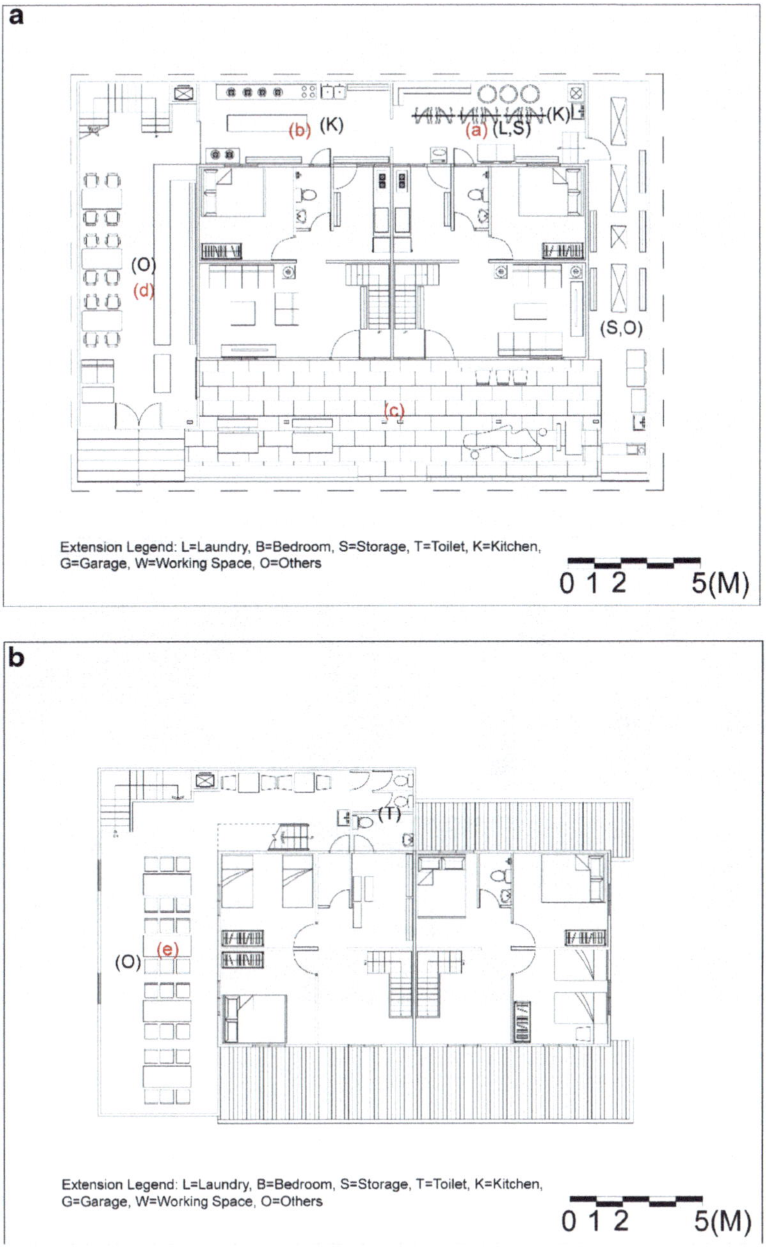

Fig. 3.7 (**a–c**) L.J.'s housing plan. *Source* Author. (**d-i**) Photos of L.J.'s house. *Source* Author

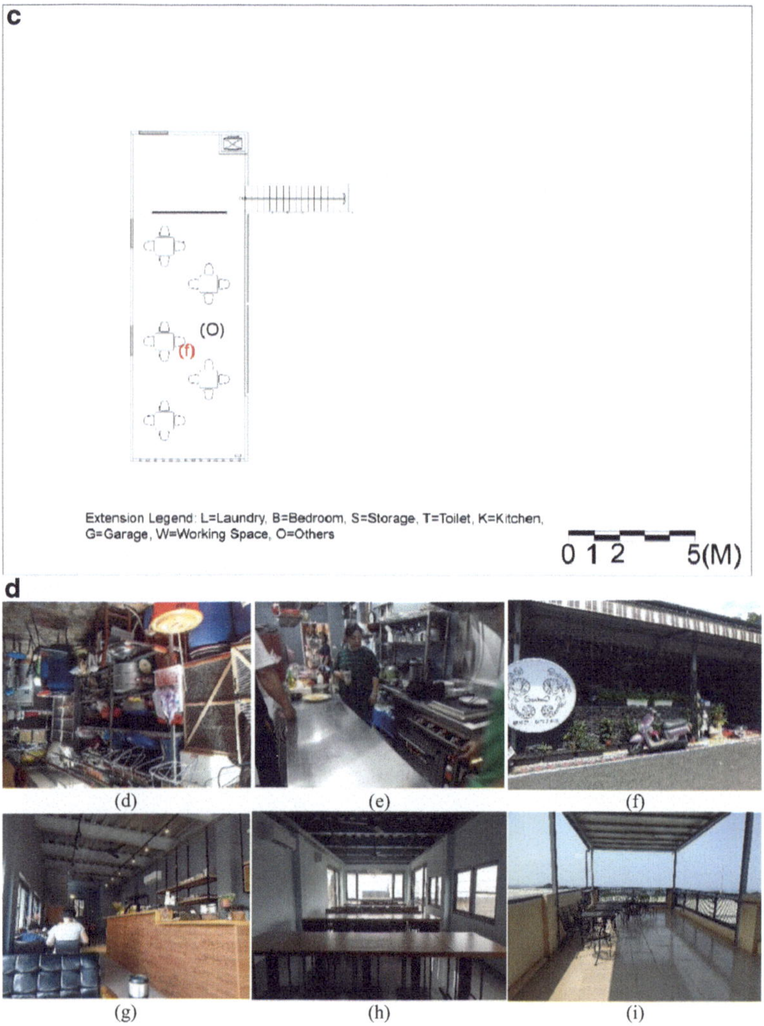

Fig. 3.7 (continued)

tableware and a water cooler. Regarding the income generation, the house owner extended the left side of the house to include an indigenous craft classroom. The classroom served as the base for the local elderly to create handicrafts and sell them to tourists (Fig. 8f). The house owner stated that these activities helped the elderly to socialize and, in this way, were essential for their mental health. On the second floor, to generate extra income, the house owner and her husband lived in one room, allocating another room on this floor as guest accommodation for tourists who want to stay overnight in the Rinari settlement (Fig. 8g). Expecting future expansion of the tourism industry, the household had already extended the second floor to further increase the number of guest rooms (Fig. 8h).

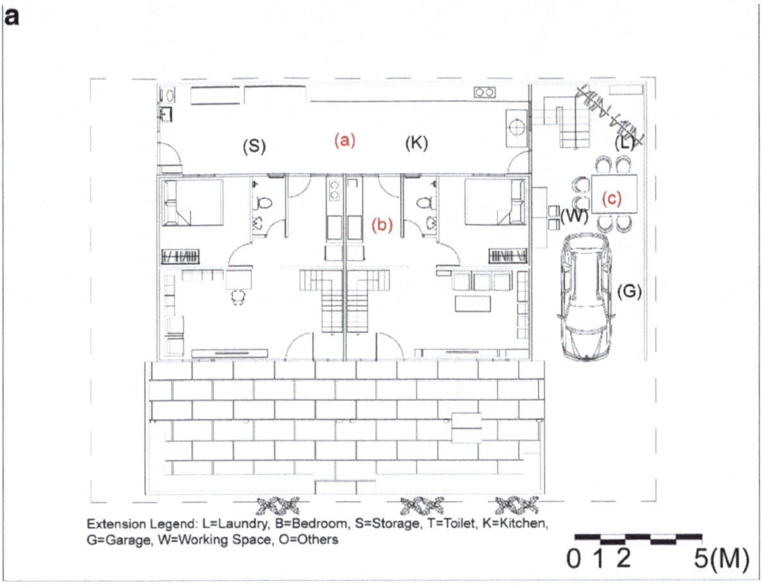

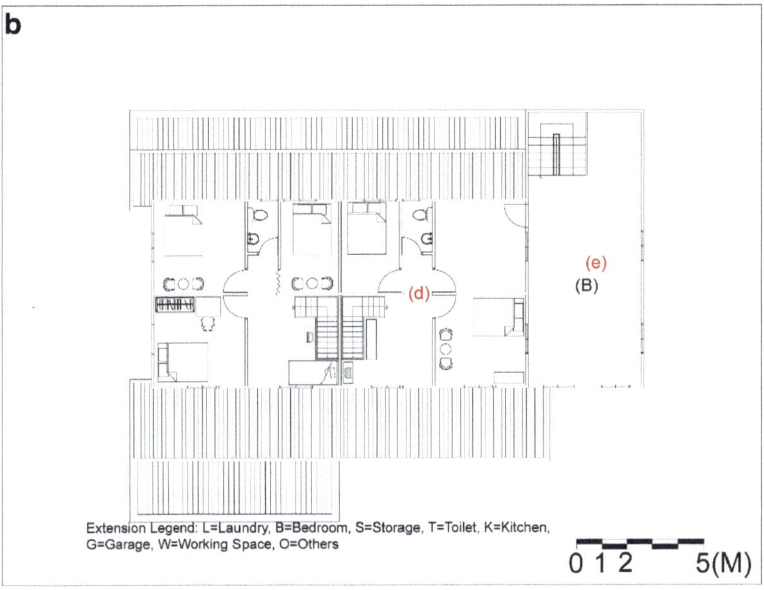

Fig. 3.8 **a** L.M.'s housing plan (First floor). *Source* Author. **b** L.M.'s housing plan (Second floor). *Source* Author. **c–h** Photos of L.M.'s house. *Source* Author

Fig. 3.8 (continued)

By using three case studies, this chapter illustrates the relationship of housing extension to the built environment and households' pre-and post-disaster culture and livelihood, including providing a satisfactory living space (built environmental factor), catering to traditional livelihood (pre-disaster livelihood), expressing one's identity and indigeneity (cultural factor), and catering to post-disaster livelihood (post-disaster livelihood). Table 3.5 summarizes the findings from the perspective of housing extension and the spatial characteristics of permanent housing. Moreover, since the extension behavior was household-initiated, the violation of the building code spurred a dispute between the government and the community, causing residents to feel insecure about their housing.

3.6 Discussion and Conclusion

This chapter discusses and analyzes the post-disaster housing modification issue thoroughly. First, the author reviews the transition of the focus of disaster-related studies from short-term to long-term recovery. Second, two important elements impacting PDR achievement—the built environmental and socioeconomic factors—are introduced. Specifically, from the socioeconomic perspective, this chapter underscored the importance of looking at a PDR project in the indigenous context, as indigenous peoples often suffer more hardship after a disaster compared to other citizens. Lastly, the indigenous context in Taiwan and studies regarding housing extension are introduced.

This chapter makes three scholarly contributions. First, it compares the indigenous permanent housing settlements, indigenous-Chinese-mixed permanent housing settlements, and Chinese permanent housing settlements. The results indicate that

Table 3.5 Multiple perspectives that influenced post-disaster housing spatial characteristics

Perspective	Spatial characteristics
Providing a satisfactory living space	1. Increased housing size to accommodate more household members 2. Diversified housing functions to boost livability
Catering to traditional livelihood	1. Space catered to agriculture activities conducted for self-sufficiency (e.g., space for millet barn) 2. Space for perpetuating traditional food culture (e.g., space for the steam case and an extended kitchen) 3. Extended front yard for gatherings (i.e., socializing)
Expressing one's identity and indigeneity	1. Slate pavement, table, and chairs in the extended front yard 2. Family stories portrayed on the slate parapet
Catering to post-disaster livelihood	1. Housing features catered to income-generating activities (e.g., restaurants, hosting guests, and shops) 2. Note that disputes between the government and residents, and disparities among households regarding extensions have occurred

indigenous households tend to extend their permanent housing to a greater extent than the Chinese groups. This implies that the special livelihood and cultural characteristics of indigenous households prompt them to build housing extensions of a greater area than those by the Chinese groups. Second, this chapter differs from other similar studies, as it categorizes the housing extension in the Haocha community into five different patterns. This chapter confirms that these five different patterns had a corresponding relationship with a household's economic capability. Third, by looking at three case studies, the built environmental factor (providing a satisfactory living space), pre-disaster livelihood factor (catering to traditional livelihood), cultural factor (expressing one's identity and indigeneity), and post-disaster livelihood factor (catering to post-disaster livelihood) have been identified as factors that can motivate a household to implement a housing extension.

The finding of this chapter aligns with Rapoport's (1999) argument. The author argues that housing extension is the method employed by disaster-impacted households to fulfill their wants and needs by modifying the living space. In most PDR projects, the engineer plays a dominant role in conducting the housing construction, while the sociologist and anthropologist usually have no stake in it. However, especially in indigenous communities, the author suggests that some socioeconomic or anthropology studies should be conducted before the disaster to understand the pre-disaster livelihood, culture, and architecture of tribal communities. This information can be helpful in post-disaster design and planning decision-making. This process should be considered critical at the disaster preparedness stage.

In the design and planning stage, though the NGO that had earned the trust of indigenous peoples attempted to involve the indigenous community in the decision-making process, the pressure from the government and practical conditions deterred them from adopting the traditional method of constructing housing. Therefore, even in an NGO-led PDR project, understanding and support from the government are crucial. Recently, the incremental design—the housing provider provides only the core housing while the residents can fill in the functions and partitions based on their needs—has become popular and was implemented in several PDR projects (O'Brien et al. 2020). However, based on the findings of this chapter, ten years after the disaster struck, the discrepancy in the housing extension among Haocha households implies that the economic capability and livelihood opportunities after the disaster were distributed unevenly. Therefore, the author suggests that increased engagement from NGOs and the government is needed while residents attempt to extend their permanent housing.

References

Ahmed I (2011) An overview of post-disaster permanent housing reconstruction in developing countries. Int J Disaster Resil Built Environ 2(2):148–164

Ahmed I, McEvoy D (2014) Post-tsunami resettlement in Sri Lanka and India: site planning, infrastructure and services. Int J Disaster Resil Built Environ 5(1):53–65

Airriess CA, Li W, Leong KJ, Chen ACC, Keith VM (2008) Church-based social capital, networks and geographical scale: Katrina evacuation, relocation, and recovery in a new Orleans Vietnamese American community. Geoforum 39(3):1333–1346

Ali RA, Mannakkara S, Wilkinson S (2020) Factors affecting successful transition between post-disaster recovery phases: a case study of 2010 floods in Sindh, Pakistan. Int J Disaster Resil Built Environ 11(5):597–614

Carrasco S, Ochiai C, Okazaki K (2016) Impacts of resident-initiated housing modifications in resettlement sites in Cagayan de Oro, Philippines. Int J Disaster Risk Reduct 17:100–113

Chang C (2020) What you should know about being Indigenous in Taiwan. https://www.taiwannews.com.tw/en/news/4017174. Accessed 23 Jan 2022

Chen YL (2009) Report on Typhoon Morakot (0908) of 2009. Weather Forecast Center Central Weather Bureau, Taipei, Taiwan

Chuan KC (2018) A study on the post disaster village relocation at the town of Rinari in Majia Township Pingtung County. Degree thesis of department of public administration and policy, National Jinan International University, Taiwan

Dias NT, Keraminiyage K, DeSilva KK (2015) Long-term satisfaction of post-disaster resettled communities: the case of post-tsunami Sri Lanka. Disaster Prev Manag: Int J 25(5):581–594

Hsieh ZC, Fu ZC, Chen ZC, Lin WY (2012) A road farther and farther from hometown?: Rethinking the policy of rebuilding after typhoon Morakot. Natl Taiwan Univ Soc Work J 26:41–86

Hsu M, Howitt R, Miller F (2015) Procedural vulnerability and institutional capacity deficits in post-disaster recovery and reconstruction: insights from Wutai Rukai experiences of Typhoon Morakot. Hum Organ 74(4):308–318

Jha AK, Barenstein JD, Phelps PM, Pittet D, Sena S (2010) Safer Homes, Stronger Communities: A Handbook for Reconstructing After Natural Disasters. World Bank Publications, Washington, DC, US

Lee YC (2012) A study for exterior space improvement of relocated village and homeland recognition after disaster–a case study of Haocha tribe in Rinari. Unpublished master thesis, National Taiwan University, Taiwan
Lin JJ, Lin WI (2016) Cultural issues in post-disaster reconstruction: the case of Typhoon Morakot in Taiwan. Disasters 40(4):668–692
Lin PSS (2019) Building resilience through ecosystem restoration and community participation: post-disaster recovery in coastal island communities. Int J Disaster Risk Reduct 39:101249
Masterson JH, Peacock WG, Van Zandt SS, Grover H, Schwarz LF, Cooper JT (2014) Planning for Community Resilience: A Handbook for Reducing Vulnerability to Disasters, 3rd edn. Island Press, Washington, DC, US
Monteil C, Simmons P, Hicks A (2020) Post-disaster recovery and sociocultural change: rethinking social capital development for the new social fabric. Int J Disaster Risk Reduct 42:101356
O'Brien D, Carrasco S, Dovey K (2020) Incremental housing: harnessing informality at Villa Verde. Archnet-IJAR: Int J Arch Res 14(3):345–358
Quarantelli EL (1982) Sheltering and Housing after Major Community Disasters: Case Studies and General Observations. Ohio State University Press, OH, US
Rapoport A (1999) On the relationship between family and housing. In: Housing Provision and Bottom-up Approaches-Family Case Studies from Africa, Asia, and South America. Ashgate Publishing Ltd., Aldershot, UK
Schilderman T (2010) Putting People at the Centre of Reconstruction. IFRC, Practical Action Publishing, London, UK, Building Back Better, Delivering people-centred housing reconstruction at scale
Shao PJ, Lin WI, Chang LZ, Lee CJ, Juan MZ (2018) Post-disaster relief and reconstruction. Wunan, Taipei, Taiwan
Speranza CI, Wiesmann U, Rist S (2014) An indicator framework for assessing livelihood resilience in the context of social–ecological dynamics. Glob Environ Chang 28:109–119
Taiban S (2016) History of the Ha-cha tribe. Taiwan History Center Publisher, Taipei, Taiwan
Taiban S, Lin HN, Ko CC (2020) Disaster, relocation, and resilience: recovery and adaptation of Karamemedesane in Lily Tribal Community after Typhoon Morakot, Taiwan. Environ Hazards 19(2):209–222
Tsai SL, Ochiai C, Deng CZ, Tseng MH (2021) A sustainable post-disaster housing development framework for an indigenous Hao-Cha community in Taiwan: considering culture and livelihood in housing extensions. Int J Disaster Resil Built Environ 13(5):583–600
Tucker S, Gamage A, Wijeyesekera C (2014) Some design aspects of sustainable post-disaster housing. Int J Disaster Resil Built Environ 5(2):163–181
United nations general assembly (2007) Declaration on the rights of indigenous peoples. United Nations, Geneva, Switzerland

Chapter 4
Long-Term Spontaneous Resident-Driven Housing Modifications in Post-disaster Resettled Communities in the Philippines: A Longitudinal Study

Sandra Carrasco and Chiho Ochiai

Abstract This study explores the long-term impacts of reconstruction projects in the city of Cagayan de Oro in the southern Philippines, hit by typhoon Washi in December 2011. Recovery plans prioritized the resettlement of affected communities through housing projects in resettlement sites with NGOs as strategic partners in managing and building housing units. However, recovery plans largely overlooked the inputs of housing beneficiaries in the planning and implementation stages. Subsequently, families, upon receiving their homes, embarked on a progressive construction process to expand and customize their housing to better suit their needs and preferences. This research presents a longitudinal study focusing on the Mahogany village in the Calaanan resettlement site. Data was collected at two intervals, two- and seven-years post-resettlement. The study observes the process of resident-driven progressive adaptation of post-disaster housing and its relationship with residents' satisfaction levels with their living conditions. The findings underscore the diverse capacities of individuals in creating suitable living spaces for their families and highlight the necessity of incorporating a community capacity approach in post-disaster recovery and urban planning that fosters self-help initiatives and building community resilience in disadvantaged urban areas.

4.1 Introduction

The impact of disasters on housing often results in the temporary or permanent displacement of people, leading to cascading effects on affected individuals and communities. Housing reconstruction is considered the backbone of post-disaster

S. Carrasco (✉)
School of Humanities and Social Sciences, Deakin University, Melbourne, Australia
e-mail: s.carrasco@deakin.edu.au

C. Ochiai
Graduate School of Global Environmental Studies, Kyoto University, Kyoto, Japan

C. Ochiai et al. (eds.), *Disaster and Displacement*, Disaster Risk Reduction,
https://doi.org/10.1007/978-981-96-2583-3_4

recovery (Peacock et al. 2018; Kroll-Smith et al. 2015; Carrasco and O'Brien 2022a). However, its appropriateness and success are often evaluated considering short-term perspectives as agencies involved in housing recovery move somewhere else or find other priorities (Schilderman and Parker 2014). Humanitarian agencies frequently seek to complete housing projects and never come back to analyze the impacts of the housing recovery projects they led (Schilderman and Parker 2014). Remarkably, in recent years, the humanitarian sector has emphasized levering disaster-affected people's capacities to actively address their housing needs through people-centered approaches, such as self-recovery (Twigg et al. 2017). Nevertheless, challenges persist, including the short-term focus of humanitarian agencies and governments, which may undermine the long-term recovery aspirations of affected communities (Davis and Parrack 2018).

Post-disaster resettlement implies removing communities from their physical and social environments where they have evolved over decades or even centuries, where people develop crucial social networks and patterns for adapting to multiple stresses and crises (Oliver-Smith 2009; Carrasco et al. 2016c, 2023; Völker 2022). Communities' links with their homes and neighborhoods are based on economic, political, and socio-cultural factors that are severely disrupted when people are transferred to other locations (Jha et al. 2010). Thus, resettlement is considered one of the less desirable scenarios for post-disaster recovery as it impacts communities' loss of supporting social networks, sense of home and belonging, and reduction of people's livelihood opportunities (Jha et al. 2010).

Typhoon Washi, which hit the northern region of Mindanao in the Philippines in December 2011, severely impacted urban areas, with Cagayan de Oro being one of the most affected. Almost 40% of the population in Cagayan de Oro was displaced due to the disaster, which triggered a government response to minimize the hazard exposure for informal settlers along the Cagayan de Oro River. Thus, residents were resettled in sites far from the city center and their livelihoods, disrupting their social networks, financial stability, and sense of home and community. Furthermore, residents were given housing units built by contractors and NGOs considering minimum standards in place in the country, which for social and emergency housing incompatible with concerns of appropriateness such as reduced housing area and limited services provided. A 2014 study evidenced the residents' motivations to build housing extensions to fulfill needs and aspirations dismissed in the provided housing designs and community resilience consideration in the resettlement planning (Carrasco et al. 2016c).

This study focuses on analyzing the resettled people's perceptions and opinions about their housing conditions to determine their satisfaction levels and the patterns for housing modifications, observing the types of extensions built by residents. By combining qualitative information collected through questionnaires, qualitative data from residents' interviews, and extensions' mapping through observation this study aims to shed light on the types of extensions and factors that triggered their construction.

4.2 Research Design

This paper is part of a more extensive longitudinal study that considers primary data collection in 2014, 2019, and 2021. This paper examines the current housing conditions of residents in Mahogany Village, Calaanan resettlement site. The housing units were constructed by the NGO Habitat for Humanity Philippines in partnership with national and local governments, with limited involvement from the affected communities. The study investigates residents' efforts to improve their housing situations and explores the factors that motivated households to extend their houses. This is achieved by assessing residents' attitudes, as indicated by their satisfaction levels at three distinct intervals: upon moving into their NGO-constructed homes and at 2- and 7-years following occupancy.

Given the complexities of residential satisfaction and residents' motivations and decision-making in Mahogany Village, this study employs a mixed research methodology drawing on quantitative and qualitative analyses (Doyle et al. 2009). Previous studies on residential satisfaction in resettled communities following disasters in the Philippines and Chile (Carrasco et al. 2017; Carrasco and O'Brien 2022b) inform this approach. Quantitative analysis will empirically validate the situation and trends in Mahogany Village, focusing on the residents as a collective. In contrast, qualitative analysis of interview data will deepen understanding of residents' perceptions and behaviors at the individual level.

4.2.1 Sample and Participants' Selection

This study is based on a longitudinal study that considers two data collection fieldworks on site. The first data collection was conducted between July and August 2014, and the second data collection was conducted in December 2019. Additionally, semi-structured interviews were conducted through video calls with government officials in November 2021.

The initial contact with the community was through community leaders facilitated by local government staff working on community support, local parish volunteers, and direct interaction with residents facilitated by two local research assistants. The data collection considered four villages in the Calaanan site, although, for this study, only the data from Mahogany Village is considered for analysis. Mahogany Village has 160 housing units, and 159 households because one of them is used by the local government as a support and logistic base in the settlement. The fieldwork in 2014 focused on qualitative data, and 60 questionnaires were completed with support from local assistants. In 2019, 28 questionnaires were completed by residents, and 14 semi-structured interviews were conducted with local assistants who translated the questions and answers from the local language, Bisaya, into English. The questionnaire and interview participants were randomly selected. The details of the tools used for data collection are detailed below:

(1) A closed-question quantitative questionnaire, used in 2014 and 2019, seeking (a) general household data, (b) previous housing conditions, (c) initial conditions at Mahogany Village, and (d) current housing conditions.
(2) Qualitative, semi-structured interviews, used in 2019, extending data gathered by questionnaire to include (a) timelines for incremental additions, (b) difficulties encountered when undertaking incremental improvements, (c) future plans for improvements, and (d) changes to the broader neighborhood.
(3) Mapping of settlement conditions using architectural research tactics, including architectural drawings, photographic surveying, and physical trace analysis (Groat and Wang 2013; Yin and Campbell 2018).
(4) Secondary data from government reports, books, and academic papers.

4.2.2 Data Analysis

The collected information underwent a comprehensive analysis focusing on illustrating the various modification patterns and assessing the constructed spaces' current functionality. Additionally, an examination of residents' opinions and perceptions was collected through interviews to explain the underlying reasons driving these modifications, considering the unique challenges and circumstances encountered at the resettlement site. The observation of the housing extensions in Mahogany Village was also processed as a map of the settlement and compared with the state of housing extensions in 2014 and 2019. This holistic approach aimed to provide a deeper understanding of how the spaces are being adapted and utilized in response to the specific needs and conditions of the community.

4.3 Post-Disaster Resettlement and Housing Design

The typhoon Washi, which hit the southern region of Mindanao in December 2011, caused unprecedented destruction in the cities of Cagayan de Oro and Iligan. Washi displaced almost 40% of the urban population in Cagayan de Oro, who gradually returned to their pre-disaster homes. However, approximately 85% of the affected households belonged to the informal housing sector, many located in highly vulnerable areas (OCD 2012). The disaster resulted in permanent displacement for residents of informal settlements along the Cagayan de Oro River, as the government declared the former informal settlements as "no-built zones" and prioritized the resettlement of communities to the outskirts of the city (Fig. 4.1). This controversial measure unequally targeted informal settlers and prevented them from returning to their pre-disaster homes close to the city center and some of the most important commercial areas where many displaced communities found their livelihood opportunities (Carrasco and Egbelakin 2022). Figure 4.1 (left) shows the location of the resettlement sites concerning the former informal settlements ranging from 5 to 19 km.

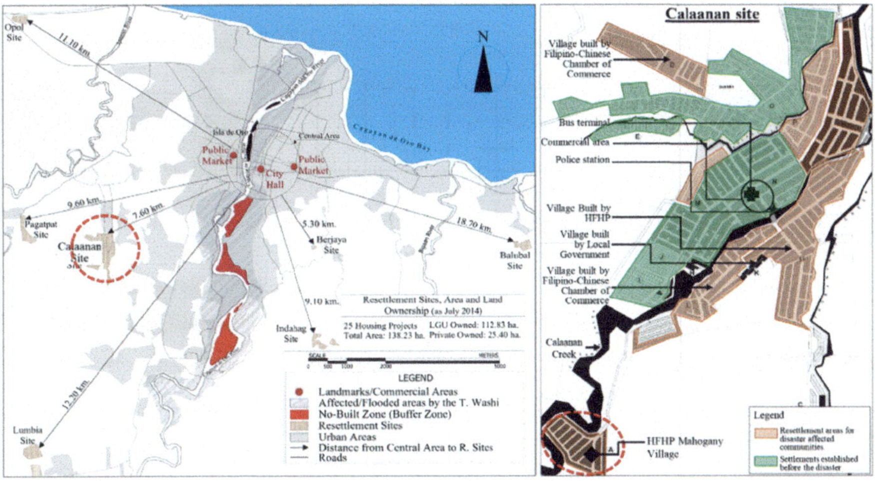

Fig. 4.1 Resettlement sites in Cagayan de Oro (left) and location of Mahogany Village in Calaanan site (right)

The resettlement sites are expected to minimize risks, especially flooding from future typhoons and recurrent rainy seasons that have affected the residents due to increased Cagayan de Oro River levels.

The government recovery plans prioritized partnerships with agencies such as NGOs and other stakeholders, which acted as contractors to build housing units in local government-owned land purchased prior to the disaster and intended to be used for social and emergency housing (Carrasco et al. 2016a). Figure 4.1 also shows the location of the Calaanan resettlement site, which was the target of this research. Figure 4.1 (right) shows the different neighborhoods or villages built by different agencies in Calaanan, where social housing settlements were built before the disaster. One of the key government partners for housing reconstruction was Habitat for Humanity Philippines (HFHP), which built houses at various resettlement sites in Cagayan de Oro. HFHP built the houses in Mahogany Village in partnership with the international volunteer organization "All Hands Volunteers" (now "All Hands and Hearts"). Builders included local masons and local and international volunteers.

The neighborhood analyzed in this study, Mahogany Village, was built by HFHP, as shown in Fig. 4.1 (right). Mahogany Village is the most isolated of the housing areas built for resettled communities. The local government, which owned the land, was responsible for assessing the suitability of resettlement, coordinating the design of the settlement layout, developing the land, and providing infrastructure, services, and additional facilities. In Mahogany Village, 160 housing units were built following a quadruplex design (Fig. 4.2), which implies a building divided into four housing units. This particular design responds to concerns to maximize the use of the limited land available. However, the quadruplex design implemented in Mahogany and other settlements built by HFHP responds to political decisions from government officials

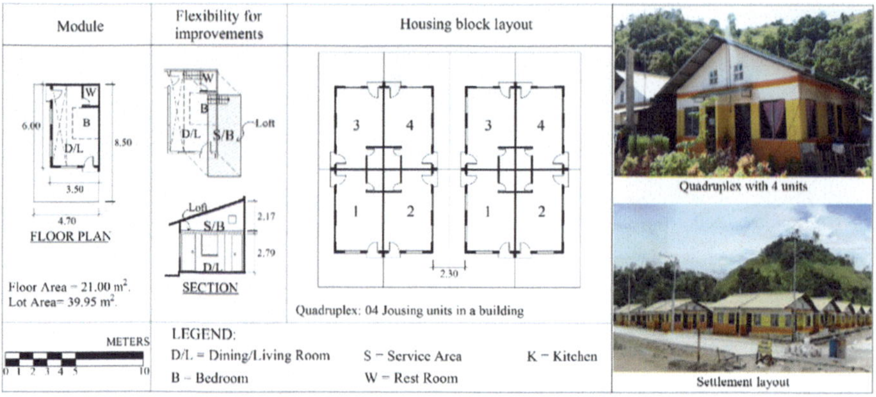

Fig. 4.2 Housing layout in Mahogany village

rather than technical reasons or households' preferences. Each housing unit includes spaces to be used as a dining/living room, bedroom, and a small restroom built in one corner, all of which are contained in a single-story one-bedroom. The internal divisions of the space were not included in the provided house, and the residents were expected to add them.

Similarly, the housing design considered the construction of a mezzanine or loft inside the house, which could add more livable areas to be performed by the residents based on their needs and resources. The covered floor area is 21.00 m^2, and the entire lot area for each housing unit is 39.95 m^2. The minimalist house design and the reduced livable areas were commonly used in social housing in the Philippines, and they followed social and emergency housing regulations by the National Housing Authority and the Department of Social Welfare and Development (Carrasco et al. 2016c).

4.4 Long-Term Analysis of Post-Occupancy Housing Modification

Agencies involved in reconstruction, such as NGOs and other government partners in the Philippines, have an understanding of recovery in discrete stages, such as relief, recovery, and reconstruction. However, this consecutive stage vision contradicts the perceptions of disaster-affected people, who see housing as a continuous change from emergency shelter to permanent housing (Davis and Parrack 2018). Therefore, it is crucial to comprehend the pathways leading to a gradual process of self-recovery and the implications of disaster recovery programs over the long term. Examining the people-driven transformations in housing and neighborhoods and their influence on social and community structures yields crucial insights into the impacts of various shelter and settlement reconstruction initiatives on community recovery (Davis and Parrack 2018; Flinn et al. 2017).

4.4.1 Housing Satisfaction as a Driver for Modification

Households regularly undergo a self-assessment process in which they contrast their residential environment and its conditions with their ideal housing attributes based on their aspirations and social norms in their environment (Bunster and Bustamante, 2019). The adaptation of residential conditions also plays a crucial role in restoring disrupted relationships to physical and social environments resulting from processes of post-disaster resettlement (Snarr and Brown 1980). Resident-driven housing customization or modifications are triggered when households' perceived conditions do not align with their expectations to address changes to the household structure, socio-economic conditions, or cultural appropriateness (Snarr and Brown 1980; Carrasco et al. 2016c).

Examining residents' perspectives about their houses both before and after occupancy can reveal the successes and challenges of a resettlement project that may not have been anticipated prior to establishing daily routines and activities in their new houses (Oliver-Smith 1991). Furthermore, shifts in housing satisfaction serve as a predictive factor for residents' attitudes toward housing adjustments (Manalang et al. 2003; Carrasco et al. 2017). In this study, residents' opinions about their houses are assessed by considering three points in time: at the moment of receiving their houses in mid-2012, after two years of occupancy in 2014, and after seven years of occupancy in 2019.

In order to assess residential satisfaction, ten indicators were considered, with residents rating them on a five-point satisfaction scale: (1) Location of the settlement; (2) Size of the lot; (3) Size of the house; (4) Strength of the house; (5) Quality of construction; (6) Construction quality; (7) Thermal comfort; (8) Acoustic comfort; (9) Fulfils residents' needs; (10) Tenure conditions; (11) Cultural appropriateness and (10) Safety within the settlement. These indicators were defined based on an understanding of the physical environment, residents' socio-economic circumstances (Lizarralde and Bouraoui 2010; Barenstein 2012; Perera et al. 2012), and experiences inherent in their daily routines (Ahrentzen 1992; Francescato et al. 1989) and living conditions (Carrasco et al. 2017; Carrasco and O'Brien 2022b). The analysis of satisfaction indicators involved comparing residents' perceptions of their housing conditions and environment before resettlement and after four years of living in the houses, having undergone multiple housing adaptations. Importantly, the three last indicators, Tenure conditions, Cultural appropriateness, and Safety, were only considered in the 2019 survey as these were added to the survey based on the residents' comments from the study conducted in 2014.

The data from the resident questionnaire surveys conducted in 2014 and 2019 was processed according to Eq. (4.1) to obtain the satisfaction scores 'x' per indicator:

$$x = \frac{R_1(i) + R_2(i) + R_3(i) + R_4(i) + R_5(i)}{R} \tag{4.1}$$

R = Total number of respondents
R1 = Highly satisfied
R2 = Satisfied
R3 = Neutral
R4 = Dissatisfied
R5 = Highly dissatisfied

For the analysis or comparison of the initial and post-occupancy satisfaction per indicator, the answers obtained were assigned a scoring index 'i', detailed as follows:

Highly satisfied = + 2, satisfied = + 1, neutral = 0, dissatisfied = −1, highly dissatisfied = −2.

Additionally, the quantitative data was processed following the score mentioned above, and qualitative data was introduced to the analysis to provide a rationale for the changes in residents' levels of satisfaction based on their opinions when asked about the advantages and disadvantages of living in agency-built houses in Calaanan, and in particular in Mahogany Village, and how these impact their living conditions.

Figure 4.3 shows the satisfaction analysis results, broadly evidence of a substantial positive change in residents' perspectives about their housing conditions from the initial housing without improvements in 2012 to 2014. However, the survey conducted in 2019 revealed that residents' satisfaction has decreased at various degrees depending on each indicator.

One of the significant changes in residents' satisfaction levels corresponds to the first indicator related to the location of the settlement. In the 2014 survey, resettled

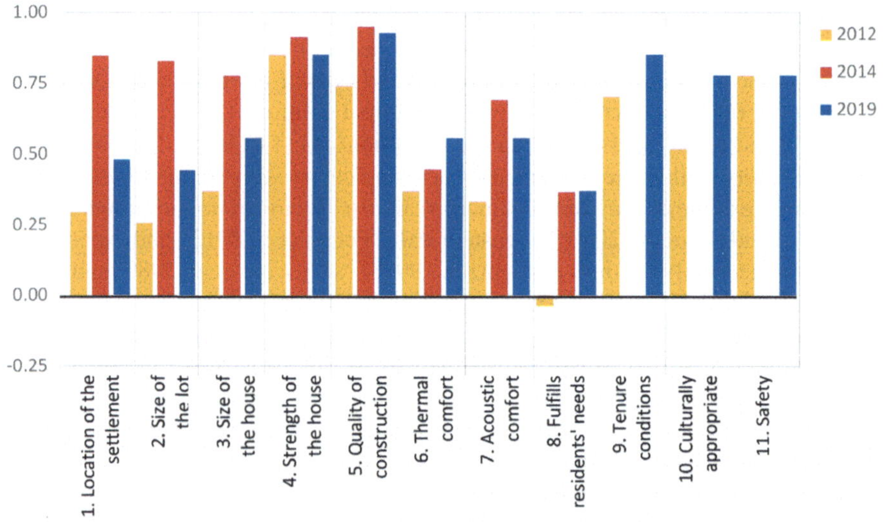

Fig. 4.3 Comparison of residents' satisfaction levels in 2012, 2014, and 2019

people had a relatively positive opinion because they compared the proximity to the city of Mahogany Village in comparison with other resettlement sites (Carrasco et al. 2017). However, the 2019 survey contrasted with previous perceptions as it revealed the negative long-term impacts of accessibility difficulties and the effects on their efforts to rebuild their livelihoods. For instance, two residents mentioned, *"It's difficult to go to work because it's far and the fare is high. But we can't do anything about it because we are already living here."* (Female, 34, MV 3); *"It's far from our workplace and [public transport] fare is expensive. The people have no working opportunities here. Also, it is difficult to find transportation at night."* (Female 49, MV 20). Similarly, residents expressed a decrease in their satisfaction with the size of the lot, which is closely linked to the space available to build extensions, compared with other housing typologies in Calaanan. A resident mentioned, *"In Mahogany, the houses are close together, unlike in the other areas in Calaanan where they have a little space in between [to build extensions] and even some are single detached."* (Female, 35, MV 28).

Indicators related to the housing units revealed different levels of households' satisfaction. Residents' satisfaction with the size of the house at the beginning of inhabitation was low and reached its peak in 2014. However, residents' satisfaction further decreased despite the added area as residents performed extensions to their houses. Residents compare their houses in Mahogany Village with their pre-disaster houses, which they consider bigger. However, the quality of the construction and the stability of the tenure conditions play a significant role in how people appreciate their current houses, *"Our house before [the disaster] was bigger, but our house now is nicer. We are ok here because I have siblings living near the affected areas, and they struggle because they need to rent a place there"* (Female, 34, MV 3). The satisfaction with the strength and the quality of construction still remains high as the materials used are perceived to provide stable structures *"This house was given to us. We could not have afforded a house if the typhoon did not happen. Our previous house was not made of concrete"* (Female, 34, MV 3); *"This house is solid, even better than the others built in Calaanan. It is solid, made of concrete hollow blocks"* (Male, 38, MV 34).

Thermal comfort relates to the already perceived mismatches of the local conditions of high temperatures and the houses' design that do not have proper ventilation (Carrasco et al. 2017). Thus, satisfaction is low, and this became the trigger for residents to build extensions to have shaded spaces with better ventilation, *"When my children come to visit me, they have to sleep in the extension outside because it is too hot inside the house"* (Female, 48, MV 16). Acoustic comfort is related to the residents' privacy as the housing typology is not adequately insulated. Residents' perceptions remain low as neighbors' noise often disturbs them. As they mentioned, *"If neighbors have fights or loud music, we can hear them. We cannot complain because we're on the same boat. Maybe if the houses were separated by at least one meter, it would be better"* (Female, 35, MV 28); *"If our neighbor fights, we can hear it because [the house] is quadruplex"* (Female, 34, MV 3).

Residents' perceptions about the appropriateness of the houses based on how they fulfill their needs were the lowest when they moved in. However, it gradually increased as they built extensions that followed their lifestyles and cultural concerns. For instance, the houses did not include kitchens, so residents-built spaces for "dirty kitchens", which are common in low-income groups in the Philippines as they use firewood collected from the proximities. Although these kitchens may have a negative impact on residents' health, as one resident mentioned, *"The materials used were good, but the space was just small. We could have still made a dirty kitchen if it had more space. We had to build it, but our neighbor complained because of the smoke"* (Male, 40, MV 26).

Remarkably, residents positively perceive the changes in the tenure conditions. In previous studies, residents expressed their concerns about the construction of extensions, which are not allowed. However, in 2019, local government officers sought alternatives to transfer the ownership of the houses and land to residents. Thus, households felt relieved of the pressure to fulfill their increasing space needs, which contradicted the tenure conditions. Additionally, residents feel satisfied about the safety of their homes as the potential risks to natural hazards are minimized. However, it remains uncertain how the incremental construction of extensions could be detrimental to residents' safety, as it was already warned in a previous study (Carrasco et al. 2016b).

4.4.2 Patterns of Housing Modification

This section centers on analyzing the types of modifications built by residents throughout the seven years since the agency-driven houses were turned over to them. It aims to analyze the changes made to the original houses and shed light on the factors that might have driven the residents' efforts in performing certain modifications, in contrast to the state of the houses and settlements mapped in 2014 and 2019.

The housing design by Habitat for Humanity Philippines considered only internal partitions and the construction of a mezzanine or a loft. Therefore, any construction outside the housing unit is considered informal or unplanned extensions, either horizontal or vertical, adding upper stories on top of the original houses. The usufruct conditions by which residents were granted the right to live in the houses limit the construction of extensions and could be considered a violation of the occupancy conditions threatening residents' permanence in these houses (Carrasco et al. 2016c).

This study mainly focuses on the informal housing extensions that might be located within the property line or the provided lot and the extensions that are built beyond the property line. Figure 4.4 presents the comparison of the housing extensions in Mahogany Village mapped in 2014 and 2019. Unplanned extensions cannot be considered permanent as these can be removed by the government if required. Thus, considering the materials used, there are two types of extensions, durable and precarious. Extensions built with durable materials, such as concrete, reinforced concrete, and hollow blocks, can be considered permanent or semi-permanent constructions. Precarious extensions refer to constructions using local materials such as wood, palm leaves or nipa, corrugated iron roof sheets, and other makeshift materials that are considered non-permanent constructions, as these can be removed or upgraded to permanent constructions.

Figure 4.4 shows precarious extensions that are used as intermediate spaces between indoor and outdoor areas (see Fig. 4e and i) that are motivated by the extremely hot temperatures inside the provided house. Figure 4e and i show the

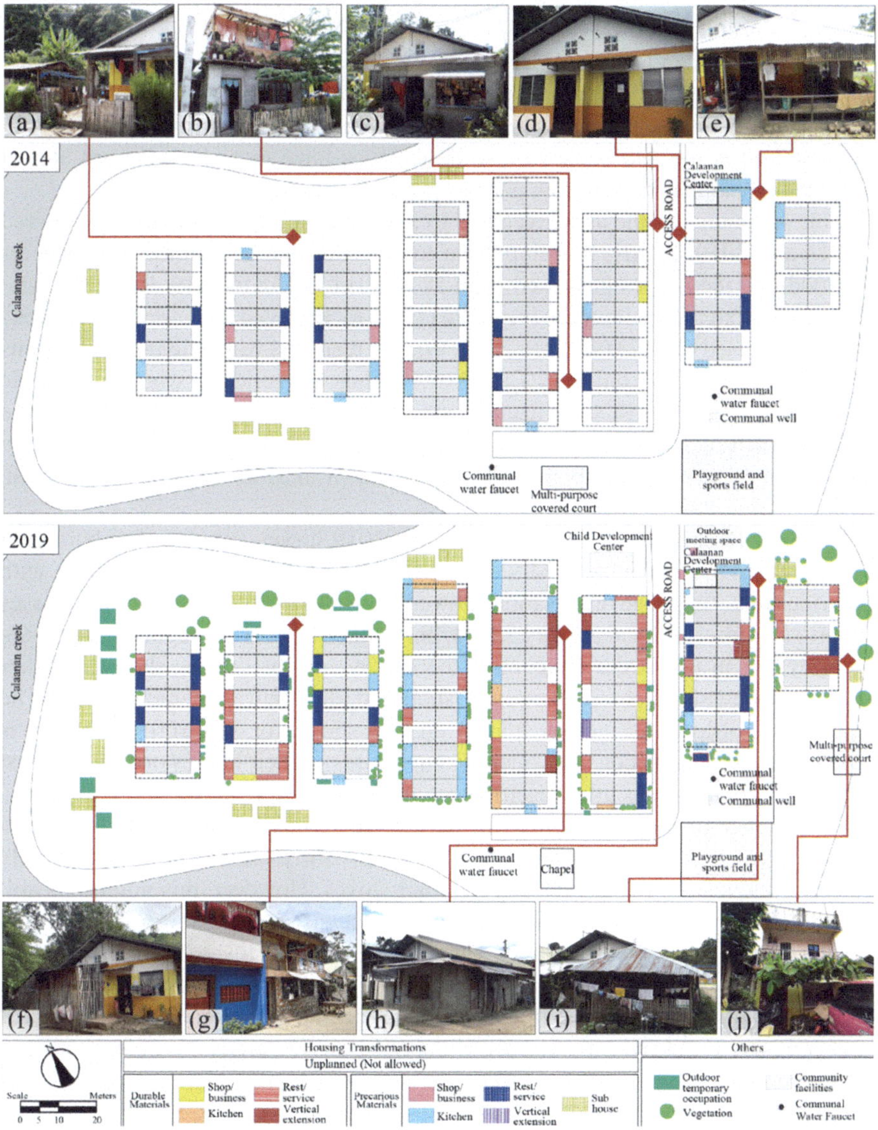

Fig. 4.4 Comparison of 2014 and 2019 modifications

minimum changes made to a large precarious extension between 2014 and 2019. Figure 4d is an unmodified house, contrasting with an extensively modified house in Fig. 4j that depicts a frontal and vertical extension built with durable materials. In Fig. 4c, residents added a frontal extension to be used as a local grocery shop or sari-sari store. The same house in 2019 presents the extension of the roofs to provide shaded spaces for customers to protect themselves from the sun and stay in the shop for more extended periods of time, as shops are also used as spaces for social interaction among neighbors.

Figure 4b and g shows vertical extensions and how these have become more structurally stable and stylized over time. Extensions outside the provided lot are called sub-houses (Carrasco et al. 2016c) because they are multipurpose spaces that can be used as bedrooms, service areas, and resting areas. Figure 4a and f shows the same house and sub-house whose dimensions have increased and are almost connected, separated by a narrow passage.

Table 4.1 compares the uses given to the extensions built by residents, as observed in 2014 and 2019. Firstly, it is important to consider that the number of residents who modified the houses increased from 57 in 2014 to 132 in 2019. Residents who built durable extensions increased from 26 (17%) to 36 (23%), precarious extensions slightly increased from 73 (46%) to 75 (47%), and residents who built both precarious and durable decreased from 11 (7%) to 9 (6%). Thus, the residents who built durable extensions were the most significant group, which can be related to the transition from more affordable and quick constructions, which are later upgraded to durable structures as the residents obtain more resources (Makachia 2005; Carrasco et al. 2016b).

A total of 18 shops were counted during the 2019 survey, most of them built with durable materials. In contrast, 14 were counted in 2014, and most were built with precarious materials. The observation showed that 13 of the shops or local businesses are sari-sari stores or local grocery shops, one of the shops is not regularly open, ten shops are not in use at the moment of the survey, and two of the shops are other kinds of businesses such as bike repair shop and electronics technician shop. The high number of local businesses, mainly grocery shops, for only 160 households is evidence that residents consider opening businesses in their homes one of the most viable livelihood options after resettlement and in the longer term.

The number of kitchens built with durable materials has increased from zero to five, meaning that the precariously built small, "dirty kitchens" that use locally sourced firewood are being replaced by other stoves. However, this number is still limited to five in the settlement. Precarious kitchens are still the most common cooking spaces, mainly small stands containing firewood stoves. Similarly, service or rest spaces used to perform daily house chores or to rest are the most common extensions built in Mahogany Village. There has been a considerable increase in the number of service and rest spaces built with precarious materials. Vertical extensions have increased to seven, and six of them are built with durable materials that provide more stable and safer extensions.

Extensions built beyond the property line of the provided lot increased from 10 to 14, and all were built with precarious materials as these represent the informal

Table 4.1 Patterns of housing modification

Materials	Uses	Year surveyed	
		2014	2019
		No (%*)	No (%‡)
Durable	Shop/business	4 (7)	13 (10)
	Kitchen	0 (0)	5 (4)
	Rest/service	6 (11)	42 (32)
	Vertical extension	0 (0)	6 (5)
Precarious	Shop/business	10 (18)	5 (4)
	Kitchen	16 (28)	30 (23)
	Rest/service	15 (26)	23 (17)
	Vertical extension	1 (2)	1 (1)
	Sub house	10 (18)	14 (11)
	Outdoor temporary※	0 (0)	14 (11)

Note *Percentages are based on the number of modified houses surveyed each year. * = 57 modified in 2014. ‡ = 132 modified in 2019.*
※ = Informal occupation of common areas that do not add living spaces.

occupation of common areas in the settlement, and there are fewer considerations before their eventual removal by the government or neighbors themselves. Interestingly, residents also built other precarious structures occupying the common areas, especially alongside the houses at the end of the settlement. The 14 outdoor temporary structures are used to define cultivating spaces that are not covered and used for residents to grow their vegetables or for raising chickens and other small animals.

Most residents undertook housing extensions in 2019, with a notable 83% doing so, compared to just 36% in 2014. Table 4.2 illustrates the areas extended in relation to the original houses. For instance, in 2014, residents, on average, added 10.11 m^2 to their homes, marking a 46% increase over the original area. By contrast, in 2019, the average extension grew to 18.42 m^2, reflecting an impressive 84% increase in housing space. Additionally, while in 2014, households that extended their homes the most nearly doubled the original size, by 2019, the maximum extension increased houses to 3.3 times their original size. Despite the high number and diversity of housing extensions built by residents, for 74% of them, it is not enough, and they are planning the construction of more extensions, 15% are unsure, and only 11% of the residents surveyed expressed that they do not need to build further extensions. Remarkably, these perceptions might have changed as the 2021 interviews with local government officials evidenced their position favoring the formal transfer of the house and lot ownership to the residents. Although few ownership transfers were made, it is expected that most, if not all, will be done progressively in the following years.

Table 4.2 Increased area of extensions

Surveyed year	Original house area (m^2)	Average increased area			Max. area increased	
		Increased area (m^2)	Original + increased area (m^2)	%	Original + increased area (m^2)	%
2014	21.00	10.11	31.11	146	43.00	196
2019		18.42	39.42	184	74.00	337

4.5 Discussion

The research findings from this study shed light on the dynamic process of post-disaster resettlement and housing modification in Mahogany Village, Calaanan resettlement site, following the devastating impact of typhoon Washi in 2011. The longitudinal approach adopted in this study provides valuable insights into residents' housing satisfaction levels and the factors influencing their decisions to extend their homes over time.

One significant finding is the substantial positive change in residents' perspectives about their housing conditions from the initial housing without improvements in 2012 to 2014. However, the survey conducted in 2019 revealed a decrease in residents' satisfaction, particularly regarding the location of the settlement and the size of the lot. Accessibility difficulties and limited livelihood opportunities were highlighted as major concerns, indicating the long-term impact of resettlement on residents' daily lives.

Moreover, the study identified patterns of housing modification driven by residents' needs and aspirations. The majority of residents undertook housing extensions, with a notable increase observed in 2019 compared to 2014. These extensions ranged from durable structures to precarious constructions, reflecting residents' resource availability and evolving priorities. Notably, the increase in the average extension area signifies residents' efforts to improve their living conditions despite challenges.

The findings also reveal the emergence of local businesses, mainly grocery shops, within the settlement, highlighting residents' entrepreneurial spirit and adaptation strategies post-resettlement. However, the real impact of shops on households' financial conditions is still unknown, as the large number of shops in Mahogany also means more competition among them to attract customers. Resident-built extensions are still motivated by the need to add living spaces in a housing unit with reduced area for large families. Thermal comfort also motivates the extensions due to the original houses' lack of proper ventilation. The construction of extensions outside of the provided lot also evidences the need for space and the appropriation of common areas, which could cause community conflicts as evidenced in social housing and resettlement studies (Sepúlveda et al. 1994; Carrasco and O'Brien 2022b). Interestingly, the study identified informal occupation of common areas for cultivating spaces, indicating residents' utilization of available resources to meet their needs.

4.6 Conclusions

This study contributes to the understanding of long-term post-disaster resettlement dynamics and housing modification processes in Mahogany Village. The findings underscore the importance of considering residents' perspectives and experiences in shaping effective resettlement and housing policies.

The decrease in residents' satisfaction levels over time highlights the need for ongoing support and interventions to address evolving challenges and improve livelihood opportunities in resettlement areas. Moreover, the prevalence of housing extensions and the emergence of local businesses underscore residents' resilience and adaptive capacity in the face of adversity.

Moving forward, policy interventions should prioritize residents' participation and empowerment, ensuring that resettlement initiatives align with their aspirations and socio-economic needs. Additionally, efforts to formalize land tenure and provide adequate infrastructure and services are essential to enhance the long-term sustainability and well-being of resettled communities. However, the impacts of the construction of extensions in the structural stability and the settlement design are required to minimize future risks that might emerge from an uncontrolled incremental housing construction process.

Finally, this study emphasizes the complexity of post-disaster resettlement processes and the importance of adopting a holistic and participatory approach to promote sustainable recovery and resilience among affected populations.

References

Ahrentzen SB (1992) Home as a workplace in the lives of women. In: Altman I, Low SM (eds) Place attachment. Springer

Barenstein JD (2012) Communities' perspectives on housing reconstruction in Gujarat following the earthquake of 2001. In: Barenstein JD, Leemann E (eds) Post-Disaster Reconstruction and Change Communities' Perspectives. Boca Raton, FL: CRC Press, Taylor and Francis Group

Bunster V, Bustamante W (2019) Structuring a residential satisfaction model for predictive personalization in mass social housing. Sustainability 11(14):3943

Carrasco S, Egbelakin T (2022) Unravelling the challenges for long-term planning post-disaster resettlement in Cagayan de Oro, Philippines. IOP Conf Ser: Earth Environ Sci 1101(2):022038

Carrasco S, Egbelakin T, Dangol N (2023) Fostering recovery through stakeholders-community collaboration in post-earthquake recovery in Nepal. Int J Disaster Risk Reduct 88(103619)

Carrasco S, O'Brien D (2022a) Incremental pathways of post-disaster housing self-recovery in villa Verde Chile. Architecture 2(3):544–561

Carrasco S, O'Brien D (2022b) Re-thinking elemental's incremental housing: Residential Satisfaction and resident-driven adaptations in Villa Verde, Chile. urbe. Revista Brasileira de Gestão Urbana 14:e20210171

Carrasco S, Ochiai C, Okazaki K (2016a) Disaster induced resettlement: multi-stakeholder interactions and decision making following tropical storm Washi in Cagayan de Oro, Philippines. Procedia Soc Behav Sci 218:p35-49

Carrasco S, Ochiai C, Okazaki K (2016b) Impacts of resident-initiated housing modifications in resettlement sites in Cagayan de Oro, Philippines. Int J Disaster Risk Reduct 17:p100-113

Carrasco S, Ochiai C, Okazaki K (2016c) A study on housing modifications in resettlement sites in Cagayan de Oro, Philippines. J Asian Arch Build Eng 15(1):25–32
Carrasco S, Ochiai C, Okazaki K (2017) Residential satisfaction and housing modifications: a study in disaster-induced resettlement sites in Cagayan de Oro, Philippines. Int J Disaster Resil Built Environ 8(2):175–189
Davis I, Parrack C (2018) Taking the long view. In: Global Shelter Cluster (ed) The state of humanitarian shelter and settlements 2018. Global Shelter Cluster, Geneva, Switzerland
Doyle L, Brady A-M, Byrne G (2009) An overview of mixed methods research. J Res Nurs 14(2):175–185
Flinn B, Schofield H, Morel LM (2017) The case for self-recovery. Forced Migr Rev 55:12
Francescato G, Weidemann S, Anderson J R (1989) Evaluating the built environment from the users' point of view: an attitudinal model of residential satisfaction. In: Preiser WFE (ed) Building evaluation. Springer Science+Buisiness Media, LLC, Albuquerque, New Mexico
Groat LN, Wang D (2013) Architectural Research Methods. John Wiley and Sons, Hoboken, New Jersey
Jha AK, Barenstein JD, Phelps P, Pittet D, Sena S (2010) Safer homes, stronger communities: a handbook for reconstructing after natural disasters. World Bank Publications
Kroll-Smith S, Baxter V, Jenkins P (2015) Left to chance: Hurricane Katrina and the story of two New Orleans neighborhoods. University of Texas Press
Lizarralde G, Bouraoui D (2010) User's participation and satisfaction in post-disaster reconstruction. In: Participatory design and appropriate technology for disaster reconstruction. Conference proceedings
Makachia PA (2005) Influence of house form on dweller-initiated transformations in urban housing. 33rd IAHS World Congress on Housing. Pretoria, South Africa
Manalang RT, Munemoto J, Yoshida T, Espina C (2003) A study on residents' self-built improvements as a predictor of their intentions on residential mobility at MRB dwelling units in metro manila. J Asian Arch Build Eng 2(2):b117–b122
OCD (2012) Tropical storm sendong post-disaster needs assessment (PDNA), final report. Cagayan de Oro, Philippines: Office of civil defense, regional disaster risk reduction and management council–region 10
Oliver-Smith A (1991) Successes and failures in post-disaster resettlement. Disasters 15(1):12–23. https://doi.org/10.1111/disa.1991.15.issue-1, https://doi.org/10.1111/j.1467-7717.1991.tb00423.x
Oliver-Smith A (2009) Development and dispossession: the crisis of forced displacement and resettlement: School for Advanced Research Press
Peacock WG, Dash N, Zhang Y, Zandt SV (2018) Post-disaster sheltering, temporary housing and permanent housing recovery. Handbook of Disaster Research, pp 569–594
Perera T, Weerasoori I, Karunarathne H (2012) An evaluation of success and failures in Hambantota, Siribopura resettlement housing program: lessons learned. Sri Lanka J R Estate 6
Schilderman T, Parker E (2014) Still Standing?: Looking Back at Reconstruction and Disaster Risk Reduction in Housing. Practical Action Publishing
Sepúlveda R, De la Puente P, Torres E, Muñoz P (1994) Desarrollo Progresivo en conjuntos de lotes con servicios. Análisis Preliminar. Revista INVI 8(20)
Snarr DN, Brown EL (1980) User satisfaction with permanent post-disaster housing: two years after hurricane fifi in Honduras. Disasters 4(1):83–91
Twigg J, Lovell E, Schofield H, Miranda Morel L, Flinn B, Sargeant S, Finlayson A, Dijkstra T, Stephenson V, Albuerne A (2017) Self-recovery from disasters: an interdisciplinary perspective. ODI–Overseas Development Institute
Völker B (2022) Disaster recovery via social capital. Nat Sustain 5(2):96–97
Yin RK, Campbell DT (2018) Case Study Research and Applications : Design and Methods

Chapter 5
Transmigrant Communities in Indonesia: Two Case Study Sites in Central Sumatra

Chiho Ochiai and Ricelli Laplace

Abstract This study examines a community resettled in Sumatra, Indonesia, by the Indonesian Transmigration program, an official government-sponsored resettlement approach of the Suharto era aimed at reducing population pressures in Java and boosting economic development in the outer islands. Since transmigration projects in the 1980s, little research has been conducted on its outcomes. This chapter explores the two case studies of the Transmigration project, namely Sitiung and Kumpeh. Although these two resettlement projects have different backgrounds, they were provided with the same standard of housing by the government. Still, there are many differences in the living environment, especially flood damage, on the land where the residential land was developed, the provision of agricultural land, and the development of infrastructure. It was also revealed that these factors greatly influenced today's living environment. At the same time, this study revealed that social relationships and social mechanisms had a major impact on the improvement of the living environment in the newly settled transmigration sites, given the different living environments and inadequate infrastructure.

5.1 Introduction

For over 75 years since its independence from Dutch colonization in 1945, the government has been promoting an internal migration and resettlement project based on agrarian rural reform, capital-intensive development, large-scale migration, and the intensified commercialization of agriculture. Transmigration is a government-led relocation and resettlement program that focuses on moving low-income and landless populations from densely populated areas like Java and Bali Island to less developed areas like Papua, Kalimantan, Sulawesi, and Sumatra Island in Indonesia. Transmigration aimed to solve these problems, but other problems, such as floods or landslides and social conflicts, are often mentioned to occur instead. In this study, the real consequences are presented.

C. Ochiai (✉) · R. Laplace
Graduate School of Global Environmental Studies, Kyoto University, Kyoto, Japan
e-mail: ochiai.chiho.2x@kyoto-u.ac.jp

C. Ochiai et al. (eds.), *Disaster and Displacement*, Disaster Risk Reduction,
https://doi.org/10.1007/978-981-96-2583-3_5

The Indonesian transmigration project (ITP), or Transmigration in Indonesia, is a government-sponsored response to these phenomena; it had been sponsored by massive international agencies such as the World Bank and the Asian Development Bank (World Bank 1988). The program resettled almost 2.5 million people, around 535,000 families, between 1979 and 1984 (Benoit and Pain 1989). This process had huge environmental and social impacts on the people and the land, resulting in many controversies. Inaccessibility and isolation of resettlement locations or unfertile land, deforestation, and conflicts between resettled "transmigrants" and local people, often indigenous people, are common critics of the ITP. The Indonesian transmigration program is the world's oldest and largest sponsored internal migration project. The case of transmigration in Indonesia offers unique insight into the long-term outcomes of resettlement projects and possibilities for the future role of resettlement projects. However, the main interest of past research has focused on governmental policy, economic development, or environmental impacts, and the majority of these studies were conducted in the 1980s and 1990s (Arndt 1983; Gondowarsito 1990; Hardjono 1986; Suratman and Guinness 1977). Considering the trends of human mobility and displacement, it is highly relevant to revisit the ITP and analyze its impacts in the long term.

In any resettlement project, housing and basic infrastructure are among the most important elements for the good quality of life and sustainability of new communities. The Transmigration program often includes residential land, infrastructure, and housing (Republic of Indonesia 1982). These physical structures directly influence living conditions and are a key factor in the adaptation process of these communities. To improve living conditions and adapt to resettlement, dwellers usually make the best efforts to improve and transform their housing and surroundings, changing the quality of basic infrastructures and housing using their own resources and help from families and neighbors.

Therefore, this chapter explores how physical transformations in their living environment have been made by dwellers who play an important role in the environment and communities' adaptation process. The two main objectives of the study are (1) To understand physical transformations made by transmigrants to cope with challenges and adapt to the new environment and (2) To understand the current condition of living conditions in transmigration resettlement projects by showing two case study sites, namely Sitiung and Kumpeh located in Sumatra Island.

5.2 Study Sites and Methodology

5.2.1 Study Sites

Two case study sites, Sitiung and Kumpeh, were selected from Sumatra Island (see Fig. 5.1). Sumatra was chosen as the targeted island since it is the largest receiving region for transmigration projects. For comparison reasons, the sites were chosen

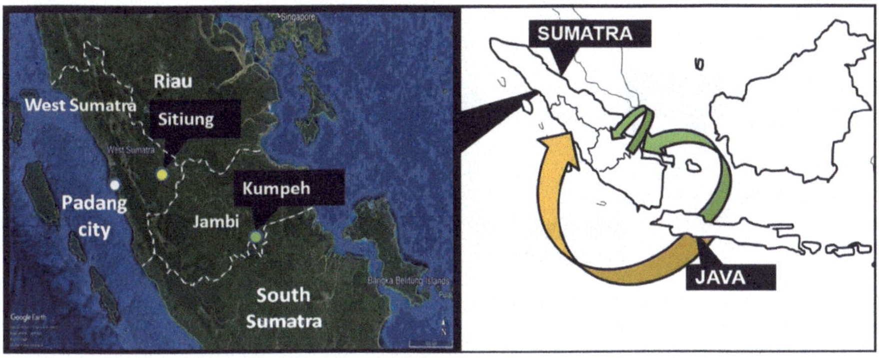

Fig. 5.1 Study site locations. *Source* Google earth, modified by author

based on relevant representational differences: size of resettlement, economic level, and migrant's background and original place.

Sitiung is located in West-Sumatra, Dharmasraya Regency, 200 km from Padang City. It is the biggest transmigration resettlement in West Sumatra (as the biggest performed by the Indonesian government). In 1976, 3,000 families were moved from Java Island (Wonogiri) due to a dam-driven displacement resettlement project.

Kumpeh is located in Jambi, Muaro Jambi regency, 64 km away from Jambi City. It is a smaller resettlement of 500 families, comprised of Javanese and local transmigrants (people who have already transmigrated before and settled in nearby locations), both groups coming from different parts of Java and Jambi. Most of the dwellers were resettled because there was not enough land in their original village, there were high rates of poverty, and people were living in cramped conditions. Unlike the residents in Sitiung who were forcibly displaced, the people resettled in Kumpeh voluntarily moved to this new location.

5.2.2 *Methodology*

Field surveys were conducted at both sites between July 2015 and November 2017. Analysis drawn from two primary surveys, a questionnaire survey and a measurement survey, are complemented by in-depth interviews, quantitative data analyses, and field observations. In total, 102 households were surveyed in Sitiung, and 109 households were surveyed in Kumpeh. Figure 5.2 shows the location and village layout of the study sites. In addition to field surveys, interviews were conducted with governmental representatives from the Jakarta transmigration office and local governmental offices in Sumatra and resettlement's community leaders.

The surveyed area in both sites targeted the area where the last remaining original transmigrant house and land are located (Fig. 5.3). Measurement data was collected on housing materials and size, construction quality, and structure durability. Changes

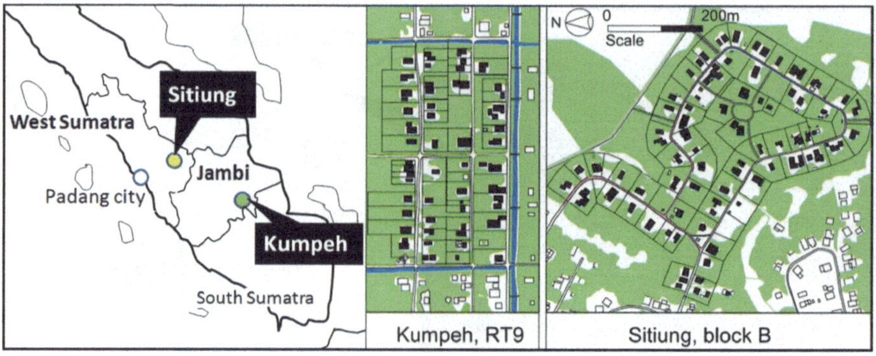

Fig. 5.2 Study sites location in Kumpeh (L) and Sitiung (R). *Source* Authors

Fig. 5.3 Original transmigrant's house in Kumpeh (L) and Sitiung (R). *Source* Authors

in residential land were studied, as well as access to infrastructure (access to electricity, water, roads, cooking fuel, sewage, and telecommunications). Infrastructure and housing durability were categorized according to Indonesian building construction guidelines and later utilized to calculate the living conditions of resettlement. Living conditions are analyzed as a result of housing durability and infrastructure combined, resulting in temporary, semi-permanent, or permanent living conditions (Republic of Indonesia, technical guidelines of household development of 2002, building regulation review of 2005, housing rental regulations 1963).

5.2.3 Resettled Conditions

(1) Sitiung

The Sitiung transmigration project (1976/1978) was planned by massive dam-displaced resettlement established during the Government's agricultural strategy of

Repelita II. It represents a new type of resettlement in which entire communities are transferred together, called Bedol Desa. The construction of the Wonogiri dam (completed in 1981) in central Java affected 6 sub-districts, moving (between 1976 and 1980) a total of 10,139 families (41,365 people).

In total, 3,201 families were relocated to two villages, Sitiung 1 and 2, under the procedures specified by the Indonesian transmigration project guidelines, and other resettlements in south Sumatra, Jambi, and Bengkulu. Between 1976 and 1977, 2,001 families were relocated from Wonogiri Regency of Central Java Province to Sitiung 1, and in the following years, 1977–78, 1,200 more were relocated to Sitiung 2, comprising a total of 13,581 people. The Wonogiri Betol Desa (village) was the biggest transmigration resettlement ever carried out in the country. Since then, no other projects have attempted to move such a large group of people (data from the official local government of Sitiung collected by the author, 2016).

The first and biggest project of its kind, as well as President Sukarno's "home area," the Sitiung project received much support from governmental agencies, media attention, and publicity. This Betol Desa project worked as an "experiment" to create an ideal resettlement scheme that could be adopted elsewhere (Gondowarsito 1990).

In Sitiung, the guidelines specified by the government promised each transmigrant family to have a house of 36m2 in a 2,500 m2 residential plot land and 2 hectares of cropland (some residents did not receive the promised amount of cropland) but in reality, the average land holding was 1.25 hectares.

Interviewees revealed that living conditions were very poor when they first resettled compared to the former settlement in Central Java. The land condition was not suitable for farming due to deforestation and floods. The government initially promised to provide one year of assistance (monthly ration of food, seeds, and working materials), but later extended it to three years due to the hard living conditions of the resettlement site.85% of the interviewees responded that the main issue was the lack of infrastructure when they first resettled. Thus, transmigrants and their community efforts to improve their living conditions were crucial to ensure a better living environment and the process of resettlement. Additionally, the government provided one year of assistance in the way of food rations, seeds and working materials which was later extended to three years considering the challenging living conditions at the beginning of resettlement.

(2) Kumpeh

Between 1975 and 1984, during the last years of the Government's agricultural strategy of Repelita III, in Jambi province in east Sumatra, more than 13,000 families (58,096 people) were settled in new transmigration sites, most of them in swamp areas and isolated locations. Swamp areas represent a big challenge for sustainable settlements. Limitations to soil management and agricultural production of these sites sometimes result in settlers living at a subsistence level. It was clear that this became a huge problem, with the response from the World Bank's last project officially related to transmigration: the Integrated Swamp Development Project (ISDP from 1994–2000), where they tried to mitigate poverty and improve development in swamp areas of Jami, Riau and West Kalimantan (Adhiati and Bobsien 2001).

Mekar Sari village, in Kumpeh, was one of these ITP, located in Muaro Jambi regency, and 76 km from the central city administration of Jambi. Located in a swampy area, it is a palm plantation-oriented transmigration project. The migrants in this project are constituted of indigenous Melayu people and Javanese and were resettled within the same area. Resettlement occurred in 1986, with 500 transmigrant households (2,430 people). Dwellers come from west Java (54 households), Central Java (70 households), and east Java (70 households), totalizing 194 transmigrant families from Java, and 306 local transmigrant families from Jambi, in a total of 806 families for the entire resettlement. Today, the resettlement has 2,629 people and 761 households (local government data of 2014/2015, obtained by the author during a field survey, 2017).

Locals and Javanese transmigrants came to the site for the same reasons of poverty and lack of productive land in their original village. As promised by the government, they received a house in a plot land of 2,500m2 and 2 hectares of "productive" land. However, the village was built on a swamp area with frequent flooding. Many families lost their houses and land in big floods that occurred in the early 90 s and 2000, forcing them to evacuate to other places or abandon the site.

A mix of locals and Javanese migrants under very different vulnerable environmental conditions, the Kumpeh project outcomes and dwellers' living conditions differ greatly from those of the Sitiung.

5.3 Transformation of Land, Infrastructures and Houses

5.3.1 Background Differences

Before understanding the main differences and outcomes of the Sitiung and Kumpeh transmigration projects, it is important to mention the five main characteristics that most differ between them are the date each resettlement site was established, environment and exposure to hazards, government design and implementation, project size, and transmigrants' backgrounds.

First, Sitiung is almost 10 years older than Kumpeh. In 2017, Sitiung completed 40 years of resettlement, while Kumpeh completed 31. Therefore, this should be taken into consideration when comparing outcomes since Sitiung had more time for adaptation and changes. Second, although both sites had flooding problems during the initial phase, Sitiung's floods were properly managed, and risks were mitigated. On the other hand, Kumpeh has suffered from chronic floods until today. The surrounding land of the Sitiung site hosts a diversity of crops, while Kumpeh only has palm oil plantations. As explained in the study site, the Sitiung project was given more government attention than other projects, and for this reason, it has been considered by some researchers as a political project (Gondowarsito 1990). Not only was the allocated budget for the project higher than in Kumpeh, but assistance and follow-up by the central government were closely implemented. Fourthly, the resettlement sizes were different. While Sitiung is a dam-forced resettlement inside the Betol desa scheme that involved moving an entire community, Kumpeh is a small and less ambitious project to move poor and landless people from different communities and develop a new settlement focused on mono-crop production. Finally, in Sitiung, all the migrants came from Wonogiri and belonged to the same background, culture, traditions, and even the same family. When migration happened, the surroundings and environment changed, but community ties, roles, and relationships continued. Also, friends and families shared the same experience while moving and shared the same hardships, strengthening ties and a sense of belongingness. On the other hand, transmigrants in Kumpeh come from different backgrounds and places, including people from Java and Jambi province. The survey showed that even those who came from inside Jambi originated from at least four different villages.

5.3.2 Resettlement Impacts on Land

Significantly, in Sitiung, land traditionally belonged to local indigenous Minangkabau people, while Kumpeh land had no traditional ownership. As a result, the land allocation conflicts and certification delays impacted the land distribution in Sitiung. Residents in Sitiung received around 1 to 1.25 hectares of cropland and cattle as compensation, while in Kumpeh, most of the transmigrants received 2 hectares as designated by the transmigration land allocation law. In Sitiung, cropland was also limited, while in Kumpeh, the forest could be cleaned and opened by residents themselves, in a process described by dwellers as "clean and take.“ However, due to flooding problems, many residents in Kumpeh lost their land, and many bought land from residents who gave up on the project. Therefore, landholding is not regular and can vary from residents with no access to land at all to residents owning 3.6 or more hectares. This creates a great land access disparity, which may lead to income disparities and social inequality. This imbalance was found to be more present in Kumpeh than in Sitiung.

In Sitiung, 10% of transmigrants in the surveyed area gave up on resettlement and returned to Java or sold their land to spontaneous migrants who came later in the first 20 years. Today, 23% of dwellers are spontaneous migrants. The residential land transformations that happened to accommodate them represent 17% of total modifications.

In Kumpeh, a total of 40% of transmigrants in the surveyed area left the resettlement site and went back to their original village or somewhere else. This number is greater than in Sitiung because of the flood problem and the difficulty of improving their physical living environment. There was a large quantity of cheap land available to sell at reasonable prices from transmigrants who wanted to move out due to the large quantity of cheap land available to sell from transmigrants moving out, spontaneously due to transmigrants moved into the resettlement, and it was found that they comprised 21% of residents by now. To accommodate new migrants, it represents 30% of total modifications by dividing land to sell or rent land and houses.

5.3.3 Transformation of Residential Land

As a result of the transformation analysis in residential land, particular patterns can be found in each resettlement. Six transformation patterns of residential land were observed, namely: (1) a single house on the residential land, (2) additional houses by a family member on the residential land, (3) different houses and families on the residential land, which was divided by ownerships, (4) additional houses for renting on the residential land, (5) a new house on the new land, and (6) communal land. Although land transformations were made for different reasons, there is not much difference in their outcomes (see Fig. 5.4). Most of the dwellers live on the same residential land received by the government with their nuclear family or extended family with other houses inside the same residential land. It counts for 76% in Sitiung and 70% in Kumpeh. When it comes to divided land (sold or given to another family to use), the land division was mainly made to be sold to spontaneous migrants that came after resettlement in Sitiung. Kumpeh has a higher percentage of it due to the families that lost their lands by the flood and moved to resettlement in different areas. While new residential land was made available in Sitiung for spontaneous migrants, this was not found in Kumpeh. Interestingly, compared to the strictly designed Sitiung project, land borders and restrictions in Kumpeh were more flexible. With many empty plots, new land was expected to be opened and used by spontaneous migrants or even by original dwellers. That did not happen could be an indicator that all the available residential land was already occupied. It could also be attributed to the process of land transfers, in which land that the original residents had already established was more attractive than uncertain land parcels. There was also a disparity in land use practices.

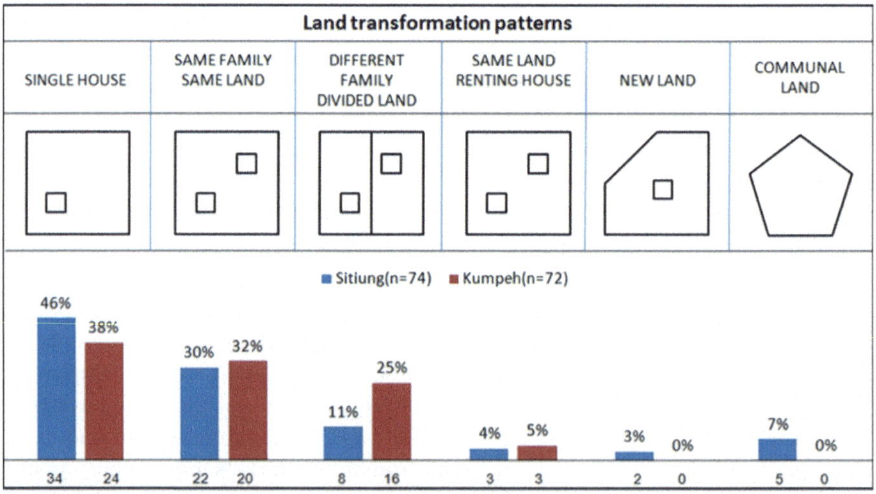

Fig. 5.4 Transformation patterns of residential land of resettlements. *Source* Authors

5.3.4 Transformation of Infrastructure

In both the Sitiung and Kumpeh sites, basic infrastructure, such as drainage systems, paved roads, rainwater canalization, and access to drinkable water was not provided. After some years, the government provided electricity and some institutional buildings like schools and governmental buildings after some years. Therefore, most of the basic infrastructures related to their living were made by the transmigrants themselves.

Sitiung was a special case in which the entire community was resettled together from Wonogiri of the dam project, therefore, community ties have been kept very strong. Also, their original region was known for helping each other concept of "Gotong Royong" to improve their own living environment, instead of waiting for external assistance. One of the first infrastructural improvements they made was building rainwater collection and canalization along with the road improvement (1979). It was urgently necessary since their residential area suffered from floods, and dwellers said that during the rainy season, water would enter their houses and cause damage. The community members also built community facilities like watching/gathering posts (for community safety), as well as storage facilities to store common materials for ceremonies like weddings or festivals. Also, digging the wells was also one of the first improvements during the first years.

At the time of the survey, 100% of households had access to electricity, 84% had access to piped water, 93% had septic tanks, 92% had access to paved roads, and 85% of dwellers had access to good infrastructure according to the categories explained in the methodology. As a result of improvements made by the dwellers and the installation of infrastructure later by the government, living conditions were much improved such as electricity covers the whole area, and more than half of the dwellers have access to gas, septic tanks, telecommunications, piped water, and paved streets.

On the other hand, in Kumpeh, the area lacked sufficient infrastructure and access to electricity, drainage systems, paved roads, rainwater canalization, and drinkable water. Improvements in the living conditions were made by transmigrants themselves individually, and there were no community-based construction works conducted since many dwellers gave up on the resettlement, and dwellers changed rather quickly. Also, came from different backgrounds and places, making it more difficult to create and build community relationships and ties. After many years, the village today has access to piped water (59%), septic tanks (74%), and electricity (100%). However, only 18% have access to paved roads, mainly due to the flooding issue still today and not mitigated. Also, this study revealed that 77% of dwellers live with poor infrastructure like access to only oil for fuel and electricity (see Fig. 5.5).

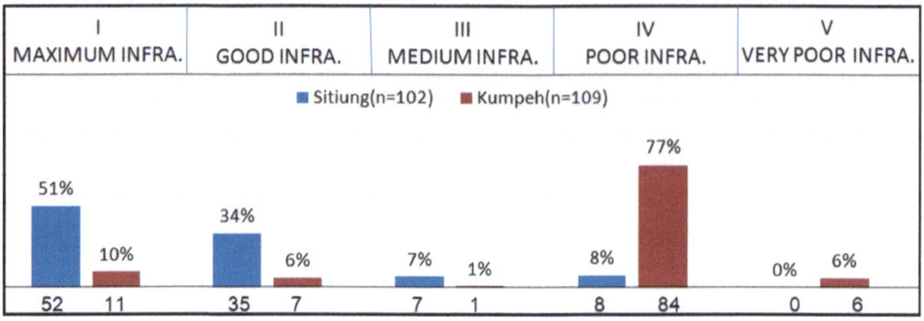

Fig. 5.5 Infrastructure patterns in 2017. *Source* Authors

5.3.5 Housing Transformation

In both resettlement sites, dwellers received the same type of housing and residential land according to transmigration guidelines (Republic Indonesia 1982). One difference is that in Kumpeh, the house was elevated one meter from the ground, showing that designers had already predicted the flooding problem. Also, in Sitiung, most houses had toilets and wells, which were not provided in the Kumpeh. Interviews revealed that for both projects, the main sources of dissatisfaction with the houses were similar: the temperature inside was very hot, there was poor ventilation, the small size didn't fit the family's needs, poor materials and structure, and water entering the house during floods. To cope with these problems, dwellers improved their houses, and now the resettlements have a wide range of sizes, shapes, and housing materials.

(1) Sitiung Case

In Sitiung, house size increased from 36 m^2 to an average of 130 m^2 to accommodate family needs (see Fig. 5.6). The material of the structure of the houses also changed with time most houses are now built by of a confined structure (64%) often confined concrete block, and only 18% still contain the elements of the original wooden structures, while 19% are built with mixed materials of wood, brick and concrete block. In fact, 82% of the houses are completely reconstructed. Only one original house that was provided at that time remains in the area. Modifications were performed by dwellers themselves with help from the community members during the first two decades. However, in the past two decades, dwellers paid construction workers to build their houses, improving the quality of the house. In the last two decades, dwellers paid construction workers to build and improve the quality of houses and structures.

Today, 76% of the dwellers live on the same land that they received from the government with their nuclear family or extended family. Many have other houses built in the same cluster. Each land cluster is often divided by a fence, wall, or plantation.

Most of the spontaneous migrants (76%) came between 1985–1995 (around 8–18 years after resettlement) which was made possible by increasing the number of

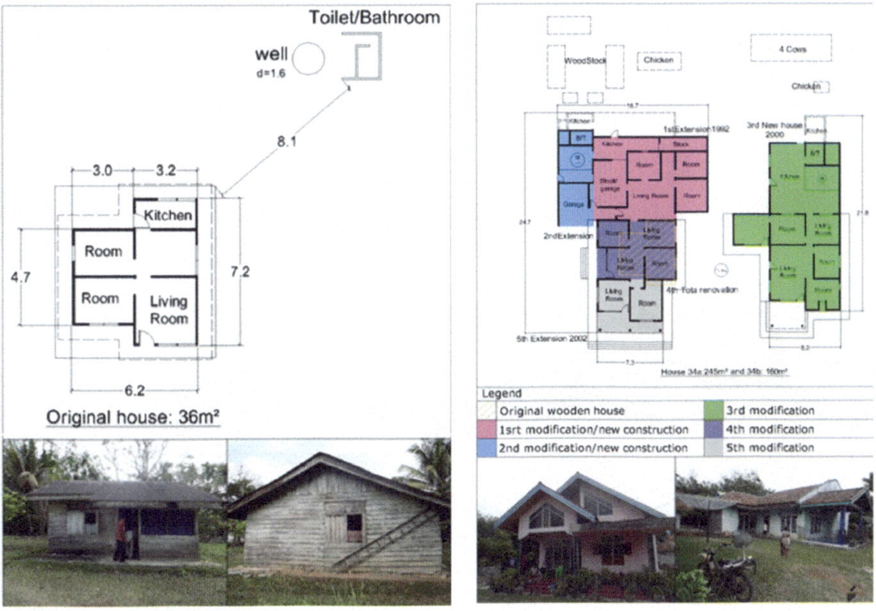

Fig. 5.6 Example housing transformation in Sitiung (example case). *Source* Authors

plots. This indicated that resettlement probably took around 13 years to reach a certain level of stability, as it started to become attractive for new migrants to move in. In summary, out of the total 59 original transmigrant landholders, 6 owners (10%) gave up living there and sold the land (4 sold the entire land, 2 sold part of their land first, and the rest after some years), 1 owner passed away, and 52 original inhabitants still remaining.

(2) Kumpeh Case

In Kumpeh, houses increased from 36 m2 to an average of 86 m2 (but varieties from 14 m^2 to 230m^2), and dwellers have their houses now made of confined brick (35%), non-confined brick (10%), wooden structures (29%), and mix structure (26%). Due to the flooding problem and the conditions of original transmigrant housing, today, 88% of the houses are completely reconstructed, and only one original house remains intact, while 11% of houses can still see traces of the original house. No presence of a concrete block was found. Instead, dwellers use brick. Most of the houses are made of confined brick, followed by wood and a mix of both. Partition is mainly made with brick and a mixture of brick and wood. Corrugated iron is the material most used for roofing, followed by a tile roof and a mix between them. Modifications were performed by dwellers themselves with help from neighbors in most cases, only for the last decade (concerning quality brick houses) were built by construction workers; therefore, structural quality is not as high as in Sitiung (see houses in Fig. 5.7).

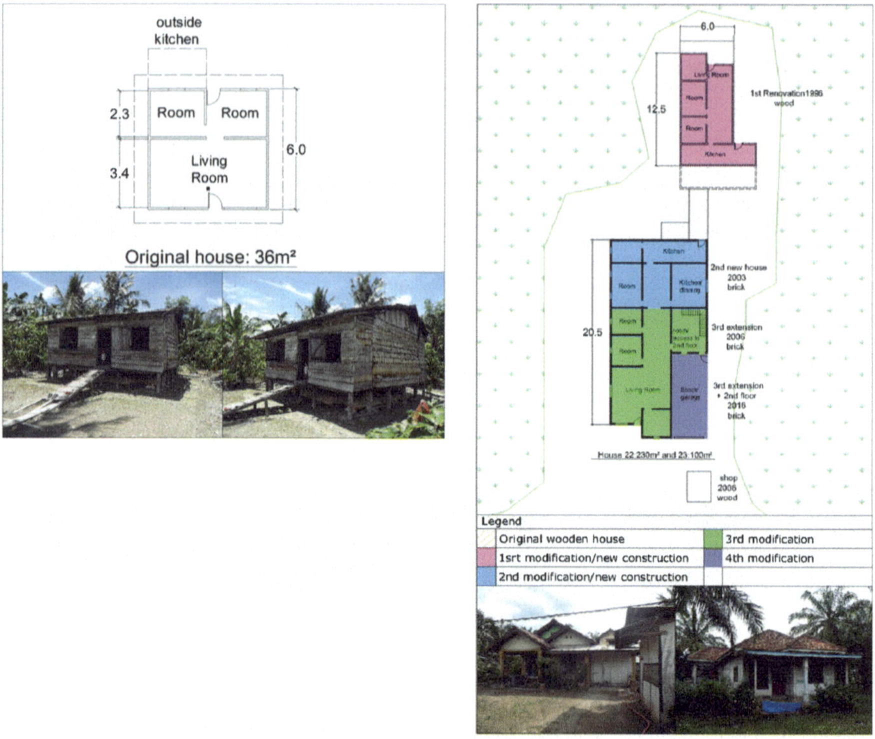

Fig. 5.7 Example housing transformation in Kumpeh (example case). *Source* Authors

Comparing the type of structure, Sitiung has the majority of structures confined (64%), with no presence of only non-confined structures. Wood and mixture between confined and wooden structure or confined and non-confined structure. In Kumpeh, 35% is confined, 29% is wood, 26% is a mixture, and 10% is a non-confined brick structure.

(3) Housing Durability Comparison

After all these transformations, dwellers improved their living conditions compared to the initial conditions of resettlement. The durability of the houses differs considerably. Kumpeh houses have around the same ratio of durability I (permanent), II (Semi-Permanent), and III (Temporary), while Sitiung durability I (permanent) is the most prevalent. In Sitiung, 59% of dwellers live in the permanent category, meaning they have good access to infrastructure and high durability of their houses. In contrast, 83% of dwellers of Kumpeh are still found in temporary living conditions, meaning that there is a lack of infrastructure, and houses are still built with wood. Therefore, even after 30 years of resettlement, parts of the village are still found almost in the same situation as in 1986. Figure 5.8 shows the comparison of road conditions and housing durability comparison.

Fig. 5.8 Road condition and housing durability comparison. *Source* Authors

(4) Living Satisfaction

Most dwellers already reached the fulfillment stage in Sitiung since their living conditions are better than in the original village, but not in Kumpeh. These modifications include enlarging houses to receive friends and family for long periods of time during vacations, building one's own business for an increase in income, or hobbies. Houses in Sitiung are bigger, with spaces to receive visitors, while houses in Kumpeh are smaller and usually just enough to accommodate the family. However, transformations regarding building their own business were found in both resettlements at almost the same rate, with 11 families in Sitiung and 13 families in Kumpeh making this kind of modification.

It was surprising when data showed the high level of satisfaction of dwellers in both sites regarding joining the resettlement and life in general. This happened because transmigrants who passed through the adaptation process and survived in the resettlement until today, in one way or another, could improve their living conditions, land holdings, and housing in comparison with their situation before the resettlement. Especially in Kumpeh, where transmigrants didn't have much land before and lived in cramped conditions with their families. Many said they were satisfied with the resettlement because they had the opportunity to be independent from their parents and have their own house and land.

5.4 Community Efforts to Improve Their Living Environment

Although environmental factors have affected the adaptation process of the resettlements differently, as explained that Kumpeh is located in a swamp area and suffers from chronic floods during the rainy season (six months) every year, comparatively, the Sitiung community had problems with flooding in the beginning but became relatively stable after transformations made by own community and government's irrigation project. These differences are influenced by the social structures of the community relationship and are rooted in the original community.

5.4.1 Motivation for Transformation

Motivations for transformations in both resettlements showed some differences and similarities. Regarding similarities, it was found that both villages made similar changes during the survival phase of adaptation. These changes focused on overcoming flooding problems, cramped conditions, and access to water. Motivations may be the same, but strategies to cope with similar problems differ a little. In Sitiung, the flooding problem was mitigated by community building construction of drainage systems as well as individual housing modifications (like elevating the house level and paving the floor). In Kumpeh, only individual housing modifications were made, although neighbors also helped with the construction. Cramped conditions were solved by households by enlarging their own houses. Access to water and building water wells was also mainly made by the community in Sitiung, while in Kumpeh it was made individually.

5.4.2 Goton Royong and Community

If the community is healthy and well-established, cultural and social activities are very present. Mosques and prayer houses are very common, and in the case of Sitiung, local schools hold festivals with traditional dancing and music. The community manages marriages, religious festivals, and other activities using its own resources. In some other cases, they also have the help of credit associations to deal with public goods (Benoit and Pain 1989). Another important practice held by communities is called "Gotong Royong" (mutual cooperation), where dwellers engage in construction, organization, maintenance, and building activities to improve the resettlement. Some researchers call attention to the fact that the government took advantage of this "traditional" community labor practice to impose and secure free labor for public infrastructure projects (Bowen 1986). In the case of Sitiung, it was observed that both cases were present. Transmigrants were managed by the government to build irrigation systems and land clearing during the first years. However, they also cooperated with each other as a community to build housing for the elderly or poorest ones, wells, septic tanks, and drainage systems to avoid floods.

Usually, a good social network happens in large-scale resettlements when people come together from the same village and share the same background. When resettlement is made by people with different backgrounds, community bonds can be weakened. In Kumpeh's study, people with different backgrounds who had no previous relationship with community-based construction works, for example, were found to have less impact on the community than in the Sitiung case. This was also influenced by the fact that many transmigrants moved due to the floods, making social bonds harder to maintain with spontaneous transmigrants coming and going.

On a smaller scale, although Kumpeh residents said there was no organized community-based construction work, gotong royong was cited by Kumpeh dwellers as an activity to help build new houses for the poorest newcomers or those who lost their houses during floods.

5.5 Conclusion

This chapter analyzed the long-term adaptation of transmigration resettlements in Sitiung and Kumpeh, Sumatra Island, Indonesia. After reviewing the resettlement provision process, this study focused its analysis on the physical transformations and adaptation of houses, infrastructure, and land made by beneficiaries in two main settlements. Under the premise that transformation should be understood as an important aspect of adaptation to life in a resettlement resulting from the unsatisfied needs of residents in a new environment, analysis in this chapter identified transformation patterns of transmigrant dwellers who made physical changes to government provisions, first as a matter of survival to cope with challenges in the resettlement, and then as part of socio-economic evolution. The study also explored the transformations with a focus on how physical transformations and the adaptation process of dwellers are linked, influencing their adaptation level.

The main differences were found in environmental conditions (since Kumpeh is a swamp area suffering from floods) affecting the infrastructure conditions; therefore, the infrastructure in Sitiung is in better condition. Floods also influenced housing quality, forcing dwellers in Kumpeh to reconstruct their houses from scratch more than once. While in Sitiung, they made gradual transformations and improvements.

Resettlement deeply affects the lives of transmigrants. More attention should be paid to transformations in built capital. Assistance in improvements of these structures is essential to improve living conditions of resettlements, with assistance for relocation and site location with proper infrastructure and services, and assistance for rehabilitation to maintain and/or improve physical conditions, therefore improving standards of living in the post resettlement period. Consideration of local conditions of resettlement should be taken more seriously. Identifying patterns in transmigrants' physical needs to sustain their daily activities, family structures, and livelihood under conditions of the local climate is critical for the provision of suitable resettlement and construction of local sensitive solutions.

References

Arndt HW (1983) Transmigration: achievements, problems, prospects. Bull Indones Econ Stud 19(3):50–73

Adhiati M, Bobsien A (2001) Indonesia's transmigration programme–an update: a report prepared for down to earth. International campaign for Ecological Justice in Indonesia. http://www.downtoearth-indonesia.org. Accessed June 2018

Benoit D, Pain M, ORSTOM (Agency : France), Indonesia (eds) (1989) Transmigration et migrations spontanées en Indonésie: Propinsi Lampung. Jakarta Selatan, Indonesia: Bondy, France: Departemen Transmigrasi ; ORSTOM

Bowen JR (1986) On the political construction of tradition: gotong royong in Indonesia. J Asian Stud 45(3):545

Gondowarsito R (1990) Transmigrasi Bedol Desa: Inter-Island Village Resettlement From Wonogiri To Bengkulu. Bull Indones Econ Stud 26(1):48–68

Hardjono J (1986) Transmigration: looking to the future. Bull Indones Econ Stud 22(2):28–53

Republic of Indonesia (1963) Hubungan sewa menyewa perumahan, peraturan pemerintah republik indonesia nomor 49 (Rental regulations)

Republic of Indonesia (1982) Guidelines for transmigration settlement planning: introduction, phase I, phase II., Republic Indonesia, department of public works and electric power, directorate general of housing, building, planning and urban development

Republic of Indonesia (403ad, kpts/m/2002) Tentang pedoman teknis pembangunan rumah sederhana sehat (rs sehat), keputusan menteri permukiman dan prasarana wilayah. Menteri permukiman dan prasarana wilayah (Temporary health housing regulation)

Republic of Indonesia (2002) Bangunan gedung, peraturan pelaksanaan undang-undang nomor 28,peraturan pemerintah (Building regulation)

Suratman GP (1977) The changing focus of transmigration. Bull Indones Econ Stud 13(2):78–101

The World Bank (1988) Indonesia–The Transmigration Program in Perspective. A World Bank country study, Washington, DC

Chapter 6
Evolution of Traditional Mayangna Housing in Nicaragua: Effects of Resettlements and Cultural Transformations on Its Habitability Through Comfort Simulations

Nathalia González Meixueiro and Guido Cimadomo

Abstract The Sumu-Mayangna ethnic group had its first settlements in Central America; however, due to a lack of archaeological research in this area of Nicaragua, there is little information about the prehispanic history of the community. Currently, most of the Sumu-Mayangna population inhabits the Bosawás Biosphere Reserve in northeastern Nicaragua. As a consequence of a series of historical events, the community has occupied a region separated geographically, politically, and economically from the rest of the country. Disregarding the multiple foreign interventions that the Sumu-Mayangna has suffered, they have maintained within their cosmovision the harmonious order between nature and human; even today, they are still very much related to their natural environment. During the Nicaraguan Revolution, which started in 1979, this community was expelled by force from its ancestral land, villages being burnt, and land devastated. Around 25% of the Sumu-Mayangna community had to escape from their homes, obliged to resettle in refugee camps in Honduras, close to the border with Nicaragua, while the rest were captured to work in concentration camps settled in a land known as Tasba Pri. Sumu-Mayangna people were able to return to their country only at the end of the conflict, being resettled in new camps close to the village of Bonanza and forced to adapt to a new style of life. This research focuses on the influence of historical and social events on the Sumu-Mayangna dwelling. It pivots on three relevant historical periods: the era of colonization and the impact of foreign cultures up until the mid-twentieth century; the Nicaraguan Revolution period between 1979 and 1989; and the last three decades, meaning from 1990 to the present day. Through archival research, interviews, architectural surveys and bio-climatic simulations, this chapter will explore the interrelation between their

N. G. Meixueiro (✉) · G. Cimadomo
Department of Art and Architecture, University of Malaga, Málaga, Spain
e-mail: nathaliameixueiro@gmail.com

G. Cimadomo
e-mail: cimadomo@uma.es

C. Ochiai et al. (eds.), *Disaster and Displacement*, Disaster Risk Reduction,
https://doi.org/10.1007/978-981-96-2583-3_6

cosmological worldview, natural environment, political-social events, and the transformations of their homes. In this research, thermal comfort will be used as a way to establish the degree of repercussion, serving as a measure of human well-being in their interactions with buildings.

6.1 Introduction

The Sumu-Mayangna is a transboundary civilization that has historically inhabited the Atlantic Coast region of Nicaragua, currently with settlements in less than ten localities related to the Bosawas Biosphere Reserve in the north of the country. The Sumu-Mayangna community is also located in southern Honduras, in the departments of Olancho and Gracias a Dios. The Sumu-Mayangna indigenous manner of living is the result of historical processes involving violence, forced migrations, colonization, and conquest. Although diminishing in contemporary times, they are strongly linked to nature. Their vital source is the forest, establishing a harmonious order between nature and humans. It is within these forested domains that individuals procure essential resources such as firewood, sustenance, medicinal plants, and construction materials (Cordón Pedregosa 2007, von Houwald 2003).

For part of the Sumu-Mayangna indigenous population, the Bosawás Biosphere is a resource provider. The Bosawás Biosphere Reserve (BO: Bocay River, SA: Saslaya Hill, and WAS: Waspuk River) is located in northwestern Nicaragua and has long been the natural habitat of the Sumu-Mayangnas and other indigenous people (Fig. 6.1), where they currently have the freedom to live their traditions outside the influence of mestizos and their culture–an ethno-racial classification to refer globally to a group of people of mixed European and Indigenous American ancestry, who have the highest representation in positions of power–or by contact with foreigners who come to their lands attracted by the numerous natural resources (PNUD 2005: 129). According to the last census conducted by the National Institute of Statistics and Census (INEC) in 2005, it is currently inhabited by seven indigenous communities: Mayangna Sauni As, Mayangna Sauni Bas, Mayangna Sauni Bu, and four other indigenous communities that do not belong to the Sumu-Mayangna group.

Over the past few decades, the territory of the Sumu-Mayangna has been threatened by factors such as the exploitation of natural resources and the unlawful conversion of forested areas for agricultural purposes. Furthermore, the adverse effects of climate change have exacerbated the vulnerability of their ecosystem, subsequently impacting Sumu-Mayangna's lifestyles and the integrity of their habitats. These multifaceted challenges have had profound consequences for their socio-cultural practices and the sustenance of their dwellings.

Despite Sumu-Mayangna's understanding of nature, insertion into modern society has been inevitable for this ethnic group, either by proximity to mestizo communities. Before being introduced to the neoliberal economic model, the community shared resources in proportion to individual production capacity. Although they are aware that money cannot compensate for the sale of their resources, trees, or land and that

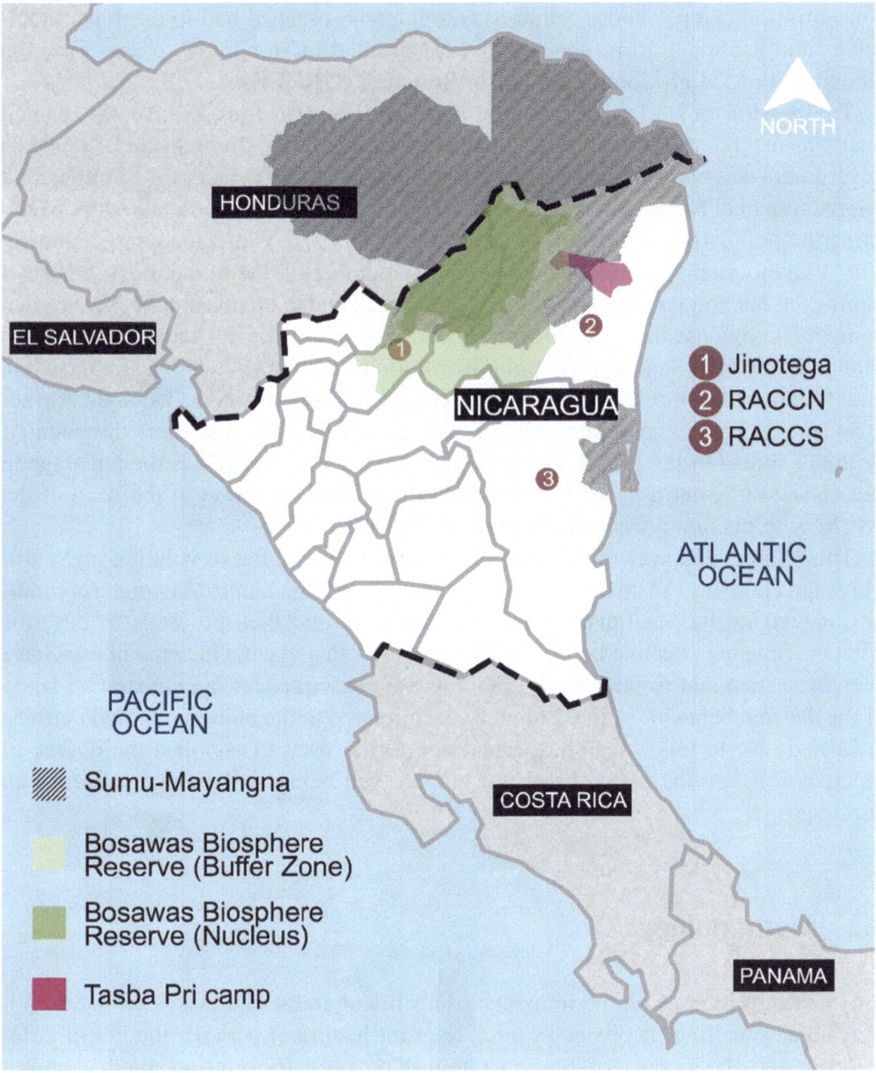

Fig. 6.1 Contemporary location of the Sumu-Mayangna communities in and around the Bosawas biosphere reserve in the north of Nicaragua, north Caribbean coast autonomous region (RACCN) (Nathalia González Meixueiro 2023)

their needs are frugal and satisfied without depending on almost any foreign product, western concepts have increasingly taken hold in the community.

Certainly, one of the significant historical occurrences in Nicaragua that profoundly influenced Sumu-Mayangna society was the Sandinista Popular Revolution (1979–1990). In this period, a considerable number of indigenous community members were involuntarily and aggressively transferred to government-organized concentration camps. Those Sumu-Mayangna who escaped had to establish themselves and acclimate to refugee camps constructed in Honduras, overseen by the United Nations High Commissioner for Refugees (UNHCR).

The Sandinista National Liberation Front (FSLN) was founded in 1961, named after the insurgent Sandino. The movement was gaining strength, and Somoza's government was being affected. This resulted in widespread persecution, murder, and imprisonment of Nicaraguans who were allegedly enemies of the dictatorship, weakening the popularity of the current government. In 1978, Pedro Joaquín Chamorro, who was a journalist opposing Somoza and a member of one of the most important families in Nicaragua, was assassinated, a fact that ended up turning the Nicaraguan bourgeoisie against the government, weakening Somoza even more (BBC News Mundo 2019). The Sandinista Popular Revolution began in 1979 in Nicaragua, when after four decades of rule by the Somoza regime, Anastasio Somoza Debayle, former president of Nicaragua, fled to Miami, and the FSLN took control of the country. With the spread of the FSLN, the opposition also arose, which was the Nicaraguan Resistance or "Contras". In 1979, Daniel Ortega came to power at the head of the FSLN, with his first government lasting until 1990.

This chapter analyzes the effects of the Sandinista Popular Revolution on Sumu-Mayangna housing. To do so, it first details the role that the Sumu-Mayangna community played in the Sandinista Popular Revolution, and then proceeds to describe what the housing was like before, during and after that event. Once the houses have been described and to objectively quantify the consequences, a comparative study of the thermal behavior of these models, each related to the aforementioned periods, is carried out. In this research, thermal comfort is used to establish the degree of repercussion, serving as a measure of human well-being in their interactions with buildings.

6.2 Methodology

This research focuses on the influence of historical and social events on the Sumu-Mayangna dwelling. It pivots on three relevant historical periods: the era of colonization and the impact of foreign cultures up until the mid-twentieth century; the Nicaraguan Revolution period between 1979 and 1989; and the last three decades, meaning from 1990 to the present day. This chapter will show, through archival research, interviews, architectural surveys, and bio-climatic simulations, the interrelation between their cosmological worldview, their natural environment, political-social events, and the transformations of their homes.

This research is based on primary sources for documentation on original and contemporary housing, in detail through fieldwork surveys and conversations with the inhabitants of the Bosawas Biosphere Reserve (Fig. 6.1), and on secondary bibliographic sources for the history of the Sumu-Mayangna community and for the reconstruction of housing during the Revolution period.

This analysis considers aspects related to thermal behavior and humidity with the support of dedicated software such as OpenStudio, Autodesk CFD, and Revit. Each of the three housing models was developed based on bibliographic descriptions and field visits carried out under the University Volunteer and Cooperation for Development program, organized by the University of Malaga and the Andalusian Agency for International Development Cooperation in Nicaragua, when it was possible to find and survey case study houses similar to those described in the bibliographic sources.

6.3 The Sumu-Mayangna and the Popular Sandinista Revolution

On November 11, 1974, the National Association of Sumu-Mayangna Communities (ANCS, Asociación Nacional de Comunidades Sumu-Mayangnas) was created with the support of religious institutions and local associations. Sumu-Mayangna representatives from fourteen villages met to discuss their collective development. Their purpose was to obtain titles to their land, make their rights and obligations as Nicaraguans known, bring education to all, improve sanitation and health infrastructures, and promote their culture. In all the history of the community, there had never been a supra-regional organization, on the contrary, the links among small groups were lost in proportion to the advance of the conquerors in their land until they remained in small, isolated groups. At the same time as the Sumu-Mayangna's constitution was getting stronger, the political conflicts in Nicaragua worsened, and the indigenous people ended up involved in political conflicts alien to them. Their territory spread through actual North and South Caribbean Coast Autonomous Regions (RACCN and RACCS), poorly integrated economically and politically into the national state. Since 1894, this area had been abandoned–it was basically a colony–with concessions to logging, mining, or fishing companies or with new settlers coming from the Pacific region who settled near the rivers, expelling the indigenous people from their places. The indigenous inhabitants of the coast had not benefited from the same development as the rest of the country. This population was distinguished from the rest of Nicaragua; they maintained many of their ancestral customs, and their language was their traditional tongue.

In the third ANCS assembly held in 1980, thirty-two communities and government representatives met; however, there was no exchange of opinions. When issues such as the dispossession of their lands by settlers were brought up, government representatives answered, 'That will not happen anymore because now the land belongs to the people' (von Houwald 2003:570). The Sumu were recently strengthening their individual consciousness and developing as a community it was the first time that high officials had visited them. Although there were some skeptics, everyone in the community was excited and confident in the new doctrine. The Sandinistas, by this time already established in the power of the national government, were taking their ideology to every corner of Nicaragua and used every possible means to approach the

Sumu-Mayangna, including a "Literacy Crusade." Their goal was to instill Sandinista values. All these coercive measures were at first disguised as good deeds, but later, their true intentions came to light, being contrary to Sumu-Mayangna customs and therefore difficult to accept: "The administration of land by the state (adding them to the production of state farms) and increasing production: that was the only thing that interested the regime!" (von Houwald 2003: 574).

The recent creation of the ANCS and the lack of political experience produced a situation of confusion and disorientation among the Sumu-Mayangna. They tried to stay out of political conflicts, and during the fourth ANCS in 1981, the vast majority voted to maintain a neutral profile. However, their reserved attitude was interpreted as hostility by the Sandinistas, who responded with violence. This first phase of indoctrination was followed by a second, much more violent phase. On January 14, 1981, forced relocations began, a decision taken by the General Staff to protect the civilian population against invasions from the Contras.[1] The truth was that the General Staff's decision was economically motivated. The Sandinistas arrived in Sumu villages to recruit young people, give them military training, and then return them to their villages so they could later be responsible for military actions (Bataillon 2001). The initial contingent was carried out by evacuating different villages between 1982 and 1983, displacing the indigenous people in remote areas of the jungle. Thousands of indigenous people were forced to abandon their homes violently: the regime burned houses, cut down trees, and killed their domestic animals, forcing them to march through the jungle to work camps in an area called "Tasba Pri" (Free Land) (Fig. 6.1). The affected people wandered into the mountains because the authorities did not allow them to create new settlements; in consequence, many Sumu-Mayangna died from starvation and disease (von Houwald 2003: 596). These camps had an economic and political advantage for the government, as it was easier to maintain control over the population. They served as collective work camps in the cultivation of coffee plantations to support the national economy. This method resembled the *Kolkhozes*, a stimulus program to increase agricultural production devised by the government of Joseph Stalin in the early twentieth-century Soviet Union. The displacements caused the division of indigenous peoples: those who stayed, those recruited for the labor camps, and those who, confused, fled to Honduras and Costa Rica. Until then, the population of the Atlantic Coast had not had strong participation in the Nicaraguan economic system, since any production was difficult to sell due to the lack of communication routes.

Upon the arrival of the Sandinista army in the Sumu-Mayangna villages, many decided to flee, leaving all their belongings behind. Six hundred twenty-one Sumu were reported to have set foot on Honduran soil on July 25, 1982, fleeing the regime's conscription, followed by five hundred twenty-five on August 2, 1982. Another indefinite number moved to Costa Rica. Although it is impossible to know exactly how many, it is estimated that at least a quarter of the Sumu-Mayangna population left the country during this period (Dodds 1989). The refugees who arrived in Honduras

[1] Colloquial is the name given to the U.S.-funded opposition from Honduras that tried to overthrow the FSLN (some groups are known as MISURA, KISAN, or YATASMA).

were taken by the United Nations High Commissioner for Refugees (UNHCH) to the Mocorón camp in the eastern part of the country. Such a drastic difference between their native land and their new settlements resulted in the death of many indigenous people due to the change in food, camp life, and topographic and climatic conditions alien to their customs (Desenclos et al. 1990).

Local UNHCR leaders did not hide the fact that they felt identified with the Sandinistas and even declared themselves willing to undertake repatriation actions by force if necessary. It should be noted that, in addition to the Sumu and the rest of the population fleeing the conflict, Honduras also hosted refugees from the former Somocista political party, housed in independent camps, such as El Paraíso (Gómez García 2021). For the Sandinistas, it was part of their propaganda to get the refugees from the Atlantic Coast to return to Nicaragua. UNHCR used coercion to encourage the Sumu to return voluntarily, reducing food supplements and medicine and offering them economic incentives. When some finally returned to Nicaragua, they were unable to return to their hometown and were relocated to camps near the town of Bonanza. UNHCR admitted that they were returning to desolate conditions, to barracks camps in the mining regions. Naturally, the Sumu-Mayangna were used as part of Sandinista proselytizing, making it appear that they were taking the multi-ethnic character seriously. It was not until 1984 that President Daniel Ortega recognized that the approach by the Sandinista front had been misguided. In early 1985, negotiations began between the Sandinistas and the Sumu, concluding in a pact that recognized that the revolution depended on guaranteeing the people of the Atlantic Coast their political and cultural autonomy, thus restoring the confidence of the Sumu (Morris 2014).

Beginning in 1990, after the end of the armed conflict, the National State was restructured, and with it, the policies of the newly created Autonomous Regions (Fig. 6.2). The Autonomy of the Atlantic Coast was approved in 2003, resulting, among other things, in the power of the indigenous people to decide on the approval of contracts for the exploitation of natural resources, and independent decision-making in education, promotion, development, and conservation of cultural heritage. The war also strengthened indigenous peoples' sentiment towards land title rights. Although investments in the Atlantic Coast have increased over time, most ethnic communities continue to live in extreme poverty (PNUD 2005: 55). The Sandinista conflict ended up destroying what was left of Sumu's history. They were involved in a conflict they were not part of; they went through a process that, again, after the Spanish colonization of the sixteenth century, marked their culture and forced them to change, directly or indirectly, customs and ideologies.

Fig. 6.2 Most relevant events and policy changes (Nathalia González Meixueiro 2023)

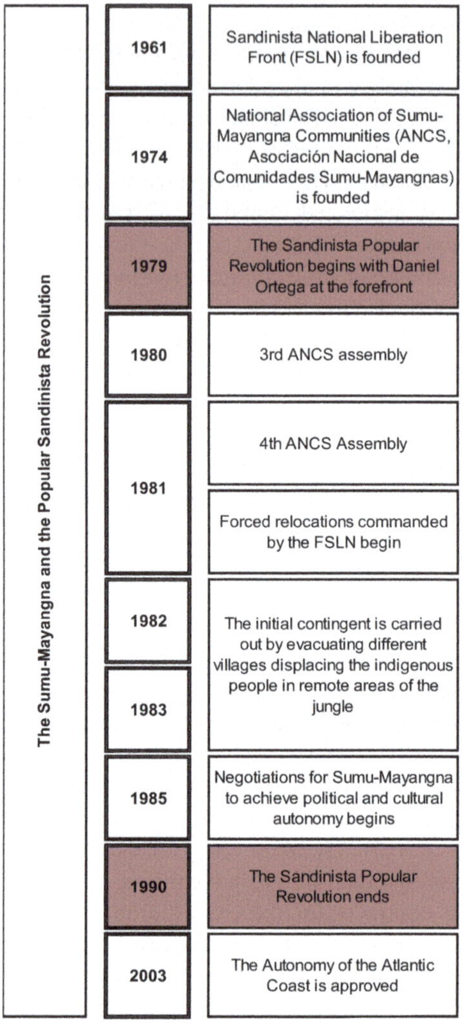

6.4 Consequences of the Sandinista People's Revolution on Sumu-Mayangna Traditional Housing

6.4.1 Sumu-Mayangna Traditional Housing

Prior to the Sandinista Revolution, and as a result of the transformations brought about by sixteenth century European colonization, Sumu-Mayangna housing was constructed one meter above the ground. While it was common to find rooms elevated on pillars on the Atlantic Coast due to seasonal flooding, in the interior of the region this was not such a common solution. The Sumu have traditionally inhabited mountainous areas, and this method of elevating their homes likely served to prevent flooding caused by heavy rainfall and runoff. It may have also been a measure to keep crawling insects out of the houses and protect the wooden floors. In any case, this practice allowed for constant ventilation in the lower part of the structure, which helped regulate the temperature inside the house. By raising the house, a thermal cushion was created with the ground. The primary structure of these houses was built using wooden logs, and the walls of the houses were made of rammed bamboo. This technique involves cutting bamboo stalks in half and then ramming them to create a fibrous panel, which is then attached to the main structure. Bamboo has historically been an abundant material in the area, and resulted in a thin, porous envelope that allows for the constant renewal of air. This makes it ideal for the local climate, as it provides well-ventilated walls. Another widely used material in the region is the Suita Palm (*Calyptrogyne Sarapiquensis*), whose leaves were commonly used for constructing roofs.

Inside the houses, there was very little furniture, among them the "Kubus" stove—a low wooden table with a drawer filled with white clay. On top of it, an inverted "U" of clay, 10 to 15 cm high, which worked as a cooking plane under which the firewood was placed . Their austere lifestyle was evident also in their vocabulary. In Lehmann's Central America dictionary of 1920 (von Houwald 2003), there are only sixty words in the chapter on "House and Objects", while there are more than sixty names for types of birds.

Today, few houses are built with bamboo walls and Suita roofs. In 2019, a study was carried out in the Sumu-Mayangna communities of Mukuswas and Santa Maria (Cimadomo et al. 2020), where it was still possible to find among the settlements some dwellings replicating the postcolonial housing model described in the bibliographic sources (Fig. 6.3). These were surveyed and photographically documented. Some huts consisted of a single volume that served as kitchen and living quarters; others were composed of two volumes that did not necessarily have a specific use. In some houses, the Kubus stove was still used.

6.4.2 Sumu-Mayangna Housing During the Sandinista Era

6.4.2.1 The Tasba-Pri Labor Camps

Under the pretext of protecting the civilian population from Contra invasions, the indigenous people recruited by the Sandinista regime were distributed in camps in an area called "Tasba Pri," located along the road that crossed the region from East to West approximately 60 km south of the Coco River, on the Rosita-Puerto Cabezas Highway. These labor camps had 53,543 hectares of fertile land, schools, and health services. Jorge Jenkins described these settlements as modern urbanization, standardized and equidistant housing, comparing them to a Romanian program of production and urbanism called *sistematizare*, which aimed to eliminate rural settlements that

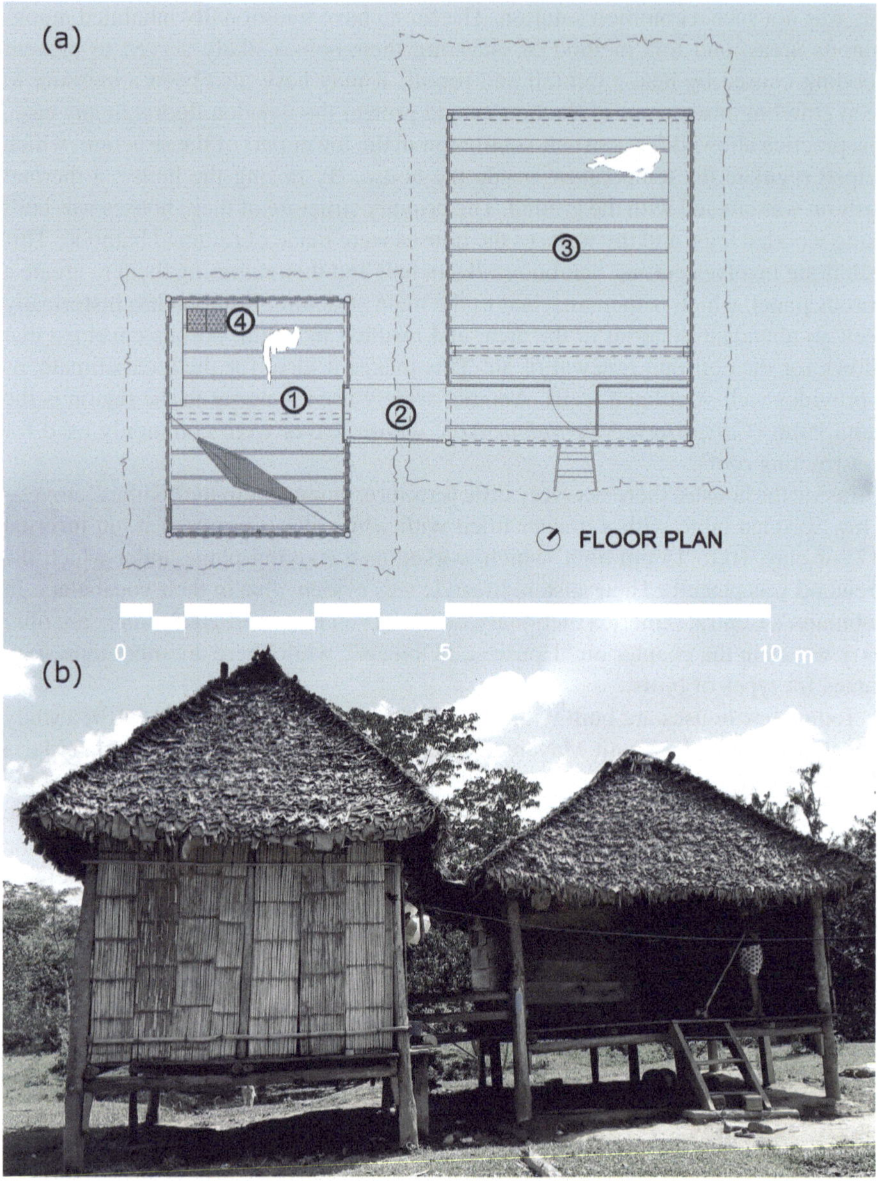

Fig. 6.3 Pre-Sandinista Sumu-Mayangna housing. Room dedicated to kitchen open air connection free space room with single shelf Kubus stove (Nathalia González Meixueiro 2019)

represented a burden to the Romanian government (von Houwald 2003: 584). In Tasba Pri, there was a similar intention to replace the indigenous villages in order to integrate their population into the national production system.

Due to their barn shape, Von Houwald called the Tasba Pri dwellings "Animal Farm" type prefabricated houses (von Houwald 2003: 591), but besides the fact that the construction of houses was serially arranged in a straight line, there is not much documentation available that accurately describes urbanistic intentions on the part of the government, nor are there many accounts of how the indigenous people found the houses in Tasba Pri, the distribution of the properties or how many people lived inside each one. However, some texts that relate life in general in the camps, such as the book "Nicaragua: The people speak (Levie 1985) and again "Mayangna: Notes on the History of the Sumu Indians in Central America" (von Houwald 2003), include photographic documentation from where their construction system can be deduced. These photographs (Fig. 6.4) show gabled zinc roofs over the prefabricated houses with fiber cement walls. The constructions in Tasba Pri followed the traditional logic of the elevated structural system, probably to protect the dwellings from the humid soil, with only a small window in the main façade, lateral access, and a porch. The latter represents half of the dwelling as if they had intentionally left them half built so that the indigenous people could expand them according to their possibilities. These constructions, at first sight, deficient in terms of thermal comfort, were as foreign to the Sumu-Mayangna culture as the vocabulary inculcated by the literacy crusade with words such as Marxism. The strangeness of their new dwellings led the indigenous people themselves to begin to adapt and personalize the constructions.

6.4.2.2 Refugee Camps on the Border with Honduras

Most of the Sumu population that escaped from the Sandinista military, reached Honduras, welcomed by the humanitarian association UNHCR. It soon became known that the conditions in the refugee camps were not pleasant, and many Sumu preferred to find themselves a place to settle near the border. In the refugee camp in Mocorón (Honduras), they lived in houses similar to those they had in their villages, elevated above the ground, with wooden partitions and Suita roofs. Although no documentation available that describes in detail the living conditions in the refugee camps, analysis of some secondary sources, as it occurs in Tasba Pri, allows us to deduce that the Sumu adapted to this form of construction, adjusting it as much as possible to their customs. María R. Cordón Pedregosa (2007) explains that during their stay in Honduras, the Sumu practiced wood gathering to build their houses in the refugee camps, but they also continued to use traditional materials such as Suita and bamboo.

Partly motivated by the uncomfortable conditions of Mocorón, alien to their context and partially manipulated by the Contra, the Sumu decided to move to Tapalwas (Honduras), where they suffered even more repression from this paramilitary group, as there was a base very close to the new settlement. However, Tapalwas

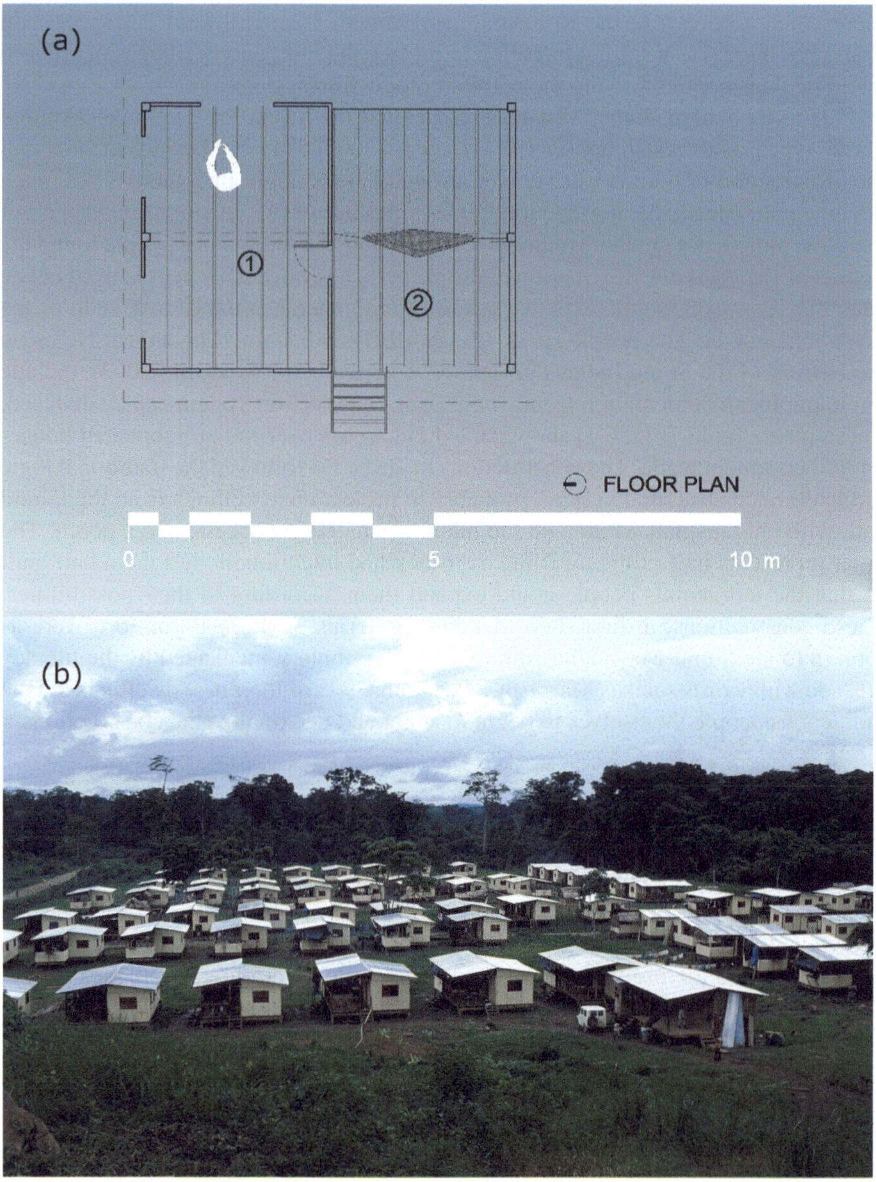

Fig. 6.4 Sandinista Tasba-Pri camp **a** Housing floor plan (Nathalia González Meixueiro 2023) **b** Panoramic view of a camp (Almasio Cavicchioni 1984)

had a similar context to that of their origin, where they were able to practice agriculture, and soon, a village that resembled a traditional Sumu village was born. On the other hand, due to political and propaganda reasons for the Sandinistas, UNHCR began to pressure the indigenous people to be repatriated to Nicaragua, to the point of using coercion to encourage the Sumu to return. The repatriated Sumu-Mayangna were moved to camps around the present municipality of Bonanza built with UNHCR support along the Pispis River; such is the case of Sakalwas.

6.4.3 Sumu-Mayangna Housing After the Revolution

The Sandinista conflict of the 1980s marked a before and after for traditional Sumu-Mayangna housing due to the uprooting of the population when they returned to their place of origin. When the Sumu displaced by the war returned to their territory, some rebuilt their villages, and others, as already mentioned, were housed in UNHCR camps. Documentation of Sumu-Mayangna household culture, carried out in the village of Sakalwas (Cimadomo et al. 2020), presents evidence of how the dwellings stopped using traditional construction methods passed down from generation to generation to adopt Western influences, with the use of industrialized materials and standardized systems alien to their traditions. Although in 2019, it is still possible to find houses that resemble those built before the Sandinista Popular Revolution, the mestizo influence, together with the imposed cultural uprooting, has resulted in the oblivion of construction traditions.

The process of building their houses begins with preparing the land and cutting the undergrowth and trees that could pose a threat in case of storms. Afterward, a wooden slat structure is built, which is supported by a grid that will become the base for the floor of the house, the latter raised on four posts buried in the ground. The roof of Suita leaves has been replaced by sheets of zinc, a symbol of prestige, easy to obtain, although costly. In part, it is justified by the scarcity of the palm for the protection of natural resources against the aggression produced by the settlers who intended to increase their crops. The result of using zinc sheets without any insulation, directly on the vertical walls and without sufficient ventilation, produces a poor level of comfort inside the house. The construction process ends with the envelope, which is made with wooden planks. Wood does not have the same porosity as rammed bamboo, worsening the necessary ventilation of the house, in addition to the fact that the resulting interstices in the arrangement of the planks are the perfect hiding place for insects and crawlers. Intuitively, gaps are left in the envelope as windows, which can be closed with folding shutters. The dwellings continue to have a free interior layout, being more like a shelter that adapts to the different immediate needs (Fig. 6.5).

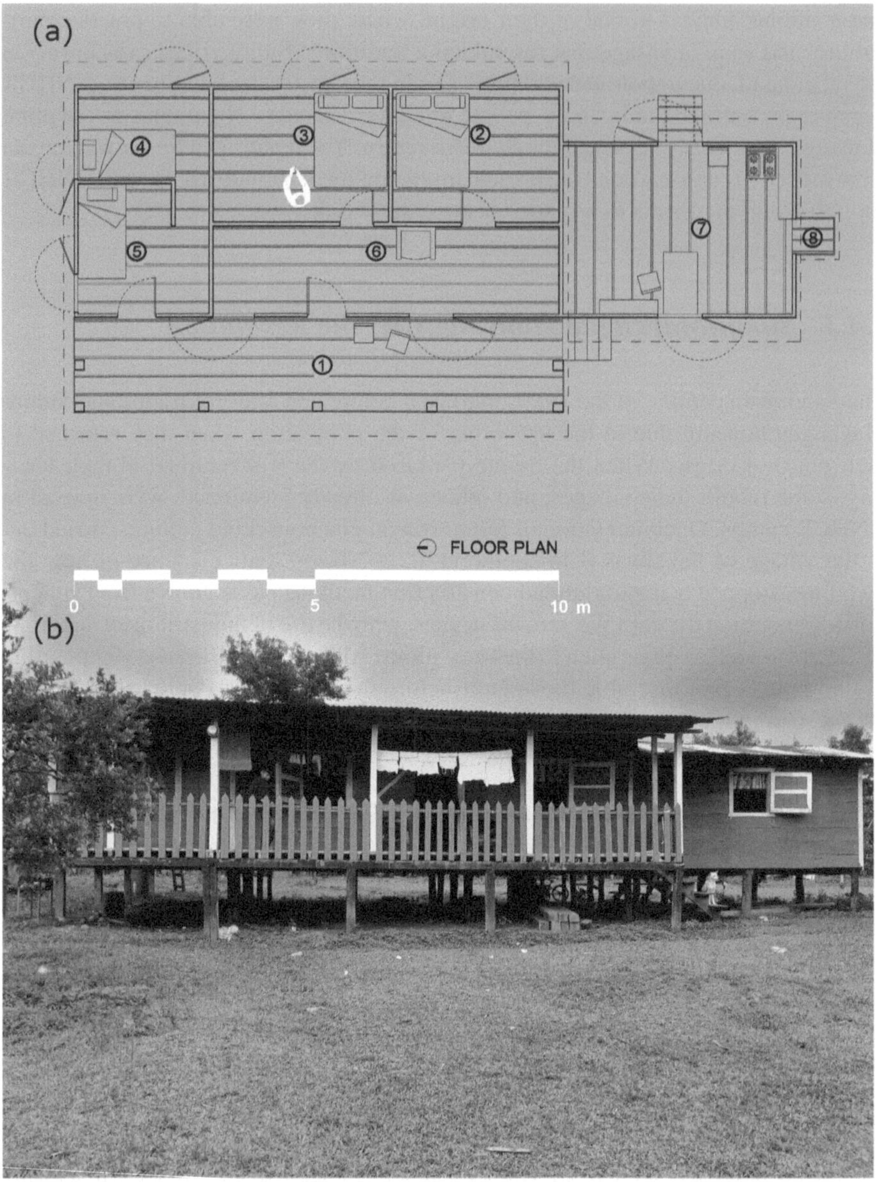

Fig. 6.5 Current Sumu-Mayangna housing. **a** Plan: 1) porch, (2) bedroom, (3) bedroom, (4) bedroom, (5) bedroom, (6) living room, (7) kitchen. **b** Main façade (Nathalia González Meixueiro 2019)

6.5 Comparative Study Based on Bioclimatic Simulation

As in any conflict, it is complex to measure the socio-cultural effects that the Sandinista war left on the Sumu population (Darke 1987), but it is possible to quantify the degree of repercussions in certain aspects, such as the impact that the revolution had on the thermal comfort inside the house, due to the new houses built for the refugees. Transformations in the comfort conditions of a dwelling have a direct influence on the way of life of its inhabitants; thus, this analysis can offer a complementary approach to understanding the effects of the revolution on a vulnerable community.

With the support of software such as Autodesk Revit for modeling and OpenStudio for simulation, an analysis of the thermal comfort of the Sumu house before, during and after the Sandinista Popular Revolution is carried out. For this study, a pre-revolutionary house, a house during the revolution emulating the "Tasba Pri" camps, and a post-revolutionary house as the Sumu live nowadays were modeled, resulting in three different typological models based on the descriptions previously outlined.

6.5.1 Sumu-Mayangna Traditional Housing

The first typology–pre-Sandinista revolution housing–was modeled after an existing house in the Sumu-Mayangna community called Santa Maria and visited in 2019 as part of the UMA-AACID University Cooperation for Development program. This is one of the rare cases where the traditional construction system is preserved. Its envelope is made of bamboo, wood, and a suite roof. The volume is made up of two blocks; one of them has a stove similar to a barbecue in its interior, while the other has only a shelf, which, in turn, was part of the structure of the house. Structurally, a platform resting on equidistant posts and beams was elevated. Outside this platform, higher posts supported the roof and, in turn, served as a base for the envelope, as shown in Fig. 6.3.

6.5.2 Sumu-Mayangna Housing During the Sandinista Era

The second model is based on a hypothetical reconstruction of the houses in Tasba Pri, modeled with the support of historical photographs of the camp. The result is a single volume elevated one meter above the ground with a fiber cement envelope and a zinc gable roof (Fig. 6.4).

6.5.3 Sumu-Mayangna Housing After the Sandinista Era

The third case study was collected in Sakalwas. The house has four rooms, a kitchen, and a living space, in addition to a porch along the entire length of the house (Fig. 6.5). It rises 0.80 m above the ground in a structure similar to the previous ones. The envelope is made of "Maria" wood (*Calophyllum brasiliense*) with 1.00 × 0.80 m openings that allow the wind to pass through. The roof is of zinc gable sheets, built into two modules, one corresponding to the kitchen and the other containing the porch, which leads to the main entrance, a living room, and four bedrooms.

6.5.4 Simulation Results

The simulation software, apart from modeling the case study houses, necessitates the inclusion of bioclimatic contextual data to approximate real-world conditions, in this case, those found in the Bosawas Biosphere region. During the modeling process, it is also necessary to divide the dwellings into zones (rooms) and define walls and roofing. Other important variables to obtain reliable results are the individual characteristics of the materials employed: roughness, thickness, conductivity, density, specific heat, thermal absorptance, solar absorptance, and visible absorptance. All of them have been defined according to the surveys and the hypothesis employed in the case of the second model. Finally, the parameters to be measured must be defined, in this case, the same for all three houses: thermal behavior and humidity, considered the most relevant and more precise to simulate.

To measure comfort values, this study is based on Givoni's diagram to compare each case (Fig. 6.6). Givoni's chart points out the effects on which architecture has on climate and identifies the qualities and strategies that a building in a given area must have in order to achieve thermal comfort. In this particular site (14.061728218299182 N, 84.47614037787929 W), the lowest temperature is 21° Celsius, and the highest temperature is 33° Celsius throughout the year, whereas the lowest absolute humidity percentage is 70% and the highest is 85%. With this given information, thermal comfort without passive or active strategies, according to Givoni's diagram, is ideally found between a dry bulb temperature of 24° and 30° Celsius and an absolute humidity between 17 and 80%, shown in gray in Fig. 6.6. Nevertheless, there is still an allowance for comfort between 29° and 34° Celsius and 17% and 100% of absolute humidity shown in green, provided that there is enough natural ventilation. On the other hand, according to the chart, it is possible to achieve thermal comfort at the lowest temperature as long as there is a passive heat strategy. As exhibited in Fig. 6.6, normally, any sort of passive strategies is enough to keep a building within thermal comfort levels.

The results of the bioclimatic study carried out with the aforementioned software are presented in the graphs below, where it is possible to observe the pre-Sandinista house, the Tasba Pri house, and the current house, each one distinguished by color

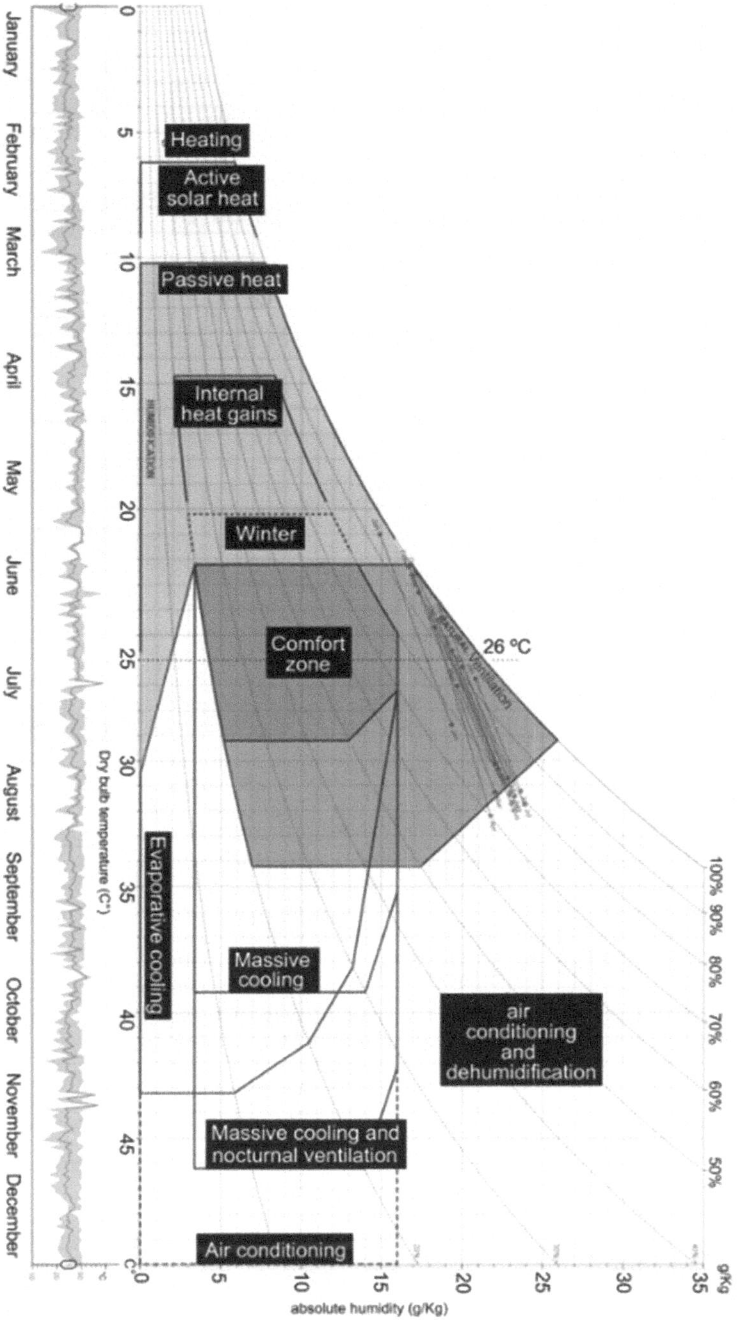

Fig. 6.6 Baruch Givoni's psychometric diagram for the selected location, Bosawas Biosphere reserve (Nathalia González Meixueiro 2023)

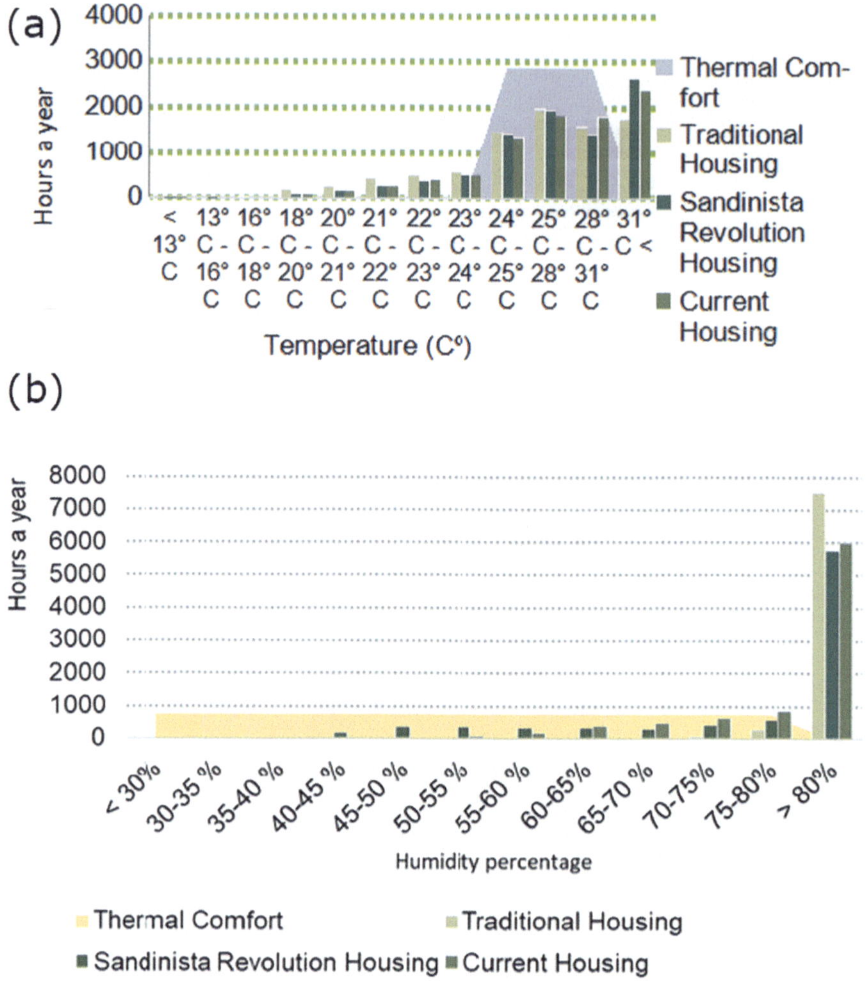

Fig. 6.7 Traditional housing, Sandinista revolution housing, and current housing thermal **a** and humidity **b** behavior during the year (Nathalia González Meixueiro 2023)

and its thermal and humidity behavior distributed in hours per year. As a reference, a fourth color is added, showing the thermal comfort zone in yellow (Fig. 6.7).

6.6 Discussion

According to the simulation results, the best thermal performance housing is the pre-Sandinista house, with 85.73% within the comfort parameters according to Givoni's psychometric diagram. Its ventilation through the rammed bamboo enclosures and windows, as well as the separation from the ground, offers the best thermal comfort. Its thick Suita roof was key in insulating the interiors from the sun's rays and high radiation, while at the same time protecting it from rain. The next level of comfort, with 55.49% throughout the year, is achieved by the current post-revolution housing. The integration of zinc is the factor that decreases the comfort inside the houses, with its high transmittance being the biggest disadvantage that this material provides, in addition to the noise generated during the long rainy seasons. Finally, the worst thermal performance is the housing introduced by the Sandinistas in the labor camps. Its industrial materials and prefabricated construction, alien to the socio-cultural and environmental context in which Sumu-Mayangna are implanted, generate evident problems in thermal performance. These results are obtained also if the typology was not formally modified, maintaining a gable roof, a rectangular volume, and a separation from the ground. There is no record that a correct orientation was considered, but the variation in its materials was enough to lower the level of comfort inside the house, which added to a foreign environment and decreased the well-being of the displaced community (Fig. 6.8).

The evolution of construction materials used in the building of Mayangna houses has ceased to be a response to the analysis of the context to be an imitation of mestizo housing, a type of construction with industrial influences, from the materials and construction systems to its furnishings. After the Sandinista revolution, housing recovered its traditional form, but its materials were no longer the Suita roof or bamboo walls. Now, especially in the villages closest to mestizo cities, most houses are built with wooden plank partitions and zinc roofing. Disregarding traditional construction systems not only decreases internal comfort but also impoverishes the community, both in its cultural identity and in the resources needed to raise their dwellings.

Leaving aside the historical facts, it is necessary to contextualize and understand in depth the reasons for such changes. Currently, an important factor that has influenced the loss of traditional models is the weight the zinc sheet has in the community's social status. It is a trend in all low-developed countries (Cimadomo et al. 2012; Naboni et al. 2013). Analysis has shown the negative effect that zinc has on the roof construction system: in addition to its high transmittance, communication within the living space becomes almost impossible since the noise of the frequent torrential rains on the metal sheet is very high.

Being aware that no single factor is causing the transformation of Sumu-Mayangna society and, therefore, of their housing, this study provides relevant information on their constructive habits to better understand the ongoing changes. Poor thermal comfort in vernacular dwellings has multifaceted implications, affecting health, well-being, productivity, energy usage, the environment, and cultural and social aspects. Sumu-Mayangna's life today is the consequence of political interests, neo-liberal influences, globalization, and the loss of their oral tradition, among many other factors. However, the foremost challenge faced by the Sumu-Mayangna community is climate change, which, in conjunction with the escalating encroachment of agricultural activities, drives the erosion of their territory and resources, and ultimately, their ancestral connection with nature.

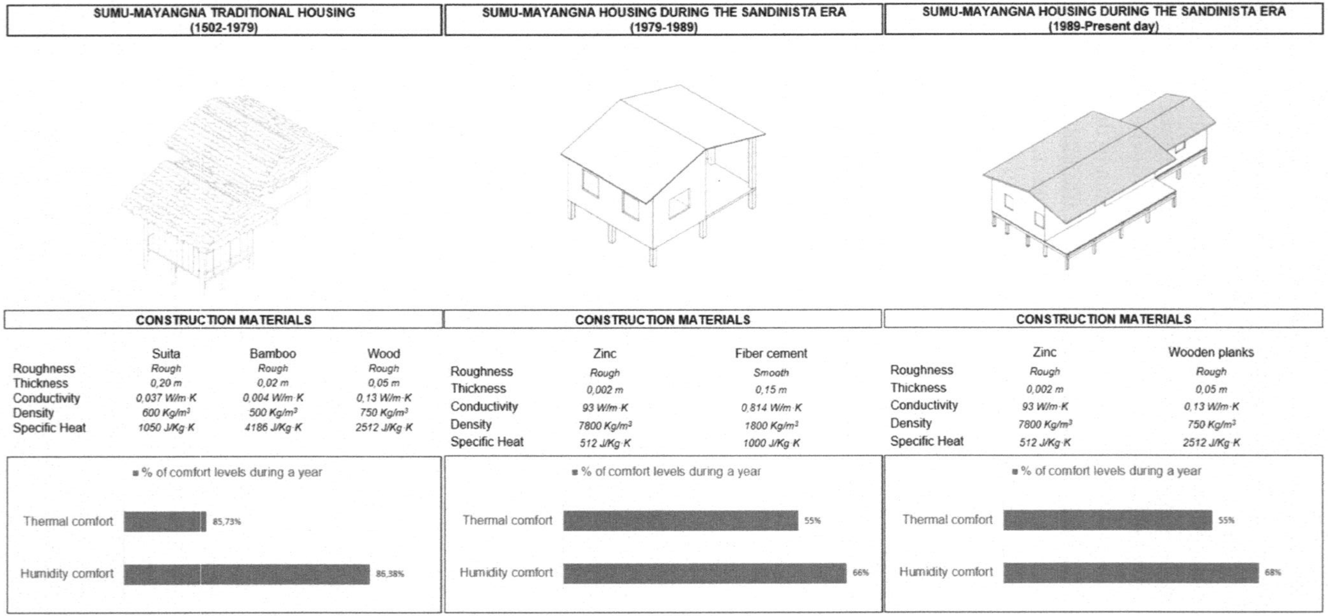

Fig. 6.8 Traditional housing, Sandinista revolution housing and current housing comparison chart (Nathalia González Meixueiro 2023)

6.7 Conclusions

Initially, the Sumu-Mayangna were a self-sufficient and sustainable society thanks to their understanding of and relationship with nature. However, as a result of the armed conflict during the Sandinista Popular Revolution in the 1980s and 1990s, their understanding and relationship with the natural environment was transformed by the cancellation of their places, their displacement to Honduras, and the deportation to the labor camps set up by the new Sandinista government. With the bioclimatic analysis of the housing models, it is possible to observe an important reduction in the level of thermal comfort of these constructions, originated by the use of new industrialized materials, directly and indirectly affecting the transformation of the population's way of life. The difference in comfort parameters between the traditional Mayangna house and the current one has been calculated as 30,24%, which falls an additional 0,84% when compared with the Sandinista Revolution house. In vernacular dwellings, inadequate thermal comfort has a wide range of consequences, impacting health, overall welfare, productivity, as well as cultural and social dimensions.

During the revolution, the government obliged the Sumu-Mayangna to abandon their dwellings and habits, forcing them to relocate to labor camps whose conditions have been a determining factor in the rejection produced towards the national institutions. The Sandinista government has not made any effort to adapt the model of life and housing for this population in these camps, an aspect that could have had a positive impact on relations with this ethnic group in a period as important as that of the revolution.

Future developments of this research should better investigate the urban and built characteristics of the Tabsa Pri refugee camps during the Sandinista revolution, as data collected is scarce and originated from secondary sources. Furthermore, data obtained through the simulations should be contrasted with in situ measuring, a task that started during the volunteer program held in the summer of 2023. These results may also help to better characterize the parameters used for the materials in the software, an aspect that is extremely complicated with traditional and natural materials.

With the disappearance of their territories, the Sumu-Mayangna ethnic group—a millennial civilization with deep respect for nature and other cultures, based on an understanding of the natural world—would become extinct. It is necessary to cooperate for the preservation of ancestral cultures such as the Sumu-Mayangna, as indigenous philosophies offer us the understanding to build and generate resilient and sustainable infrastructures. The concluding focus of this research should center on raising awareness about the relevance of conventional construction methods within the community. This endeavor would not only contribute to heritage conservation but also bolster the preservation of Sumu-Mayangna cultural identity.

Acknowledgements This work has been possible thanks to the University Volunteering Program in International Cooperation for Development Exp. 2017SEC004, funded by the Universidad de Málaga and Agencia Andaluza de Cooperación Internacional al Desarrollo.

References

Bataillon G (2001) Cambios culturales y sociopolíticos en las comunidades Mayangnas y Miskitus del río Bocay y del alto río Coco, Nicaragua (1979–2000). [Cultural and socio-political changes in the Mayangna and Miskitus communities of the Bocay and upper Río Coco rivers, Nicaragua (1979–2000)]. J Société Américanistes 87(87):376–392. https://doi.org/10.4000/jsa.1238

BBC News Mundo (2019) Revolución Sandinista: 4 claves para entender la última revolución armada de América Latina y lo que queda de su legado en Nicaragua [Sandinista Revolution: 4 keys to understanding Latin America's last armed revolution and what remains of its legacy in Nicaragua]. https://www.bbc.com/mundo/noticias-america-latina-49035196. Accessed 2 July 2023

Camacho J O (2016) Ipegüe: Intimidades en las Fuerzas Armadas Sandinistas de Nicaragua (1980–1989) [Ipegüe: Intimacies in the Sandinista Armed Forces of Nicaragua (1980–1989)]. Lulu Press, np

Cimadomo G, Simón Montesinos JA, Vacas Álvarez A (2012) Cataloging and typological study of Mgoun Valley's traditional fortified architecture, Morocco In: Mileto C, Vegas F, and Cristini V (eds) Rammed earth conservation. Taylor and Francis London

Cimadomo G, González Meixuero N, Jamauca JL, Castaño Gil C, Martín Sánchez M (2020) Documentation of traditional housing in mayangna communities. Bosawás biosphere reserve, Nicaragua. The international archives of the photogrammetry, remote sensing and spatial information sciences XLIV-M-1–2020: 203–9. https://doi.org/10.5194/isprs-archives-XLIV-M-1-2020-203-2020

Darke R (1987) Housing in Nicaragua. Int J Urban Reg Res 11(1):100–114. https://doi.org/10.1111/j.1468-2427.1987.tb00038.x

Desenclos JC, Michel D, Tholly F, Magdi I, Pecoul B, Desve G (1990) Mortality trends among refugees in Honduras, 1984–1987. Int J Epidemiol 19(2):367–373. https://doi.org/10.1093/ije/19.2.367

Dodds D (1989) Miskito and sumo refugees: caught in conflict in Honduras. Cult Surviv Q 13:3–6

Gómez García EE (2021) Refugiados Nicaragüenses y Desplazados En Honduras En La Década de Los Ochentas [Nicaraguan Refugees and Displaced Persons In Honduras In The 1980s]. Historia Contemporánea 65:163–195. https://doi.org/10.1387/hc.21013

Von Houwald GF (2003) Mayangna: apuntes sobre la historia de los indígenas Sumu en Centroamérica [Mayangna: Notes on the History of the Sumu Indians in Central America]. Imprelibros S.A.: Santander De Quilichao

Levie A (1985) Nicaragua: the people speak. Praeger, Westport, CT

Cordón Pedregos María R (2007) Estrategias indígenas, conservación y desarrollo rural sostenible en la reserva de la biosfera de Bosawas, Nicaragua [Indigenous strategies, conservation and sustainable rural development in the Bosawas biosphere reserve, Nicaragua]. PhD Dissertation, Universidad de Córdoba, Spain

Morris N (2014) Between two fires: Mayangna Indians in post-revolutionary Nicaragua, 1979–1990. Bull Lat Am Res 33(2):203–218. https://doi.org/10.1111/blar.12119

Naboni E, Bruun R, Ikonomides K, von Seilden L, Knudsen J (2013) Low-cost prototyping of comfortable and healthy homes in Tanzania. In: Stacey M (ed) Prototyping architecture: the conference papers. prototyping architecture international conference, London, March, 2013

UNPD (2005) Informe de Desarrollo Humano 2005 de las Regiones Autónomas de la Costa Caribe [Human Development Report 2005 of the Autonomous Regions of the Caribbean Coast]. INPASA, Managua

Part II
Resettlement Planning and Management

Chapter 7
From Vernacular Architecture to Temporary Housing: A Case Study of Itakura Construction in Japan

Jingying Wang, Chiho Ochiai, and Edward Ng

Abstract The provision of temporary housing after disasters encounters various challenges, including cost, sustainability, and adaptability. In light of these issues, vernacular architecture, renowned for its low environmental footprint and cultural adaptability, emerges as a promising alternative. However, there are limited studies that connect temporary housing with vernacular architecture. In Japan, a construction system was developed from vernacular storehouses and applied in temporary housing projects after the Great East Japan Earthquake and Tsunami in 2011. It is known as Itakura construction. This research aims to provide a comprehensive overview, tracing the evolution of Itakura from its origin as rural storehouses to its application in temporary housing. Specifically, three challenges associated with the application of Itakura construction are examined: legal, institutional, and technical. To overcome these challenges, a continuum of efforts bridging pre- and post-disaster contexts is observed. Achieving the legitimacy of Itakura construction and establishing a stable supply chain require decades of dedicated effort, which were undertaken prior to the disaster. Additionally, despite the prefabrication of some structural components for efficiency in temporary housing projects, engineered Itakura houses can be constructed using traditional carpentry skills during non-disaster periods. Thus, the application of Itakura construction not only mitigates the issues associated with temporary housing but also ensures the preservation of traditional craftsmanship within modern society.

J. Wang (✉) · E. Ng
School of Architecture, Chinese University of Hong Kong, Ma Liu Shui, Hong Kong SAR
e-mail: jingying@link.cuhk.edu.hk

C. Ochiai
Graduate School of Global Environmental Studies, Kyoto University, Kyoto, Japan
e-mail: ochiai.chiho.2x@kyoto-u.ac.jp

C. Ochiai et al. (eds.), *Disaster and Displacement*, Disaster Risk Reduction,
https://doi.org/10.1007/978-981-96-2583-3_7

7.1 Introduction

Post-disaster reconstruction is one of the most demanding forms of activity after a disaster, given the uncertainty of situations and stringent time constraints (Wu and Lindell 2004). Specifically, the provision of temporary housing immediately after disasters is particularly challenging and often encounters issues such as high cost, climatically or culturally inadequate design, and low recyclability, among other difficulties (Félix et al. 2013; Johnson 2007). The high cost of temporary housing is criticized for diverting resources away from permanent housing (UNDRO 1982). However, the importance of temporary housing is widely recognized. The reconstruction of permanent housing requires significant time and effort, during which temporary housing allows people to swiftly return to their normal life routines after disasters (Félix et al. 2013; Quarantelli 1995). In light of these discussions, this research presents a case study in Japan that demonstrates the possibility of integrating vernacular architecture in temporary housing.

Vernacular architecture represents the expertise of master builders across generations, who can skillfully utilize natural resources to fulfill the socio-cultural needs of communities (Oliver 2007). Thus, integrating vernacular architecture into temporary housing allows experts to adjust their designs to suit local contexts. More importantly, this approach corresponds to the concept of people-centered recovery. Unlike top-down approaches, where housing reconstruction is outsourced to construction companies external to affected communities (Barakat 2003), people-centered housing recovery prioritizes community participation, viewing housing as a process instead of a product (Barenstein and Iyengar 2010). Similarly, applying vernacular architecture to temporary housing encourages the involvement of local artisans in recovery activities, strengthening community ties through collective participation.

7.1.1 *Existing Challenges*

Worldwide, there are post-disaster projects that apply engineered traditional construction methods for housing recovery (Wang and Ng 2023). However, most of the existing projects addressed permanent housing needs. The integration of vernacular architecture in temporary housing poses significant challenges, particularly from legal, institutional, and technical perspectives, as outlined below.

Firstly, the implementation of vernacular construction often faces the issue of insufficient scientific documentation to meet existing building regulations (Schacher 2008). The lack of legitimacy can hinder its acquisition of governmental approval for reconstruction at the very beginning.

Secondly, existing projects are often based on ad hoc planning and suffer a lack of institutionalization among relevant stakeholders. It can hinder the provision of long-term support and lead to an inconsistent reconstruction approach (Barenstein and Iyengar 2010).

Thirdly, traditional construction systems are often linked with lower productivity. Proper integration with technologies like mechanization and prefabrication is crucial to enhance construction efficiency. Additionally, technical solutions are required to improve living quality and recyclability of temporary housing.

7.1.2 Research Design

Japan is chosen as a case study due to its innovative use of an engineered traditional timber construction system, Itakura, for temporary housing. The objective of this research is to understand how Itakura construction was developed from vernacular architecture and adapted to temporary housing.

Specifically, three research questions are proposed.

1. What is Itakura construction?
2. What led to the adoption of Itakura construction for temporary housing?
3. How were the legal, institutional, and technical challenges navigated when applying Itakura construction to temporary housing?

7.2 Methodology

The research methodology for data collection comprises semi-structured interviews with carpenters and a key informant that leads the research of Itakura construction, a systematic literature review of existing journal and conference articles regarding Itakura construction on the database of Architectural Institute of Japan (AIJ), and the review of related book chapters, government reports, and publications from key associations such as the NPO Forum for Wood Architecture (NPO FWA) and the Japan Itakura House Association (JIHA).

Data extracted from the transcripts and documents is analyzed following the legal, institutional, and technical perspectives mentioned above. Specifically, the study identifies solutions to overcome legal restrictions related to Itakura construction, explains the networking of main stakeholders across decades, and examines technical improvements made for temporary housing that adopted Itakura structures.

7.3 Results

7.3.1 *The Origin of Itakura Construction*

The term "itakura" in Japanese is formed by two characters, "ita" and "kura," signifying "board" and "storehouse," respectively. Thus, itakura literally translates to storehouses built with timber boards.

With a floor area of about 10 to 30 square meters, storehouses have been commonly utilized in rural families to store valuable items like grain, either for tribute or as a preparation for potential famine (Kurosaka et al. 2004). In some regions, storehouses serve as temporary evacuation spaces for family members in case of disasters (Yamada and Sato 2021; Ando and Ando Laboratory 2010). They are usually built with materials that are locally available, such as timber, earth, or stone. What sets Itakura apart from other storehouses is its exclusive use of timber materials for wall construction, although earth may also be applied to timber boards to enhance fire resistance (Aoyagi and Ohashi 2021).

Itakura storehouses are widely distributed throughout Japan, from Hokkaido and Miyagi Prefecture in the North, to Gifu, Nagano, and Toyama Prefecture in Central Japan, and Kumamoto Prefecture in the South. Meanwhile, Itakura storages can also be commonly found in temples and shrines, such as Ise Shrine, Kasuga Taisha, Shosoin of Todaiji Temple (Fig. 7.1).

Fig. 7.1 An Itakura storehouse of Kasuga Taisha in Nara prefecture

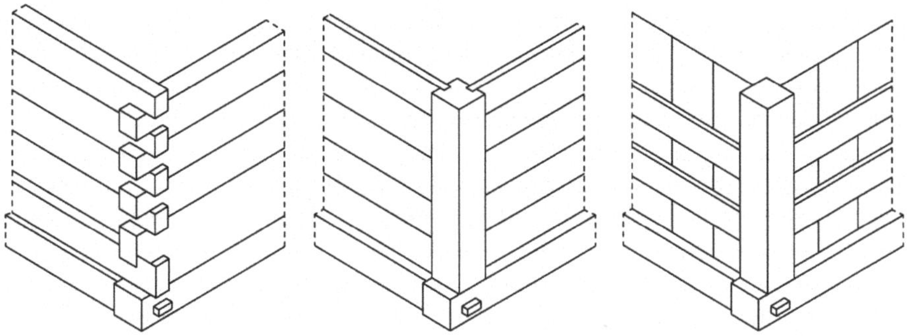

Fig. 7.2 Three structural typologies (left to right: Seiro-gura, Otoshi-itakura, Nuki-itakura)

7.3.1.1 Structural Typology

Despite the significant range of regional variations, three main structural typologies of Itakura storehouses can be identified: 1. Seiro-gura; 2. Otoshi-itakura; 3. Nuki-itakura (Ando 1995).

Similar to log construction, Seiro-gura is assumed as the most primitive construction system of wooden storehouses among the three typologies.[1] It is formed by horizontally stacking timber boards with ends interlocked by notches, usually without using any pillars (Fig. 7.2). In addition to the notches, boards are also joined with wooden dowel pins, with board thickness varying from 60 to 120 mm.

Otoshi-itakura introduces pillars and beams into the Seiro-gura system, where grooves are carved into the pillars to insert either horizontal or vertical timber boards. The term "Otoshi" translates to "drop," giving rise to the alternative translation of Itakura as the "cedar board drop method" (Iwata et al. 2023). With the help of pillars, the dimensions of timber boards can be reduced compared to those of Seiro-gura, with board thickness ranging from around 40 to 90 mm.

Nuki-itakura further introduces lintels (known as "Nuki" in Japanese) to insert between the pillars and timber boards are nailed from inside to the lintels. In this case, timber boards may not necessarily serve as load-bearing structures like Seiro-gura and Otoshi-itakura. The board thickness can be further reduced to around 30 to 60 mm.

7.3.1.2 Applicability to Housing Recovery

Based on experiment results, Itakura storehouses demonstrate a commendable ability to regulate humidity and temperature thanks to their generous usage of timber boards

[1] Seiro-gura exhibits notable similarities to Azekura, a construction system observed in monumental architecture like the treasure house Shoso-in, built for the Todai-ji temple around the eighth century. In general, Azekura demonstrates more sophistication regarding the processing of timber logs and the techniques of timber joinery compared to Seiro-gura.

(Higuchi et al. 2007, 2008). Furthermore, Itakura structures, like other vernacular buildings, are designed with traditional joinery techniques for easy relocation to serve secondary usages (Li et al. 2009). The self-regulating properties and recyclable nature of Itakura make it suitable for application in temporary housing.

Among the three typologies, Otoshi-itakura stands out with the greatest potential. Seiro-gura requires large dimensions of timber boards and faces challenges in creating openings for windows and doors. Nuki-itakura involves a more complex construction process due to its usage of lintels. Thus, Otoshi-itakura offers a balance between design flexibility and construction complexity, making it the preferred choice for further experiments. Detailed technical improvements are elaborated in Sects. 3.3.1 and 3.3.3.

7.3.2 Supply of Temporary Housing

7.3.2.1 Increasing Diversity of Temporary Housing Solutions

In Japan, the responsibility for providing temporary housing typically falls to the Standardized Committee of the Japan Prefabricated Construction Suppliers and Manufacturers Association (JPCSMA). Employing a top-down approach, JPCSMA prioritizes the swift supply of prefabricated steel-framed structures for temporary housing after disasters (Goda 2017). The design of these steel structures is based on temporary offices built at construction sites (Seike et al. 2018).

However, following the Great East Japan Earthquake and Tsunami (GEJET) on March 11th, 2011, the capacity of the Standardized Committee alone was unable to meet the enormous demand for temporary housing for around 50,000 household units, primarily in the three prefectures of Iwate, Miyagi, and Fukushima (Seike et al. 2018). Consequently, the construction of temporary housing was further delegated to the Housing Committee of JPCSMA,[2] as well as local construction companies selected through solicitations conducted by the prefectures (Seike et al. 2014). In contrast to JPCSMA, local companies were more familiar with the use of timber structures (Fig. 7.3). While some participation of local constructors in temporary housing can be observed after the 1991 eruption at Unzen Volcano (Takahashi et al. 1998) and the 2004 Chuetsu Earthquake (Kimura 2006), the GEJET marked their first systematic involvement in meeting massive temporary housing needs in Japan, totaling around 9,800 units.

Considering the contamination caused by nuclear plants in Fukushima Prefecture, it was anticipated that temporary housing would be used for a relatively long period. Thus, the Fukushima Prefectural Government aimed to select housing designs that better suited local livelihoods with the assistance of local contractors (Fukushima

[2] Temporary housing constructed by the Housing Committee generally showed higher quality compared to those by the Standardized Committee. However, housing manufacturers relied on resources from their supply chain prior to disasters that may not be connected to local contexts.

Fig. 7.3 Steel framed temporary housing offered by JPCSMA (left) and timber temporary housing offered by local constructors (right)

Table 7.1 Number of household units living in temporary housing (Adapted from Fukushima Prefecture Civil Engineering Department 2020 and Seike et al. 2014)

Supply source	Iwate	Miyagi	Fukushima
Standardized committee (JPCSMA)	7,702	14,364	6,326
Housing committee (JPCSMA)	3,677	7,208	3,655
Local solicitations	2,605	423	6,819
Total	13,984	21,995	16,800
Percentage of local solicitations	18.6%	1.9%	40.6%

Prefecture Civil Engineering Department 2020). Despite the significant increase in the workload of preparing additional documents, Fukushima prefecture showed high motivation in engaging local companies for temporary housing. About 41% of temporary housing in Fukushima was procured through solicitations (Table 7.1). In Iwate and Miyagi prefectures, however, the supply of temporary housing was primarily covered by JPCSMA, with only 18.6% and 1.9%, respectively, sourced from local solicitations.[3]

7.3.2.2 Applying Itakura to Temporary Housing

Two rounds of solicitations were launched in Fukushima, targeting construction companies located within the prefecture. In total, about 6,800 units were allocated to 24 local companies and associations, with each constructor responsible for between

[3] Miyagi Prefecture showed the least interest in involving local construction companies in temporary housing. While the exact reasons remain unclear, it is assumed that the level of urbanization plays a role. The Miyagi prefectural government, situated in the metropolitan area of Sendai city, maintains close ties with JPCSMA and major housing manufacturers. In contrast, Fukushima is less urbanized, with the authority of the prefectural government scattered across districts, resulting in a greater reliance on local constructors.

Table 7.2 Reuse of Itakura temporary housing for various purposes (Adapted from Urabe et al. 2020)

Original sites (household units)	New sites (transport distance)	Purposes (recycled household units)
Aizuwakamatsu City, Fukushima prefecture (36 units)	Aizuwakamatsu City, Fukushima prefecture (0 km)	Post-disaster permanent housing (32 units)
Iwaki City, Fukushima prefecture (162 units)	Soja City, Okayama prefecture (900 km)	Post-disaster temporary housing (48 units)
	Toyonaka City, Osaka prefecture (700 km)	An office (2 units)
	Varied	New residences (19 units)

100 and 500 household units (Fukushima Prefecture Civil Engineering Department 2020; Kiwoikasu 2012; Omizu 2013).

Thanks to its high recyclability and the utilization of local resources, Itakura was selected through the solicitations in Fukushima Prefecture. Based on the number of available carpenters, estimated at around 100, the capacity for Itakura structures was capped at about 200 household units (Kiwoikasu 2012). Consequently, a total of 198 household units were constructed using the Itakura system, with 162 units in Iwaki City (76 temporary houses,[4] two community centers, and one welfare facility for the elderly) and 36 units in Aizuwakamatsu City (17 temporary houses and one community center). The construction time for these units was 73.5 days and 45 days, respectively, exceeding the 30-day limit set by the prefectural government[5] (Haganuma et al. 2013).

In 2015, an assessment survey was conducted to evaluate the recycling rates of temporary housing built with timber structures, where Itakura construction was recognized as having the highest recyclability among various typologies (Haganuma et al. 2015). In total, approximately 51% of Itakura temporary housing, equivalent to 101 household units, has been recycled for different purposes. These include a permanent housing project in Fukushima Prefecture in 2016 and two temporary housing projects in Okayama Prefecture in 2018 (Table 7.2). Specifically, around 66% of timber materials were reused in Fukushima (Omi 2018), while a higher rate of around 88% was achieved in Okayama (Hata 2019).

[4] This research uses the number of household units rather than the number of temporary houses to represent the scale of the projects. This is because the size of buildings can vary, and different typologies correspond to different numbers of household units. Therefore, household units are considered more appropriate to reflect the scale of reconstruction projects.

[5] The causes of delay were unclear, but they may relate to the construction of the relatively complex gable roof structures and the additional loft space offered in Itakura houses.

7.3.3 *Challenges of Application in Temporary Housing*

As explained in Sect. 3.1, the original Itakura storehouses refer to diverse regional varieties and may not necessarily align with the Building Standards Law (BSL) in Japan, posing challenges to their efficient application in post-disaster contexts. This section explains how the challenges in applying Itakura construction to temporary housing were navigated from legal, institutional, and technical perspectives.

7.3.3.1 Legal Perspective

The application of Itakura construction mainly encounters two legal restrictions: earthquake and fire resistance. Thanks to the revision of BSL from 1998 to 2000, performance-based standards have been introduced alongside existing prescriptive standards in Japan (Hayashi 1998). This allows for the customization of structural components, as long as they meet the required performance criteria. Under the new law, the so-called ministerial certifications can be applied after scientifically verifying the structural performance with specific agencies delegated by the Ministry of Land, Infrastructure, Transport, and Tourism in Japan. Each certification provides detailed construction guidelines that must be adhered to during the construction process to ensure compliance with building regulations.

Initially, the wall structures of Itakura storehouses, comprising a single layer of timber boards, did not meet the required seismic standard. Diagonal braces, plywood panels, and metal connectors at joineries had to be incorporated to enhance the structure's rigidity as prescribed by the regulations, significantly increasing the complexity of Itakura construction for local carpenters. Yet, after the introduction of performance-based standards in BSL in 2000, there were expectations to simplify the retrofitting techniques for engineered Itakura structures. Substantial efforts were invested in finding simple retrofitting solutions while preserving the integrity of the traditional system.

As a result, nailing a 24 mm thick layer of wood laths onto the original 30 mm thick timber boards of Itakura structures was found to be the most appropriate (Fig. 7.4). These wood laths not only enhance Itakura's rigidity required for earthquake resistance but also enable Itakura to resist the effects of fire for at least 30 min without significant damage (Taniguchi 2010). This approach also allows traditional timber joinery to be continuously applied in engineered Itakura structures. After conducting various experiments and testing different structural variations, researchers at the NPO FWA successfully secured two ministerial certifications for Itakura construction, in earthquake and fire resistance in 2005 and 2007, respectively. Thus, Itakura had been successfully certified as engineered construction before the GEJET in 2011, which played a crucial role in its inclusion in the solicitation process in the first place.

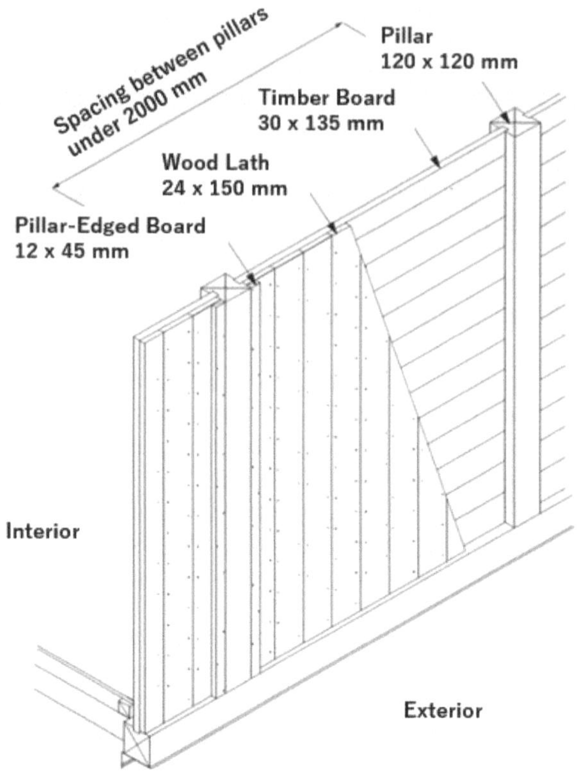

Fig. 7.4 Engineered Itakura construction

7.3.3.2 Institutional Perspective

The key stakeholders involved in the application of Itakura construction can be categorized into five groups: 1. Organizations guiding the research of Itakura construction; 2. Architectural offices or laboratories; 3. Sawmill companies or associations; 4. Construction companies or associations; 5. Prefectural governments. To better clarify the involvement of individual stakeholders, three stages are identified in chronological order (Fig. 7.5).

Stage I refers to the development of the engineered Itakura construction system. Ever since 1988, researchers from the Ando Laboratory at the University of Tsukuba have been attempting to apply Itakura structures to housing construction, although with some compromises in structures as mentioned in Sect. 3.3.1 (NIKKEI HOME BUILDER 2003). The significant potential of Itakura construction became evident during their study of vernacular Itakura storehouses across Japan. The simplicity of the material—solely timber—suggested the possibility of mass production and presented an opportunity to revitalize the forestry industry in Japan. The country has experienced a notable decline in labor forces within domestic forestry in recent decades, mainly due to substantial imports of inexpensive foreign timber (Fuchigami

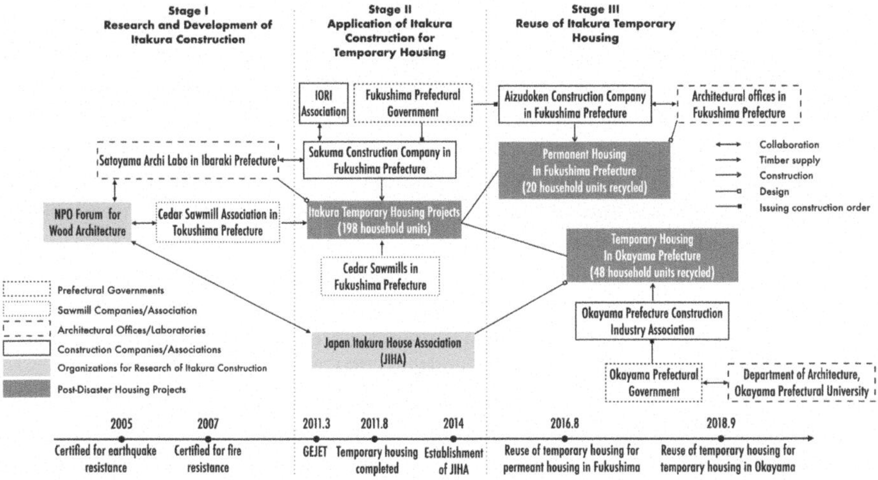

Fig. 7.5 Stakeholder network from development to application of Itakura construction

et al. 2016). Thus, promoting Itakura structures for housing construction can help increase the consumption of domestic timber materials.

Since 2003, these researchers have joined the NPO FWA and started systematic research on methods to strengthen Itakura construction. After obtaining certifications for earthquake and fire resistance, efforts were further made to create a stable supply chain for timber boards. Around 2008, cooperation was established with a cedar sawmill association in Tokushima Prefecture, ensuring a large stockpile while minimizing material costs. This stockpile eventually proved beneficial for a prompt supply of building materials after the GEJET.

Stage II refers to the application of Itakura construction for temporary housing in Fukushima Prefecture in 2011. As mentioned in Sect. 3.2, solicitation processes were initiated to involve local constructors in the supply of temporary housing. With the proposal to employ Itakura construction, Sakuma Construction Company participated in the solicitation and successfully secured two construction orders from the Fukushima Prefectural Government, in Iwaki and Aizuwakamatsu City, respectively (Kiwoikasu 2012). Sakuma Company was one of the local constructors approached by the NPO FWA prior to the disaster to promote engineered Itakura structures for housing construction. Recognizing the potential of Itakura to utilize local timber materials and preserve traditional carpentry skills, Sakuma Company established an association called IORI in 2010 with nearby local carpenters and sawmill companies to advocate for Itakura houses in Fukushima Prefecture. By February 2011, they had completed an Itakura model house that received positive feedback, bolstering their confidence in applying this method to temporary housing (Nikkei Architecture 2011). The design of Itakura temporary housing was entrusted to Satoyama Archi Labo, an architecture office founded by one of the leading researchers who developed and engineered Itakura construction.

Although the government expected the use of building materials produced within Fukushima Prefecture, not all structural components were locally available for this relatively newly developed construction system. Additionally, the drying process of cedar boards was time-consuming, making it challenging to meet the time restrictions merely with local materials. Therefore, timber supply was divided into two categories. Load-bearing structures that required drying were provided by the Cedar Sawmill Association in Tokushima Prefecture, whereas non-structural materials like substrates and finishing materials were sourced from sawmills within Fukushima Prefecture. The ratio of local to external building materials was about 4:6 (Nikkei Architecture 2011). Despite this, given the substantial timber usage in Itakura houses, the overall consumption of local timber materials per building unit remained comparable to that in other systems (Kiwoikasu 2012).

Itakura construction earned recognition after its successful application in temporary housing in 2011, attracting the interest of an increasing number of small- to medium-sized construction companies across Japan. In response, the JIHA was established in 2014 to further facilitate the research and promotion of engineered Itakura construction among interested stakeholders.

Stage III refers to the reuse of Itakura temporary housing. The prefectural government of Fukushima implemented a guideline for the gratuitous transfer of temporary housing on April 1, 2016, to control the generation of waste and reduce disposal costs during the removal of temporary housing (Fukushima Prefecture Civil Engineering Department 2020). It means that there would be no charge for the reuse of temporary housing units, but the buyers would be responsible for covering the expenses of dismantling, transportation, and construction associated with the relocation.

Itakura temporary houses in Aizuwakamatsu City were assessed as recyclable and reused as permanent houses at the same location (Fukushima Prefecture Civil Engineering Department 2020; Haganuma et al. 2015). The temporary houses were disassembled in July 2015, stored for one year in a warehouse, and then reconstructed, completing the process in August 2016 (Omi 2018; Seike et al. 2018). The demolition and reconstruction works were issued to a local construction company, Aizudoken, with support from local architectural offices. Thanks to the reuse of building materials and minimal transportation costs, the reconstruction cost per housing unit was about 18.3% lower than that of new construction (Fukushima Prefecture Civil Engineering Department 2020).

The supply of Itakura temporary houses in Iwaki City ended in March 2018, four months before heavy rainfall attacked Soya City, Okayama Prefecture (Hata 2019). Advised by the Okayama Prefectural University, Okayama Prefectural Government promptly initiated the reuse of temporary houses from Fukushima, and two reconstruction projects were completed in September and October 2018. Unlike local reuse, as seen in Aizuwakamatsu City, building components had to be transported from Iwaki City, about 900 km away from Soya City. The demolition and transportation works were entrusted to Sakuma Company, which had been responsible for the initial construction years ago. Meanwhile, the reconstruction works were assigned to the Okayama Prefecture Construction Industry Association. This arrangement not only minimized damage to the existing structures during demolition but also enabled

Okayama carpenters to acquaint themselves with the engineered Itakura construction system (Hata 2019). In addition, it is noteworthy that the swift completion of the reconstruction projects was thanks to the support from around 80 carpenters from other prefectures (Hata 2019). Due to the relatively high transportation costs, the relocation expenses were not significantly lower than those of new construction. Nevertheless, they remained within the budget allocated by the prefectural government.

In summary, the NPO FWA and the JIHA, as the primary organizations leading the research on Itakura storehouses, have played a crucial role in developing and promoting engineered Itakura construction over decades. As of the current date, they persist in enhancing the Itakura system for better performance and disseminating the knowledge to interested architects and carpenters nationwide in Japan. Moreover, the policy support from the Fukushima Prefectural Government was also crucial in enabling the application of Itakura construction for both temporary and permanent housing.

7.3.3.3 Technical Perspective

The technical challenges in implementing Itakura construction for temporary housing centered on achieving three goals: construction efficiency, living quality, and recyclability.

Construction Efficiency

Construction efficiency can be enhanced by prefabricating some structural components in advance, such as timber boards, pillars, and beams. To ensure the efficient use of domestic timber resources in Japan, the pillar dimensions were designed to be 120 mm square, aligning with the size of standard timber logs harvested from the forests. Similarly, the beams were designed in three dimensions: 120 × 180 mm, 120 × 240 mm, and 120 × 300 mm. Timber boards for wall, roof, and floor structures came in just two dimensions–30 × 135 mm and 30 × 180 mm. For wall construction, timber boards were panelized into a single piece to expedite construction (Fig. 7.6). Additionally, the traditional mortise-and-tenon timber joints were also mechanized in advance. Thanks to these modern techniques, the overall costs of Itakura construction were well managed within the allocated budget (Moriyama 2011).

Living Quality

Differing from the typically confined space in temporary housing, Itakura structures were designed to be more open for better ventilation, incorporating features from traditional Japanese farmhouses. Large gable roofs were given to offer shade during the summer. Beneath these roofs, additional space was created. It comprised lofts in the attic to serve as storage rooms and an open veranda adjacent to living rooms (Fig. 7.7). Known as "Engawa" in Japanese, the wooden veranda not only accommodated space for laundry but also enhanced communal interactions, a feature that was

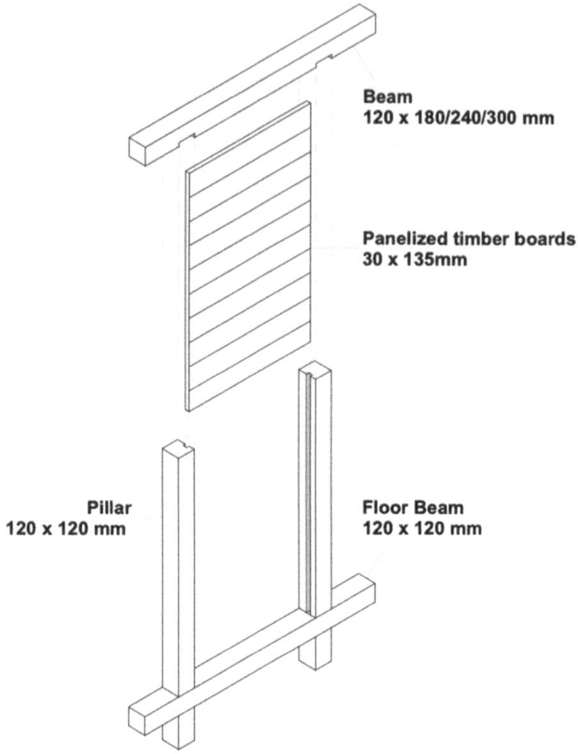

Fig. 7.6 Panelized timber boards to enhance the construction speed

not commonly observed in temporary housing (Figs. 7.8 and 7.9). Effective ventilation through Engawa space minimized the need for air conditioning, as reported in a questionnaire survey with residents (Tamamura et al. 2012).

Natural materials such as thatch and rice husks were utilized in roof and floor structures to ensure adequate insulation (Haganuma et al. 2015). In terms of wall construction, thermal insulation was achieved through four layers of timber boards totaling up to 78 cm, along with 42 cm airspace, without the need for additional

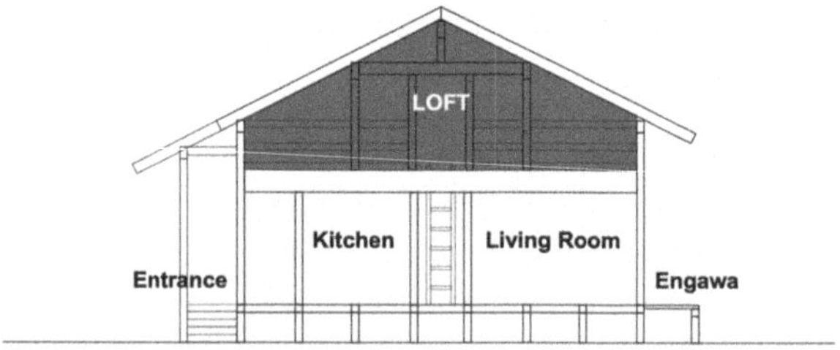

Fig. 7.7 Section of an Itakura temporary house

Fig. 7.8 The current condition of Engawa space of permanent housing recycled from Itakura temporary housing in Fukushima prefecture

Fig. 7.9 The interior of an office in Osaka recycled from Itakura temporary housing

insulation materials[6] (NPO Forum for Wood Architecture 2012). According to the measurement surveys conducted in 2012, the room temperature of Itakura houses remained relatively stable, with minimal temperature drop at night and no significant complaints of coldness reported during interviews (Kaneko and Hashimoto 2012).

Furthermore, the aroma of cedar boards evoked a sense of familiarity for the elderly populations, who constituted the majority of those affected. After three years of living in the Itakura houses, some residents expressed their willingness to continue living there, while others wished for the units to be recycled upon their departure (Ando 2014).

Recyclability

The high recyclability of Itakura houses primarily stemmed from the utilization of traditional timber joints, which can be easily disassembled (Hata 2019). Additionally, timber piles were employed for the foundation instead of concrete slab foundations, allowing for easier disassembly as the timber structures were not fixed to the foundation. The deteriorated timber piles, insulation materials, and exterior finishing

[6] Itakura temporary houses were also provided to 3 households in Tadami town of Fukushima Prefecture after a heavy rainfall in July of 2011. Recognizing Tadami as a snowy area, thermal insulation panels made from cedar barks were incorporated into the wall construction to enhance insulation performance.

materials, after years of use, were recycled as biomass. Other construction details facilitating recycling included minimizing the use of nails and employing screws for floorboards and ceiling panels to allow for easy removal, as well as exposing electrical wiring to minimize damage during the dismantling process (Fukushima Prefecture Civil Engineering Department 2020).

7.4 Discussion

The well-established legal system in Japan lays a solid foundation for the development of Itakura construction. However, this research cannot delve into the details due to the word limitations. A recommended resource for further insight is a report published by the World Bank, which explains the incremental improvements of the legal framework in Japan (World Bank Disaster Risk Management Hub (Tokyo) 2018).

While this research primarily focuses on the Itakura construction system, little discussion is given to the overall layout of housing complexes. In contrast to the complexes of steel structures, where houses were densely arranged on asphalt pavement, Itakura temporary houses were more loosely allocated on unpaved soil surfaces. The greater distance between houses reduced noise complaints and enhanced air ventilation. Additionally, the presence of unpaved grounds allowed residents to cultivate abundant vegetation, mitigating temperature increases during summers (Kaneko and Hashimoto 2012).

However, firsthand feedback from residents living in temporary housing is limited due to the dismantling of most temporary units before the start of this research. Therefore, future research aims to include opinions from residents currently residing in permanent housing using the Itakura system.

7.5 Conclusion

Itakura construction shows recyclability, adaptability to cultural and climatic contexts, as well as environmental sustainability, serving as a good alternative to temporary housing in Japan. It demonstrates how innovative solutions for temporary housing can draw upon the building traditions of communities. Nonetheless, the challenges of applying engineered Itakura construction to temporary housing are also acknowledged and analyzed from legal, institutional, and technical perspectives.

Legal restrictions of earthquake and fire resistance are solved through the issue of ministerial certifications after scientifically verifying the performance of the system through experiments. This became possible only after the introduction of performance-based standards in the Building Standards Law of Japan in 2000.

Institutional challenges are mitigated by the policy support from the Fukushima Prefectural Government, which demonstrates commendable courage in adopting innovative solutions for temporary housing and promoting its recycling after demolition. Additionally, civil societies such as NPO FWA and JIHA have also played a crucial role in developing engineered Itakura construction and establishing communication channels among interested stakeholders to facilitate knowledge transmission.

Technical challenges demand the effort of architects and engineers to further ensure construction efficiency, living quality, and recyclability of the system. It includes identifying proper structural components that can be mechanized or prefabricated in advance, understanding the climatic conditions and the living customs of affected populations, as well as paying attention to the construction details that facilitate easy recycling after demolition.

Although post-disaster research often focuses on conditions after disasters, this research examines the continuum of efforts bridging pre- and post-disaster contexts. Achieving the legitimacy of Itakura construction and establishing a stable supply chain of building materials require decades of dedicated effort, which were undertaken prior to the GEJET in 2011. In addition, during the process of legitimization, emphasis has been placed on simple modifications that maintain the integrity of traditional systems, allowing newly introduced retrofitting techniques to smoothly integrate into the practical expertise of carpenters. Thus, despite the prefabrication of some structural components for efficiency in temporary housing projects, engineered Itakura structures can be constructed using traditional carpentry skills during non-disaster periods. The application of Itakura construction not only mitigates the issues associated with temporary housing but also ensures the preservation of traditional craftsmanship within modern society.

References

Ando K, Ando Laboratory, University of Tsukuba (2010) Barns and storehouses in Japan (in Japanese). https://ci.nii.ac.jp/ncid/BB02162649

Ando K (1995) Modern wooden housing theory: dwellings of Itakura construction (in Japanese) (Vol 31). INAX. https://ci.nii.ac.jp/ncid/BN12766444

Ando K (2014) Temporary wooden housing with Itakura construction (in Japanese) Finex 26(157):19–29. https://doi.org/10.14820/finexjournal.26.157_19

Aoyagi Y, Ohashi R (2021) The Itakura construction method of traditional storehouses in the Kanbara Settlement, Tsumagoi village, Gunma prefecture (in Japanese). In: Proceeding of the architectural research meetings, Kanto Chapter, Architectural Institute of Japan

Barakat S (2003) Housing reconstruction after conflict and disaster. In (Vol 43, pp 1–40). London: Humanitarian Policy Group

Barenstein JD, Iyengar S (2010) India: from a culture of housing to a philosophy of reconstruction. In: M Lyons, T Schilderman, C Boano (eds.) Building back better (Vol 163, pp 183–186)

Félix D, Branco JM, Feio A (2013) Temporary housing after disasters: a state of the art survey. Habitat Int 40:136–141. https://doi.org/10.1016/j.habitatint.2013.03.006

Fuchigami Y, Hara K, Uwasu M, Kurimoto S (2016) Analysis of the mechanism hindering sustainable forestry operations: a case study of Japanese forest management. Forests 7(8):182

Fukushima Prefecture Civil Engineering Department (2020) Fukushima prefecture temporary housing records: emergency housing assistance for the great east Japan earthquake (in Japanese)

Goda J (2017) Construction of temporary emergency housing in times of disasters by Japan prefabricated construction suppliers and manufacturers association (in Japanese). Urban Hous Sci (98):21–27. https://www.jstage.jst.go.jp/article/uhs/2017/98/2017_21/_article/-char/ja/

Haganuma S, Hayakawa S, Ishisaka K, Urabe T (2013) A study on plan characteristic of wooden temporary housing: a study on temporary housing in Fukushima prefecture after the Tohoku Earthquake (in Japanese). AIJ J Technol Des 19(43):1043–1048. https://doi.org/10.3130/aijt.19.1043

Haganuma S, Urabe T, Ishisaka K (2015) A study on the reuse characteristic of wooden temporary housing: a study on wooden temporary housing in Fukushima prefecture after the Tohoku earthquake (in Japanese). J Arch Plan (Trans AIJ) 80(710):813–822. https://doi.org/10.3130/aija.80.813

Hata K (2019–07) A study on the reuse of temporary wooden housing using Itakura construction: a case study of transferring construction from Iwaki City, Fukushima prefecture, to Sōja City, Okayama Prefecture (in Japanese). In: Proceedings of annual research meeting Chugoku Chapter, Architectural Institute of Japan (AIJ)

Hayashi T (1998) Performance-based regulations in the building standards law and its impact on the wood industry (in Japanese). Mokuzai Hozon (Wood Protection) 24(5):335–338. https://doi.org/10.5990/jwpa.24.335

Higuchi T, Hashimoto T, Ando K (2007–07) Thermal environment in the wooden store houses in the area of Mt. Yatsugatake: results of field measurement in winter (in Japanese). AIJ summaries of technical papers of annual meeting

Higuchi T, Hashimoto T, Ando K (2008) Thermal environment in the wooden store houses in the area of Mt. Yatsugatake: results of field measurement in summer (in Japanese). In: AIJ summaries of technical papers of annual meeting (2008):93–94. https://cir.nii.ac.jp/crid/1570291227591650816

Iwata T, Harada E, Maly E (2023) Towards improving provision of wooden temporary housing: Analysis of repairs of temporary housing built by local contractors after the Great East Japan Earthquake. Int J Disaster Risk Reduct 86:103537. https://doi.org/10.1016/j.ijdrr.2023.103537

Johnson C (2007) Strategic planning for post-disaster temporary housing. Disasters 31(4):435–458. https://doi.org/10.1111/j.1467-7717.2007.01018.x

Kaneko T, Hashimoto T (2012) Residential performance survey for improving health and comfort in emergency temporary housing (in Japanese). Bull Inst Hum Environ Stud, Kanto Gakuin Univ 11:79–81

Kimura N (2006) The habitability of temporary housing (in Japanese). In Report on the Damage Caused by the Niigata-Chuetsu Earthquake

Kiwoikasu CfPAUW (2012) Case studies of emergency temporary housing utilizing wood (in Japanese)

Kurosaka T, Ando K, Okimoto T, Kariuchi K (2004) Regional characteristics and typification on the wooden storehouse construction of close-set pillars stYLE (in Japanese). J Arch Plan (Trans AIJ) 69(576):45–52. https://doi.org/10.3130/aija.69.45_1

Li L, Higuchi T, Ando K, Hashimoto T, Kobayashi H, Toyokawa S, Aoyagi Y (2009) A study on the translocation and reuse of barns at the foot of mount Yatsugatake (in Japanese) AIJ summaries of technical papers of annual meeting

Moriyama H (2011) The completion of wooden temporary housing using the Itakura construction method in Iwaki City (in Japanese). Nikkei Business Publications. https://xtech.nikkei.com/kn/article/building/news/20110906/552531/

Nikkei Architecture (2011) Iwaki City Takaku No. 10 temporary emergency housing (in Japanese). https://xtech.nikkei.com/pdf/NA/20110910/2117010/?ST=nxt_pdfdl

Nikkei Home Builder (2003) Exploring earthquake-resistant "dropped-board walls" (in Japanese). https://xtech.nikkei.com/pdf/HB/20031201/175728/?ST=nxt_pdfdl

NPO Forum for Wood Architecture (2012) ITAKURA wooden panel system: emergency temporary housing (in Japanese)

Oliver P (2007) Built to meet needs: cultural issues in vernacular architecture. Routledge

Omi Y (2018) A case study of dissolution and rebuilding from emergency temporary housing to disaster recovery housing by Itakura timber construction method (in Japanese). AIJ J Technol Des 24(57):775–778. https://www.jstage.jst.go.jp/article/aijt/24/57/24_775/_article/-char/ja/

Omizu T (2013) Empirical study: temporary housing - from the site of the great East Japan Earthquake (in Japanese). Gakugei Shuppansha. https://kuline.kulib.kyoto-u.ac.jp/opac/opac_link/bibid/BB04647342

Quarantelli EL (1995) Patterns of sheltering and housing in US disasters disaster prevention and management. An Int J 4(3):43–53. https://doi.org/10.1108/09653569510088069

Schacher T (2008) Good engineering without appropriate communication doesn't lead to Seismic Risk Reduction: some thoughts about appropriate knowledge transfer tools. In: The 14th world conference on earthquake engineering, Beijing, China. https://www.iitk.ac.in/nicee/wcee/article/14_S23-006.PDF

Seike T, Yoshiba H, Kim Y (2014) Research on construction of temporary houses in Fukushima (in Japanese). AIJ J Technol Des 20(45):503–508. https://doi.org/10.3130/aijt.20.503

Seike T, Kim Y, Hosaka Y, Ida S, Masuda T (2018) Intraregional reuse of emergency temporary housing in Japan. Sustain Cities Soc 42:650–662. https://doi.org/10.1016/j.scs.2017.11.001

Takahashi K, Nakamura Y, Shimizu Y (1998) Construction and management on living environment of temporary dwelling during the volcanic eruption of Mt. Fugen in Unzen (in Japanese). J Jpn Soc Civ Eng 1998(604):85–98. https://doi.org/10.2208/jscej.1998.604_85

Tamamura K, Hashimoto T, Toyokawa S, Ando K (2012) Indoor thermal environment in the temporary housing with wooden storehouses construction in summer (in Japanese). AIJ summaries of technical papers of annual meeting

Taniguchi R (2010) Stock-type wooden architecture in collaboration with Satoyama and carpenters. Nikkei Business Publications. https://xtech.nikkei.com/kn/article/building/news/20101201/544535/

UNDRO (1982) Shelter after disaster: guidelines for assistance. United Nations

Urabe T, Haganuma S, Watanabe M, Hayakawa S (2020) A study on the reality of relocation of emergency temporary housing: a case study on purchase method temporary housing built in Fukushima prefecture after the Great East Japan Earthquake (in Japanese). AIJ J Technol Des 26(63):667–671. https://doi.org/10.3130/aijt.26.667

Wang J, Ng YYE (2023) Post-earthquake housing recovery with traditional construction: a preliminary review. Prog Disaster Sci 18:100283. https://doi.org/10.1016/j.pdisas.2023.100283

World Bank Disaster Risk Management Hub (Tokyo) (2018) Converting disaster experience into a safer built environment: the case of Japan. World Bank. https://doi.org/10.1596/30015

Wu JY, Lindell MK (2004) Housing reconstruction after two major earthquakes: the 1994 Northridge earthquake in the United States and the 1999 Chi-Chi earthquake in Taiwan. Disasters 28(1):63–81

Yamada Y, Sato N (2021) Study on the distribution and typology of board storehouses, earthen storehouses, and compound attached buildings in the Hida region of Gifu prefecture (in Japanese). AIJ summaries of technical papers of annual meeting

Chapter 8
An Approach for Neighborhood Form Generation for Post-Disaster and Post-Conflict Temporary Housing Settlements

Nil Akdede, Bekir Özer Ay, and Ipek Gürsel Dino

Abstract Temporary housing settlements are one of the most significant post-crisis responses following detrimental disasters and forced displacement. However, the urgency to provide immediate shelter often results in the use of conventional methods that overlook fundamental spatial design considerations. Recognizing the need for a new approach to settlement generation, this study introduces a novel methodology for neighborhood form generation in post-disaster and post-conflict temporary housing settlements during the preparedness period of disaster management. In this study, a three-step methodology is adopted. First, design codes for post-disaster and post-conflict temporary housing settlements are thoroughly investigated through a comprehensive literature review. Then, the layout elements and basic spatial assumptions guiding the neighborhood form generation are identified. Finally, the requirements and constraints for neighborhood form generation in post-disaster and post-conflict temporary housing settlements are formulated, incorporating spatial design considerations. In contrast to the prevailing focus on housing design in various projects and approaches, the proposed approach, emphasizing spatial design considerations, holds promise for affected communities by enabling livable neighborhoods and for humanitarian workers involved in the provision and design of PDPC settlements. This offers a new approach to enhance shelter response efforts.

N. Akdede (✉)
Department of Architecture, Atilim University, Ankara, Turkey
e-mail: nil.akdede@atilim.edu.tr

B. Ö. Ay · I. G. Dino
Department of Architecture, Middle East Technical University, Ankara, Turkey
e-mail: ozer@metu.edu.tr

I. G. Dino
e-mail: ipekg@metu.edu.tr

C. Ochiai et al. (eds.), *Disaster and Displacement*, Disaster Risk Reduction,
https://doi.org/10.1007/978-981-96-2583-3_8

8.1 Introduction

Victims of disasters and displacement experience the impact of these crises not only at the time they occur but also in the aftermath, as the effects of such events often entail long-term challenges. Among the various physical, psychological, and social impacts of these crises, sheltering is a key area that requires coordinated responses in conjunction with logistics, public health, food, and environmental health after disasters and displacement, according to the International Federation of Red Cross and Red Crescent Societies (IFRC) in 2000.

The UNDRO (United Nations Disaster Relief Organization) (1982) defines sheltering after a crisis as fulfilling the sheltering needs of the victims with emergency, temporary, and permanent units with respect to housing policies and applications after crises. Furthermore, the concept of post-disaster and post-conflict (PDPC) sheltering has emerged, encompassing the idea of supporting people's well-being not only in the short term but also over extended periods (Alshawawreh et al. 2020). Given its importance, sheltering after crises is considered a vital yet challenging concept within the field of architecture. Sphere Association (2018), a non-profit organization focused on humanitarian action, emphasizes the significance of sheltering by stating that "Shelter is far more than a roof and four walls." In other words, it goes beyond the approach of viewing shelter merely as an object (Sanderson et al. 2014; Taheri and Mahsa 2021).

Despite its vital significance, the category of infrastructure and housing reconstruction has been identified as the one encompassing the most frequently encountered problems (Rouhanizadeh et al. 2020). The primary reasons are due to numerous barriers, such as inappropriate land use, improper physical development patterns and rules, and ignorance of traditional resources for reconstruction (Rouhanizadeh et al. 2020). In consideration of this problematic situation, the configuration of post-disaster housing has been highlighted as the "least successful form of aid when compared to other humanitarian intervention sectors" (as cited in Oxfam GB 2003).

In recent years, there has been a notable increase in the number of individuals affected by disasters and displacement, highlighting the challenges associated with providing appropriate shelters for these populations. However, conventional methods continue to be employed to generate PDPC settlements, resulting in inadequate and unsuitable solutions. Moreover, the common configuration of temporary housing settlements as a 'one size fits all' solution often results from neglecting the affected group's settlement patterns and social organization schemes (Pezzica et al. 2021). Therefore, this study aims to propose an approach for neighborhood form generation specifically designed for PDPC temporary housing settlements, with a focus on spatial design considerations.

The following part begins with a literature review on PDPC temporary housing, encompassing its characteristics and existing computational design studies. Section 8.3 covers the methodology, while Sect. 8.4 delves into the generation and representation phases of neighborhood form generation. The ensuing discussion is presented in Sects. 8.5, and 8.6 concludes the paper by addressing its limitations and outlining avenues for future research.

8.2 Background Research

8.2.1 Sheltering Response After Crisis

By acknowledging the cruciality of timing in humanitarian sheltering response, governments worldwide often apply a multi-stepped reconstruction process to meet the sheltering needs of the affected groups in four vital stages: emergency sheltering, temporary sheltering, temporary housing, and permanent housing (Quarantelli 1995). Finding the existing terminologies in the literature insufficient, IFRC (2013) presents three novel terminologies: transitional shelters, progressive shelters, and core shelters/one-room shelters. Acknowledging that the end of this path is permanent housing, the nonlinearity in the sheltering response is commonly underlined, even though these steps are presented sequentially (Bashawri et al. 2014).

The present work mainly focuses on temporary housing, which has been referred to as a key to the puzzle of housing recovery (Bektaş 2007; Di Gregorio and Soares 2017). Temporary housing is defined as a place enabling accommodation for temporary intends, approximately expected to be used for six months to three years enabling the return to daily activities (Quarantelli 1995). In reality, the usage period usually tends to extend; indeed, several cases have shown that those communities that were originally transferred to temporary housing units temporarily remain in those settlements and, in this way, turn them into permanent settlements.

Whatever the design of the temporary housing units is, they all aim to provide a temporary environment where individuals can carry on with their household responsibilities and activities in contrast to the characteristics of shelters (Johnson 2002; Carrasco and O'Brien 2019). Although game-changing approaches are increasing around the world, temporary housing units have been commonly served as prefabricated, mass-produced, and standardized units (Félix et al. 2013), based on which categorization, two types of temporary units are presented as ready-made units and kit supplies. While ready-made units are constructed in factories, the latter alternative allows for on-site assembly of the components (Félix et al. 2013). In a similar perspective, Albadra et al. (2018) categorize temporary housing as either transportable or built on-site. Regarding material, use, and construction technology, Asefi and Sirus (2012) categorize units into two forms: rigid and flexible. Depending on the construction typology, Rahmayati (2016) classifies temporary housing units into three groups: stilted, semi-stilted, and landed. Besides, container housing is presented as a type of temporary housing (Zhang et al. 2014) that can also be categorized under the above-stated groups.

Despite its cruciality, temporary housing is still one of the dilemmas in decision-making processes after crises (Fayazi & Lizarralde 2018; Akdede 2018; Perrucci et al.

2022) due to the complexity of type, duration of use, procurement, cost, location, and future utilization (Johnson et al. 2006). Therefore, design codes of temporary units and settlements should be carefully examined, as indicated in the following title.

8.2.2 Design Codes for PDPC Settlements

As a comprehensive framework in humanitarian action, the Sphere Handbook provides standards, key actions, key indicators, and guidelines for effective response in various crises, including disasters and displacement (Sphere Association 2018). The 'Shelter and settlement' chapter of the handbook encompasses seven key categories: planning, location and settlement planning, living space, household items, technical assistance, security of tenure, and environmental sustainability (Sphere Association 2018). In the context of spatial design for PDPC settlements, Table 8.1 illustrates the key indicators under the planning, location and settlement planning, living space standards, and technical assistance in the Sphere Handbook that can significantly influence the spatial design of PDPC settlements. Accordingly, key indicators under household items, security of tenure, and environmental sustainability have not been mentioned in Table 8.1 since they do not have a significant impact on the spatial design of PDPC settlements. The key indicators displayed in Table 8.1 will be the basis for the requirements and constraints for neighborhood form generation, as shown in Table 8.3.

Besides the core humanitarian standards, numerous academic studies (Chen et al. 2014; Nappi & Souza 2015; Nath et al. 2017; Albadra et al. 2017; Aburamadan & Trillo 2019, 2020; Asfour 2019; Afkhamiaghda et al. 2020) aim to address the question of how to generate temporary units and settlements after disasters or displacement. Before focusing on the criteria for such settlements, it is critical to emphasize that the main objective of generating these settlements is how temporary housing can positively impact the quality of life and help reduce the vulnerability of affected groups (Nappi & Souza 2015; Valent 2019; Pezzica et al. 2021).

In relation to this question, Félix et al. (2013) emphasize the importance of incorporating gathering spaces such as squares, parks, and gardens in temporary settlements. In a similar perspective, Carrasco and O'Brien (2019) underline the cruciality of open and communal spaces in temporary housing settlements for the recovery of the community.

Aburamadan and Trillo (2020) present 75 sub-criteria organized under eight main criteria: safety and security, comfort, social context, stability, durability/adaptability, disassembling ability, flexibility and modularity, and self-sufficiency of energy, all aligned with the same objective. Additionally, Asfour (2019) highlights the significance of compact areas, social necessities, quick assembly, portability, and health-related requirements. In another study, Albadra et al. (2017) identify security, safety, and thermal conditions as the most critical criteria for refugees.

Table 8.1 Key indicators of sheltering and settlement standards (Adapted from Sphere Association 2018)

Standards	Key indicators	Specific indicators
Planning	Determination of essential needs of the affected population by the affected people themselves and relevant authorities	–
	Maximization of the number of affected people whose needs and priorities are satisfied by providing shelter and settlement	–
Location and settlement planning	Cancelling out or minimization of natural or human-made threats, risks, and hazards in shelters and settlements	–
	Maximization of the number of shelters or settlements that provide reliable essential services within a reasonable distance	–
	Maximization of the number of affected people who feel safe about the location of shelter and settlement	–
	Maximization of settlement sites which provide sufficient space for private and public outdoor activities	45 square meters for each person in camp-type settlements, including household plots
		30 square meters for each person, including household plots, where communal services can be provided outside the planned settlement area
		Minimum ratio between covered living space and plot size is 1:2; move as soon as possible to 1:3 or more
Living space	Offering adequate living space for daily activities to as many people as possible	Minimum 3.5 square meters of living space per person, excluding cooking space, bathing area and sanitation facility
		4.5–5.5 square meters of living space per person in cold climates or urban settings where internal cooking space and bathing and/or sanitation facilities are included

(continued)

Table 8.1 (continued)

Standards	Key indicators	Specific indicators
		Internal floor-to-ceiling height of at least 2 m (2.6 m in hot climates) at the highest point
	Maximization of the number of shelters that satisfy the agreed technical and performance standards and cultural compatibility	–
	Offering secure shelters to as many people as possible	–
Technical assistance	Maximization of programs in which local authorities participate in defining the construction framework and activities	–
	Maximization of the number of affected people actively participating in the construction activities	
	Maximization of the number of shelters configured according to the agreed framework	
	Maximization of the number of affected people who receive technical assistance and guidance	

Besides the critical criteria mentioned above, it is crucial to adopt a context-sensitive approach for the design and implementation of PDPC temporary settlements, taking into account the specific short-term needs of the affected population as well as the long-term measures in each unique context.

8.2.3 Computational Design Studies on PDPC Temporary Housing Settlements

Before analyzing computational design studies on PDPC settlements and units, it is crucial to consider the use of information technology (IT) and computational tools in generating settlement and shelter designs. Despite the urgency of providing immediate shelter after disasters and displacement, a survey indicates that experts have never utilized any IT or computational tools during shelter provision (Kuchai et al. 2020). In other words, the study's findings reveal the frequent reliance on conventional methods in such scenarios. However, recognizing the advantages of IT or computational tools, such as their ease of learning and quick usability, humanitarian workers specifically emphasize the need for their application in monitoring environmental effects (Kuchai et al. 2020).

Considering this essential need in the humanitarian sector, several academic studies have explored computation-based approaches for PDPC units and settlements. Focusing on the unit scale, Benrós et al. (2011) present a rule system for PDPC units using shape grammar as a generative design technique. They conduct a two-dimensional unit design process, considering functional and spatial qualities, through a monitorable function-based generation methodology. Similarly, Gonçalves (2014) explores the generation of two-dimensional plan layouts by employing shape grammar as a rule-based generative design technique and identifying three different sizes of prefabricated modules as layout elements. Another study focusing on the unit scale, conducted by Karaoğlan and Alaçam (2018), presents a two-dimensional unit layout generation for PDPC temporary units using genetic algorithms. In this study, the objectives of generating modular post-disaster units include performance, flexibility/adaptability, ease of production/transportation/construction, and architectural quality.

From a different perspective, Torus and Şener (2015) introduce Container Post-Disaster shelters (CPoDs) as a computational tool. They utilize two different sizes of containers as layout elements to generate different configurations for a block of CPoDs based on given parameters. This approach addresses the problem neither at the unit scale nor the settlement scale but focuses specifically on the application of generating configurations for a block of containers.

Coping with the scale of settlements, Jinuntuya and Theppipit (2007) emphasize the practicality of a decision support software that specifically addresses spatial planning and cost estimation for PDPC settlements. Their approach involves integrating a 3D game engine and a Java-based CAD system to enhance the effectiveness of humanitarian support in these areas. In a similar approach, Daher et al. (2015) introduce a parametric generative design process and model for both units and settlements after displacement, collaborating with experts from the Shelter Research Unit (SRU) in IFRC. Regarding settlement design, they consider elements such as surface, configuration, spacing, orientation, sanitary/water points, and gathering spaces. These elements are categorized based on physical, contextual, and climatic parameters for settlements.

Another study, which focuses on settlement plan layout after displacement, develops a parametric and multi-criteria or multi-disciplinary design optimization-based framework for three-dimensional space layout design, using a real case in Luxembourg (Daher & Kubicki 2017). In this study, contextual and numerical constraints such as site accessibility, site orientation, site constructability, usage requirements, number of persons, required sanitary facilities, dimension of the containers, and specific requirements like sanitary containers are considered. However, the construction model for settlements in this study takes into consideration five specific parameters, namely cluster layout, entrance of clusters, the number of floors, vertical access, and structural elements, considering the aforementioned contextual and numerical constraints.

Karaoğlan & Alaçam (2019) adopt multi-objective genetic algorithms for the two-dimensional generation of post-disaster shelter neighborhoods. In this study, they establish constraints for neighborhood form generation that include avoiding collisions and maintaining a minimum distance of 1.5 m between units. The considered objective functions considered are the continuity of open areas, visual relationships between modules, daylighting in living areas, and compactness.

Addressing the scale of units and clusters, Afonso & Lu (2021) present a generative design tool that explores various two-dimensional unit designs and community plan alternatives. They use a voxel (cell) coding system where three voxel alternatives are represented as building physical compartments, non-built compartments, and constant spaces such as road access or specific site properties. The model is constructed based on connectivity rules for modules as topological rules and considerations for lighting and ventilation.

Considering the mentioned studies focusing on PDPC units and settlements from the perspective of computational design, it is important to note that most of the studies do not clearly address the requirements and constraints highlighted in the Sphere Handbook and overlook spatial considerations. This highlights the need for a new approach to neighborhood form generation that is both monitorable and incorporates the clear settlement requirements and constraints outlined in Sphere Standards. Such an approach would not only provide an approach for achieving a set of livable neighborhood alternatives for communities but also offer clarity and guidance to humanitarian workers, enabling them to design settlements that effectively meet the needs of post-disaster and post-conflict situations.

8.3 Methodology

Considering the complexity of sheltering and housing configuration, particularly in the context of post-disaster and post-conflict (PDPC) scenarios, this study aims to propose an approach for neighborhood form generation in PDPC temporary housing settlements. The goal is to present the necessary requirements and constraints for PDPC temporary settlements by considering the Sphere Handbook and incorporating spatial design considerations to develop an interactive computational model and design support tool that can offer livable neighborhoods for affected communities.

The methodology of this paper is structured based on the four components of the performance-based generation model introduced by Oxman (2006). These components include representation (R), generation (G), evaluation (E), and performance (P). Consequently, this study specifically emphasizes the representation and generation aspects of the generation process. It focuses on determining the design variables and their corresponding constraints, which are highlighted as the initial step of design optimization by Machairaset al. (2014).

To illustrate the process of neighborhood planning for PDPC temporary housing settlements, a three-step methodology was employed, as shown in Fig. 8.1. Firstly, design codes for PDPC temporary housing settlements were thoroughly examined

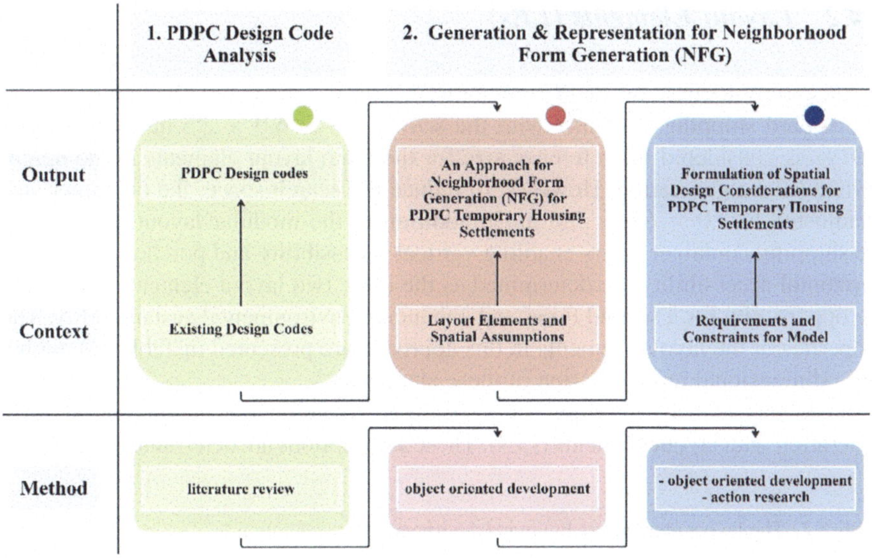

Fig. 8.1 Proposed research methodology layout

through a comprehensive literature review. Critical design codes were identified, and based on these, the layout elements (LEs) and basic spatial assumptions guiding the neighborhood form generation process were determined and represented. Subsequently, the requirements and constraints for this approach in PDPC temporary housing settlements were formulated, taking into account spatial design considerations. As a result, the exploration of diverse neighborhood and settlement layouts, considering both spatial and social determinants, can be achieved during the early design stage.

8.4 Generation and Representation of Neighborhood Form Generation Approach

8.4.1 PDPC Settlement Plan Layout

The representation of a PDPC settlement plan follows a hierarchical structure. The plan is divided into Neighborhoods (N), which are further subdivided into clusters (C). Each cluster is then broken down into its individual layout elements (LEs). The specific number of neighborhoods, clusters, and layout elements is open to discussion and can be determined based on factors such as the number of affected people, contextual considerations, available resources, and/or donations.

8.4.2 *Layout Elements (LEs)*

As this study focuses on PDPC temporary housing settlements, the common type of standard shipping container with the size of 2.5 × 6.0 × 2.5 m, as in Asfour (2019), is considered the reference size for the main layout elements in the neighborhood form generation. However, to reduce the search space, the container size is adjusted to 3.0 × 6.0 × 3.0 m. In addition to the modular layout elements of the shipping container, stairs enabling vertical accessibility and porches connecting horizontal accessibility are determined as the other two layout elements, providing the opportunity for a second floor, and enhancing environmental sustainability. The properties of the layout elements in this approach are presented in Table 8.2, with a three-dimensional representation of these elements.

Basic physical assumptions are considered to guide the neighborhood form generation with layout elements (LEs). These assumptions are determined as follows:

Table 8.2 The layout elements (LEs) in neighborhood form generation

Layout element	Description	Indoor/ Outdoor	Private/ Semi-public	3D representation of LEs for the approach
Container (A)	Container-size spaces including living and sleeping areas	Indoor	Private	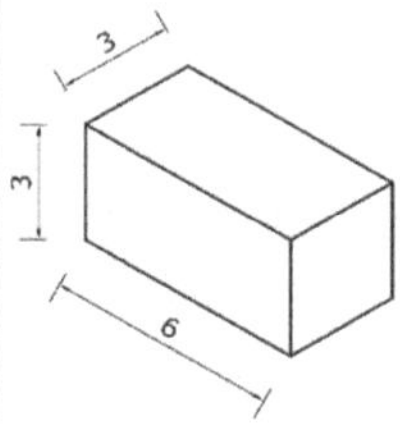
Stairs (Sta)	Open air spaces transmitting inhabitants between floors	Outdoor	Semi-public	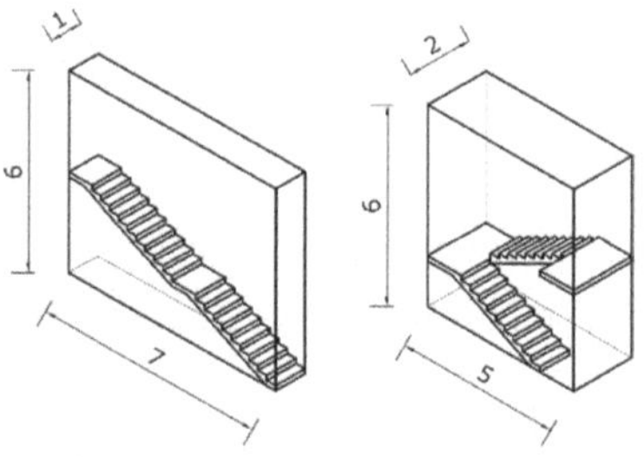
Porch (Po)	Outdoor spaces transmitting inhabitants between stairs and containers on the first floor	Outdoor	Semi-public	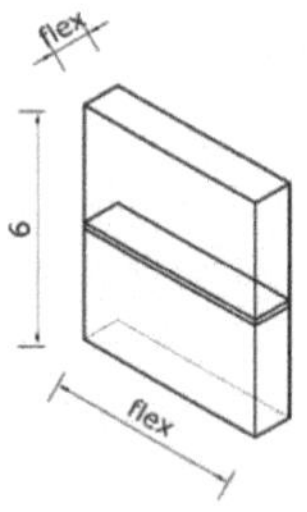

- Layout elements (LEs) in neighborhood form generation can be indoors or outdoors and can be private and semi-public spaces.
- Container (A) is determined as an indoor and private space, while Porch (Po) and Stairs (Sta) are determined as outdoor and semi-public spaces.
- Among LEs, only stairs can span multiple floors.
- Porches (Po) can only be on the first floor, aiming to facilitate the movement of inhabitants between containers (A) and stairs (Sta).
- LEs are theoretically bounded by rectangular prisms (but not necessarily physically bounded for outdoor spaces).
- The main element, the container size, is assigned as 3.0 × 6.0 × 3.0 m.
- There are two types of stairs, with sizes assigned as 1.0 × 7.0 × 6.0 m or 2.0 × 5.0 × 6.0 m.
- Sizes of porches (Po) are flexible.
- To reduce the search space, LEs are allocated onto a voxel with a size of 1.0 × 1.0 × 1.0 m, considering the predetermined standard container size.
- The maximum height of the neighborhood is limited to 6 m, meaning it can have two floors, a ground floor and a first floor.

8.4.3 Representation of Neighborhood Form Generation Approach

In the neighborhood form generation approach, the cell assignment method has been adopted, representing one of several automatic Generation of Space Layouts (GSL) methods. In this method, space geometry is first predefined and then divided into three-dimensional cells, in other words, voxels, of equal size (Du et al. 2020). This method has been utilized in various studies, including those by Yi (2016) and Gürsel Dino (2016). Through the use of this method, geometric and topological requirements can be effectively applied to achieve a feasible space layout (Du et al. 2020).

In this approach, which utilizes the cell assignment method, a matrix is used to represent the space layout, also entitled as the search space, and entities are assigned to cells within this search space. The voxel types (ranging from 0 to 7) that have been determined are presented as follows, along with the corresponding floor information and descriptions:

- Voxel type: 0 | LE:—| Ground Floor | Empty voxel,
- Voxel type: 1 | LE: A1 | First Floor | Container only on the ground floor,
- Voxel type: 2 | LE: A2 | Ground Floor & First Floor | Container only on the first floor,
- Voxel type: 3 | LE: A3 | Ground Floor & First Floor | Container both on the ground and the first floor,
- Voxel type: 4 | LE: Sta1 | Ground Floor & First Floor | Straight stairs,
- Voxel type: 5 | LE: Sta2 | Ground Floor & First Floor | U-shaped stairs,
- Voxel type: 6 | LE: Po1 | First Floor | Flying porches,
- Voxel type: 7 | LE: Po2 | First Floor | Porches on containers.

8.4.4 Generation of Neighborhood Form Generation Approach

As shown in Fig. 8.2, all cells begin with the voxel type '0' during the generation process of this approach for PDPC temporary settlements. The process starts with the assignment of a container (A), either on the ground (A1| Voxel Type:1) or on the first floor (A2 | Voxel Type:2), along with the random determination of the horizontal or vertical allocation of the layout element. After placing the first container, it checks for the availability of potential voxels for container placement. If A1 is selected as the placed object, it can only be placed on voxels labeled '0' and '2', while A2 can be placed on voxels labeled '0' and '1'.

To provide an example of the application of the cell assignment method for this approach, Fig. 7.2 illustrates the placement of two containers in the search space. As depicted in Fig. 7.2, if a container is allocated on top of another container, the overlapped voxels become voxel type '3'.

As the PDPC settlements are designed to be generated on two floors, containers can be allocated on both floors within each cluster. Therefore, Stairs (Sta1 and Sta2) enabling vertical access are necessary. Two types of staircases have been determined as layout elements: straight (Sta1 | Voxel Type: 4) and U-shape (Sta2 | Voxel Type: 5) stairs. Whenever a container is placed on the first floor, it randomly selects one of the types of staircases and positions it a maximum of 2 voxels away from the container.

The final step of placement in the approach involves placing porches (Po), which can be positioned on the available voxels between the highest point of the stairs and the nearest façade of the corresponding container. Two types of porches have been determined as layout elements: flying porches (Po1) and porches on containers (Po2). Porches can only be placed on voxel types '0' or '1'. Lastly, it checks if there

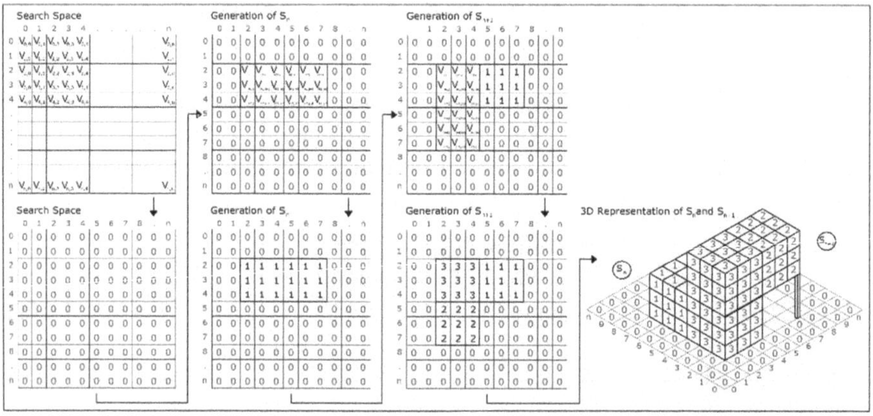

Fig. 8.2 Application of cell assignment method for two container (A) layout elements for neighborhood form generation

is only one voxel between any two porch layout elements. If this condition is met, the porches will be connected.

When the required number of container units is reached, the process stops the generation of containers. The final step is the assignment of family sizes. Due to disasters and displacement occurring in different continents around the world, the needs of affected groups vary. One crucial aspect that differs depending on the affected group is the size of the family. Recognizing the negative impact of inadequate housing designs on affected groups, three types of container plan alternatives have been determined, considering the fixed size of the modular container elements (A).

Considering the minimum living area standards specified by the Sphere Association (2018) in relation to climate conditions, the maximum number of inhabitants for the units can be determined. For example, in cold climates, container type A (use of a single standard shipping container) is planned to accommodate a maximum of 3 people, while container type B (use of two standard containers, joined one on the short side and the other on the long side, in an L-shape) and type C (use of two standard containers joined together on the long side) are designed to accommodate a minimum of 4 and a maximum of 7 inhabitants. To simplify the problem, container types B and C have not been considered as separate layout elements during the generation process.

8.4.5 Requirements and Constraints for Neighborhood Form Generation Approach

In consideration of the aforementioned standards and guidelines for PDPC temporary settlements, this study expands upon the list of requirements and constraints for neighborhood form generation, taking into account spatial design quality, equitable conditions for each inhabitant, and environmental, social, and economic sustainability. As a result, the requirements and constraints for neighborhood form generation are categorized into two main categories: neighborhood constraints and unit constraints, based on the scale of analysis.

In the approach, the category of neighborhood constraints consists of two main groups: space geometry constraints and space placement constraints, as shown in Table 7.3. Alongside the published standards at the settlement scale by the Sphere Association (2018), this study introduces four main spatial constraints under space geometry constraints for neighborhoods: balanced voxel distribution, balanced floor distribution, continuity and functionality of open areas, and centrality of open areas.

Balanced voxel distribution aims to achieve equal distribution both vertically and horizontally through the constraints of balanced voxel distribution and balanced floor distribution. The constraint of balanced voxel distribution ensures that each container in a cluster, as well as in a neighborhood, is provided with almost equal units and open areas, contrasting the scenarios shown in Fig. 8.3a. On the other hand, balanced floor distribution is determined as a constraint to maximize land use efficiency while

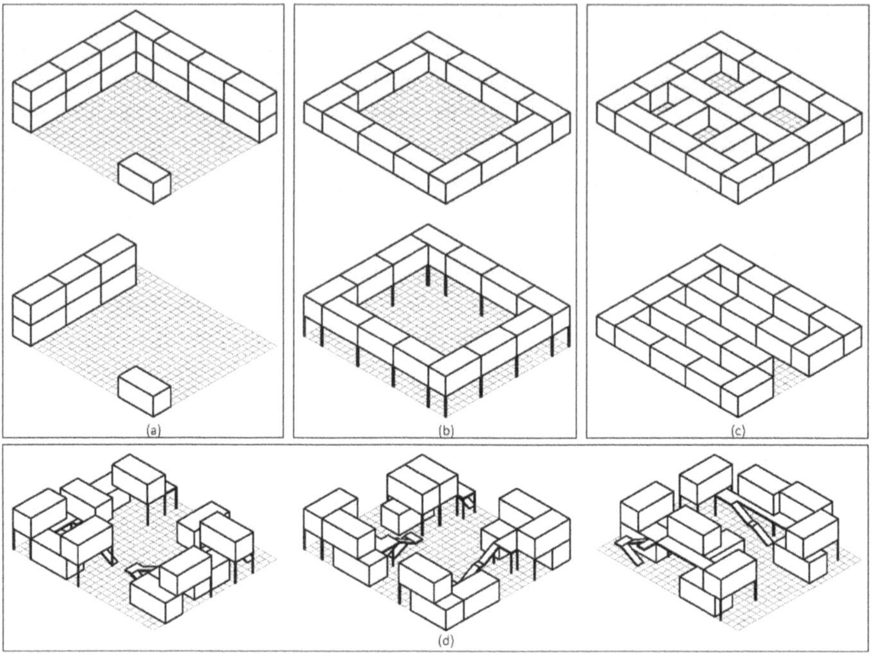

Fig. 8.3 3D Representation of the constraints of neighborhood form generation approach in cluster scale I **a** balanced voxel distribution; **b** balanced floor distribution; **c** continuity and functionality of open areas and centrality of open areas; **d** A set of alternatives for PDPC temporary housing clusters generated using the requirements and constraints in the proposed approach

maintaining spatial quality. It prevents the generation of clusters that do not achieve efficient floor utilization, as depicted in Fig. 8.3b.

Focusing on the significance of gathering spaces in post-crisis settlements, as highlighted by Félix et al. (2013) and Carrasco & O'Brien (2019), and considering safety and security issues, one critical spatial standard is the provision of functional and accessible outdoor gathering spaces for most of the containers. In light of this objective, the continuity and functionality of open areas, as well as the centrality of open areas, are determined as constraints in neighborhood form generation. This is done to prevent the creation of open spaces that resemble corridors or insecure areas, as depicted in Fig. 3c.

Table 8.3 formulates the requirements and constraints for clusters that will systematically generate a set of neighborhood alternatives using the obtained clusters.

Table 8.3 Formulation of requirements and constraints for neighborhood form generation

Category	Group	Name		Description
c1. Neighborhood constraints	c1.1. Space geometry constraints	c1.1.1. Community areas	c_{com}	Ensures a minimum of 45 square meters for each person in camp-type settlements, including household plots (Sphere Association 2018)
		c1.1.2. Plot size/ coverage	c_{plot}	Ensures that the minimum ratio between covered living space and plot size is 1:2; move as soon as possible to 1:3 or more (Sphere Association 2018)
		c1.1.3. Balanced voxel distribution	c_{cvoxel}	Ensures the balanced distribution of voxels on the land by minimizing the distance between the center point of voxels and the centroid of the land on the x and y axes
		c1.1.4. Balanced floor distribution	c_{floor}	Ensures the balanced floor distribution by minimizing the height of the center point of voxels from 3 m high
		c1.1.5. Continuity and functionality of open areas	$c_{conopen}$	Ensures the continuity of the open areas by maximizing the non-interrupted non-voxel spaces on the ground
		c1.1.6. Centrality of open areas	$c_{centrality}$	Ensures the centrality of the most continuous open area by minimizing the distance between the center point of the largest open area and the centroid of the land
	c1.2. Space placement constraints	c1.2.1. Placement of LEs	$c_{placement}$	Ensures the placement of LEs on the empty spaces or possible LEs that can be combined LEA1 can be assigned to the voxel types '0' and '2'. LEA2 can be assigned to the voxel types '0' and '1'. LEPo2 cannot be assigned to the voxel type '0' and can be assigned only to the voxel type '1'. All other voxel types can only be assigned to voxel type '0'
		c1.2.2. Placement of stairs	c_{stairs}	Ensures that if any containers are placed on the first floor, a staircase will be placed at a maximum of 2 voxels away from the container

(continued)

Table 8.3 (continued)

Category	Group	Name		Description
				Ensures that the ratio of the number of voxels of stairs to containers is less than a user-defined constant in consideration of the material capacity and budget
		c1.2.3. Horizontal accessibility on the first floor	$c_{haccess}$	Ensures that two LEs (stairs and the closest edge of containers) are connected with porches by connecting the highest point of stairs with the closest edge of the unit
		c1.2.4 Connectivity between porches	$c_{conporch}$	Ensures that if the distance between the two separated porches is one voxel, the separated porches are connected
	c1.3 Space assignment constraints	c1.3.1. Cluster size	$c_{clustersize}$	Ensures that a cluster contains the required number of people
		c1.3.2. Diversity in family sizes	$c_{diversity}$	Ensures that a cluster includes diverse family sizes by minimizing the difference between the number of diverse family sizes
c2. Unit Constraints	c2.1 Space assignment constraints	c2.1.1. Living space per person	c_{living}	Ensures that a minimum of 3.5 square meters of living space per person, excluding cooking space, bathing area, and sanitation facility, or ensures that 4.5–5.5 square meters of living space per person in cold climates or urban settings where internal cooking space and bathing and/or sanitation facilities are included (Sphere Association 2018)

8.5 Discussion

Given the current state of disaster and displacement, PDPC temporary housing settlements play a crucial role in providing shelter and supporting the well-being of affected groups. This paper emphasizes the necessity of a computer-based modular neighborhood design solution that extends beyond addressing immediate shelter needs, considering both short-term necessities and long-term measures. Thus, by centering the spatial design considerations for PDPC temporary housing settlements, this study aims to propose an applicable approach for generating neighborhood forms.

By considering the complex nature of sheltering and housing configuration, this study focuses on the generation of temporary housing settlements in PDPC scenarios. While the primary objective of this study is not to present definitive solutions for neighborhood form, it contributes to achieving a set of spatial designs by outlining requirements and constraints essential for neighborhood form generation. It recognizes the significance of gathering spaces, echoing the emphasis by Félix et al. (2013) and Carrasco & O'Brien (2019). As a result, we obtained a variety of cluster designs based on these constraints and requirements, effectively meeting the needs of post-disaster and post-conflict situations, as depicted in Fig. 7.3d. This approach provides the opportunity to generate multiple spatial design alternatives for the neighborhoods of PDPC settlements. Therefore, this study offers a valuable tool for supporting design exploration by highlighting the importance of shelter beyond a mere roof and four walls.

Although this study presents a set of requirements and constraints for the proposed approach, it's important to note that the minimum standards for settlements discussed in the literature might be inadequate to address all the necessities of the affected communities. Consequently, this study cannot comprehensively cover all the standards for PDPC temporary settlements. It is also beyond the scope of this research to evaluate the generated neighborhoods through the proposed methodology.

Overall, this research significantly contributes to the advancement of computational design approaches within the context of PDPC temporary housing settlements. It offers valuable insights into the requirements, constraints, and considerations necessary for generating the form of neighborhoods with gathering spaces. By addressing diverse aspects of spatial design directly related to social issues, this research aims to enhance the effectiveness of post-disaster and post-conflict responses. These insights can inform practical applications, offering a foundation for the development of more efficient and tailored solutions in the planning and design of PDPC temporary housing settlements.

8.6 Future Work and Conclusions

This study has introduced a computer-based approach for generating neighborhood forms in PDPC temporary housing settlements, considering the requirements and constraints defined by a set of standards and specific spatial design considerations. Acknowledging the diverse needs of affected groups, including considerations such as family dynamics, accessibility, safety and security, and privacy concerns, emphasizes the importance of a humanitarian response tailored to the specific context.

The study recognizes that standardized and one-size-fits-all approaches to settlement provision may lead to inadequate spatial design configurations. Therefore, practical models and tools are essential to provide appropriate PDPC settlements within a limited timeframe. This study highlights the significance of focusing not only on immediate crisis response but also on the preparedness period. During this phase, studies targeting suitable solutions for PDPC settlements should be developed and coordinated with relevant governmental and non-governmental authorities.

In conclusion, there is a need for ongoing research and collaboration to address the challenges in providing PDPC settlements. The integration of computational tools and practical models can enhance the design and implementation of PDPC settlements, ultimately improving the living conditions and well-being of affected communities.

Acknowledgements The authors would like to express their sincere gratitude to Dr. Cihangir Tezcan for his support in the computational implementation.

References

Aburamadan R, Trillo C (2019) A novel approach for establishing design criteria for refugees' Shelters'. In: Calabrò F, Spina L Della, Bevilacqua C (eds) New Metropolitan Perspectives. Springer

Aburamadan R, Trillo C (2020) Applying design science approach to architectural design development. Front. Archit. Res. Elsevier B.V. 9(1):216–235. https://doi.org/10.1016/j.foar.2019.07.008

Afkhamiaghda M, Elwakil E & Afsari K (2020) Creating a performance indicator for temporary housing. Can. J. Civ. Eng. (August). https://doi.org/10.1139/cjce-2020-0144

Afonso FG, Lu JC (2021) Post-disaster temporary housing system based on generative design method. Int J Struct Civil Eng Res 10(2):80–84. https://doi.org/10.18178/ijscer.10.2.80-84

Akdede N (2018) Evaluation of temporary housing units with multi-criteria decision making methods. Master Thesis. Available at: https://etd.lib.metu.edu.tr/upload/12622218/index.pdf

Albadra D et al (2017) 'Thermal comfort in desert refugee camps: an interdisciplinary approach', building and environment. Elsevier 124:460–477. https://doi.org/10.1016/j.buildenv.2017.08.016

Albadra D, Coley D & Hart J (2018) Toward healthy housing for the displaced. J Archit 23(1). https://doi.org/10.1080/13602365.2018.1424227

Alshawawreh L, et al (2020) Qualifying the sustainability of novel designs and existing solutions for post-disaster and post-conflict sheltering. Sustainability 12(890)

Asefi M, Sirus F A (2012) Transformable shelter: evaluation and new architectural design proposals. Procedia Soc Behav Sci. Elsevier B.V. 51:961–966. https://doi.org/10.1016/j.sbspro.2012.08.270

Asfour OS (2019) Learning from the past: temporary housing criteria in conflict areas with reference to thermal comfort. Int J Disaster Risk Reduct. Elsevier 38. https://doi.org/10.1016/j.ijdrr.2019.101206

Bashawri A, Garrity S, Moodley K (2014) An overview of the design of disaster relief shelters. Procedia Econ Financ. Elsevier 18:924–931. https://doi.org/10.1016/S2212-5671(14)01019-3

Bektaş E (2007) A post-disaster dilemma: temporary settlements in Düzce City, Turkey, in Post-disaster reconstruction: meeting stakeholder interests. Firenze University Press, Florence. https://doi.org/10.1400/79763

Benrós D, et al (2011) Automated design and delivery of relief housing: the case of post-earthquake Haiti. In: CAAD futures: designing together, pp 247–263

Carrasco S, O'Brien D (2019) Urbanism of emergency: use and adaptation of public open spaces in disaster-induced resettlement sites. In: Asgary A (eds) Resettlement challenges for displaced

populations and refugees. Sustainable Development Goals Series. Springer, Cham, pp 163–174. https://doi.org/10.1007/978-3-319-92498-4_12
Chen P et al (2014) A comparison of residents' satisfaction with temporary housing: a case study of two temporary housing types in southern Taiwan. J Chin Inst Eng Taylor & Francis 37(5):635–642. https://doi.org/10.1080/02533839.2014.904476
Daher E, Kubicki S (2017) 'Technologies in the planning of refugees' camps: A parametric participative framework for spatial camp planning. In: IHTC 2017—IEEE Canada international humanitarian technology conference 2017, pp 207–212. https://doi.org/10.1109/IHTC.2017.8058189
Daher E, Kubicki S & Halin G (2015) A parametric process for shelters and refugees 'Camps'. In: Proceedings of the 33rd eCAADe conference, pp 541–548
Du T, et al (2020) Gaps and requirements for automatic generation of space layouts with optimised energy performance. Autom Constr. Elsevier 116(103132). https://doi.org/10.1016/j.autcon.2020.103132
Fayazi M, Lizarralde G (2018) Conflicts between recovery objectives: the case of housing reconstruction after the 2003 earthquake in Bam Iran. Int J Disaster Risk Reduct 27:317–328. https://doi.org/10.1016/j.ijdrr.2017.10.017
Félix D, Branco J M, Feio A (2013) Temporary housing after disasters: a state of the art survey. Habitat Int. Elsevier 40:136–141. https://doi.org/10.1016/j.habitatint.2013.03.006
Gonçalves AM de CFA (2014) A grammar for shelters: an exploration of rule-based designs in prefabricated and modular shelters. In: Proceedings of the 32nd eCAADe conference, pp 327–336
Di Gregorio LT, Soares CAP (2017) Post-disaster housing recovery guidelines for development countries based on experiences in the American continent. Int J Disaster Risk Reduct. Elsevier, 24: 340–347. https://doi.org/10.1016/j.ijdrr.2017.06.027
Gürsel Dino I (2016) An evolutionary approach for 3D architectural space layout design exploration. Autom Constr Elsevier 69:131–150. https://doi.org/10.1016/j.autcon.2016.05.020
IFRC (2000) Disaster emergency needs assessment
IFRC (2013) Post-disaster shelter: ten designs. Geneva. Available at: www.ifrc.org
Jinuntuya P, Theppipit J (2007) Temporary housing design and planning software for disaster relief decision support system. In: Proceedings of the 12th international CAADRIA, pp 639–644
Johnson C, Lizarralde G, Davidson CH (2006) A systems view of temporary housing projects in post—disaster reconstruction. Constr Manag Econ 24(4):367–378. https://doi.org/10.1080/01446190600567977
Johnson C (2002) 'What's the big deal about temporary housing? planning considerations for temporary accomodation after disasters: example of the 1999 Turkish Earthquakes. In: 2002 TIEMS disaster management conference
Karaoğlan FC, Alaçam S (2019) Design of a post-disaster shelter through soft computing. Int J Archit Comput 17(2):185–205. https://doi.org/10.1177/1478077119849694
Karaoğlan F C, Alaçam S (2018) Design of a post-disaster temporary living space through the use of shape evolution. In: SIGRADI2018 technopoliticas—22th conference of the Iberoamerican society of digital graphics
Kuchai N et al (2020) The potential for computational IT tools in disaster relief and shelter design. J Int Humanit Action 5(1):1–20. https://doi.org/10.1186/s41018-020-00069-1
Machairas V, Tsangrassoulis A, Axarli K (2014) Algorithms for optimization of building design: a review. Renew Sustain Energy Rev 31(1364):101–112. https://doi.org/10.1016/j.rser.2013.11.036
Nappi MML, Souza JC (2015) Disaster management: hierarchical structuring criteria for selection and location of temporary shelters. Nat Hazards 75:2421–2436. https://doi.org/10.1007/s11069-014-1437-4
Nath R et al (2017) Investigating the key indicators for evaluating post-disaster shelter. Disasters 41(3):606–627. https://doi.org/10.1111/disa.12213
Oxfam GB (2003) Guidelines for post disaster housing reconstruction

Oxman R (2006) Theory and design in the first digital age. Des Stud 27:229–265. https://doi.org/10.1016/j.destud.2005.11.002

Perrucci DV, Hassan MM, Baroud H (2022) Planning for temporary housing through multicriteria decision analysis. J Hous Built Environ. Springer 38:541–576. https://doi.org/10.1007/s10901-022-09952-2

Pezzica C, Cutini V, de Souza CB (2021) Mind the gap: state of the art on decision-making related to post-disaster housing assistance. Int J Disaster Risk Reduct. Elsevier 53(101975). https://doi.org/10.1016/j.ijdrr.2020.101975

Quarantelli EL (1995) Patterns of sheltering and housing in US disasters. Disaster Prev Manag 4(3):43–53. https://doi.org/10.1108/09653569510088069

Rahmayati Y (2016) Post-disaster housing: translating socio-cultural findings into usable design technical inputs. Int J Disaster Risk Reduct Elsevier 17:173–184. https://doi.org/10.1016/j.ijdrr.2016.04.015

Rouhanizadeh B, Kermanshachi S, Nipa TJ (2020) Exploratory analysis of barriers to effective post-disaster recovery. Int J Disaster Risk Reduct. Elsevier 50:(101735). https://doi.org/10.1016/j.ijdrr.2020.101735

Sanderson D et al (2014) Principles, practice and lessons from haiti for urban post-disaster shelter recovery programs. Asian J Environ Disaster Manag 6(2):131–151. https://doi.org/10.3850/s1793924014000362

Sphere Association (2018) The sphere handbook: humanitarian charter and minimum standards in humanitarian response. Fourth, Sphere Association. Fourth. Geneva: Sphere Association. Available at: www.practicalactionpublishing.org/sphere

Taheri M, Mahsa T (2021) Navigating transition: post disaster housing pathways of households. J Hous Built Environ. Springer 37. https://doi.org/10.1007/s10901-021-09918-w

Torus B, Şener S M (2015) Post-disaster shelter design and CPoDS. ITU A|Z 12(1):269–282. Available at: http://www.scopus.com/inward/record.url?eid=2-s2.0-84924531879&partnerID=40&md5=7c1c92cf3e4be56318ee029687c5e321

UNDRO (1982) Shelter after disaster. New York

Valent G (2019) Disaster recovery and violence in neoliberal times: community and spatial fragmentation in L'Aquila. Disaster Prev Manag: Int J 28(4):446–459. https://doi.org/10.1108/DPM-01-2018-0032

Yi H (2016) User-driven automation for optimal thermal-zone layout during space programming phases. Archit Sci Rev 59(4):279–306. https://doi.org/10.1080/00038628.2015.1021747

Zhang G, Setunge S, van Elmpt S (2014) Using shipping containers to provide temporary housing in post-disaster recovery: social case studies. Procedia Econ Financ. Elsevier B.V. 18:618–625. https://doi.org/10.1016/S2212-5671(14)00983-6

Chapter 9
Limitations to the Use of Government-Facilitated Post-disaster Resettlement Outcomes: An Explanatory Critique of Two Landslide Resettlement Projects in Sri Lanka

Yasodhara Ranasinghe, Hannah Robertson, and Alan March

Abstract This chapter investigates the limitations of government-facilitated resettlement outcomes in Sri Lanka, particularly after the 2016 landslide event in Aranayaka region. The chapter presents a case study of two landslide resettlement projects. Both projects were delivered under the auspices of the Sri Lankan Government; However, one was built with high governmental influence, and end-users predominantly assisted the other. Notably, the government-assisted resettlement has undergone subsequent user-modifications, whereas the user-assisted resettlement has been built by adjusting the government's proposal from the outset. Based on qualitative interviews and observational data, the outcomes show significant limitations with extant government strategies, indicating that community modifications are critical to addressing those limitations. The analysis highlights that community engagement in resettlement processes is crucial to optimizing government-facilitated resettlement projects. If community engagement was considered early in the government proposals as opposed to post-implementation modifications, it would significantly improve community-identified issues being addressed in resettlement outcomes. Nevertheless, the results of this study underscore that the need for community engagement in government proposals rests on a flexible yet regulated approach.

Y. Ranasinghe (✉) · H. Robertson · A. March
Melbourne School of Design, The University of Melbourne, Melbourne, Australia
e-mail: ranasinghe.y@unimelb.edu.au

H. Robertson
e-mail: hannah.robertson@unimelb.edu.au

A. March
e-mail: alanpm@unimelb.edu.au

C. Ochiai et al. (eds.), *Disaster and Displacement*, Disaster Risk Reduction,
https://doi.org/10.1007/978-981-96-2583-3_9

9.1 Introduction

This chapter is a reflection on case study research conducted at two government-facilitated resettlement sites following the 2016 Aranayaka Landslide Event in Sri Lanka. Keeping community well-being at the forefront, this chapter explores the significant limitations of government-facilitated resettlements and community-led solutions to address those limitations. In doing so, it discusses community-inclusion as a critical element in optimizing resettlement project effectiveness.

The risk of landslides in Sri Lanka is primarily due to monsoonal rainfall on steep terrain. Landslide risk has increased in frequency in Sri Lanka due to hydro-meteorological events and unplanned land use activities such as rapid development, deforestation, and land clearing for cultivation (Ratnayake and Herath 2005; MoDM Sri Lanka et al. 2017). By 2023, approximately 30% of Sri Lanka's land and 35% of the country's population across 10 administrative districts are at risk of landslides (Konagai et al. 2023). Given the significant exposure (IFRC 2020), intensified landslide mitigation and post-disaster initiatives are significantly required nationwide. Against this background, further investigation is needed of tensions that may arise between national resettlement initiatives and the interests of resettled communities. The 2016 Aranayaka landslide (Fig. 9.1) caused by Cyclone Roanu (IFRC 2017) and subsequent resettlement projects in the area provide context for such a discussion.

Fig. 9.1 The 2016 Aranyaka landslide event (*Source* Perera et al 2018)

The 2016 Aranayaka landslide affected 11 villages causing a total of 168 deaths, with 198 families directly losing their homes. Further, nearly 2000 houses were identified as being at risk in high-risk zones, prompting the government to issue evacuation notices to those families (Sangasumana 2018; Fernando et al. 2021). In addition to damaging lives and assets, catastrophic landslide events typically cause disruption to infrastructure, lifeline facilities, resources, farmland, and places of socio-cultural importance in the area. Landslides also reduce property values and the productivity of agricultural and forest lands. Landslide disaster victims also experience reduced quality of life and emotional well-being (Cruden 1991; Dai et al. 2002; Malamud et al. 2004; Bandara and Jayasingha 2018). These direct and indirect and short- and long-term impacts following a landslide event reduce affected communities' capacity for rapid recovery, thus adding pressure on the Sri Lankan government to employ effective post-disaster resettlement solutions (Dai et al. 2002).

Resettlement tasks are often overwhelming for the Sri Lankan government, especially in light of the immediate shelter needs following the 2016 Aranayaka landslides. The challenge is further magnified because Sri Lanka is a Global South country that experiences "informal sectors, poor economic growth, rapid population growth, and infrastructure being destroyed and differentiated from the regular and transparent spaces of cities" (Arbab 2019, p. 331, 332). Moreover, in 2022, the country was under intense economic and political stress (George et al. 2022) due to its long history of centralist corruption, authoritarianism, and autocracy in politics (DeVotta 2022). In the face of such political and economic turmoil, landslide events requiring resettlement add stress to the Sri Lankan government's capacity to efficiently resettle the country's landslide victims.

In light of this context, a reflection on how efficient the government's existing resettlement strategies and built outcomes are in meeting the well-being of disaster victims allows for proposing better strategies. For that purpose, a case study (Yin 2017) analysis of two government-provided land-slide resettlements in the Aranayaka area was conducted. Research prompts inquiry to identify limitations to the use of government resettlement outcomes and the community-initiated modifications made to those resettlement outcomes to better meet the well-being needs of the resettled community members. The research asserts that a better understanding of the limitations of using government resettlement outcomes and the potential for input from the resettled community in the process will enable the proposal of appropriate alternative resettlement strategies.

9.2 The Challenge of Post-Disaster Resettlement

9.2.1 Why is Post-Disaster Resettlement so Difficult?

Post-disaster resettlement can be understood as a wicked problem, a threefold challenge where: firstly, the well-being needs and aspirations of disaster-affected individuals are diverse and complex. Secondly, implementing appropriate resettlement solutions involves contextual specificity and urgency. Thirdly, resources available for reconstruction are limited (Kruse et al. 2017; March and Kornakova 2017; Sanderson and Bruce 2020). Adding to this complexity, post-disaster resettlement programs must encompass not only short-term shelter provision but also medium- and long-term sustainable restoration of the critical infrastructures, services and housing that facilitate the livelihoods required for the full functioning of affected communities (UNDRR 2022).

A resettlement program tailored to the above-mentioned purposes, arguably, is not expected to serve in keeping the community within critical thresholds in extreme hazard conditions, absorbing their disturbances (Holling et al. 1995), or speeding up recovery from such perturbations (Adger 2000). On the contrary, its foundation lies in the necessity for "creating conditions in which people can live lives they have reason to value" (Lizarralde 2015, p. 1). Post-disaster literature suggests that the most successful resettlement approaches consider retaining pre-disaster land-ownership structures, restoring economic and social functions, adapting gradually to the hazard, preserving historical structures and the distinct identity of a place (Daly and Yenny 2012; Alexander 2013). Dai et al. (2002) suggest pluralistic approaches support effective meetings of these resettlement conditions. Such approaches embed,

the use of science in risk assessment and risk management and [disaster] studies within a sociopolitical framework, and subsumes within the decision-making process the nature of [disasters] and the nature of human activities" (Dai et al. 2002, 82).

Pluralistic approaches support the inclusion of different experiences including current and reliable knowledge and cutting-edge scientific advances, while acknowledging and incorporating invaluable wisdom and practices from traditional sources (March and Kornakova 2017, p. 246) such as locals' disaster knowledge, their building skills and locally available resources. To that end, an inclusive process that incorporates experts, academics, bureaucrats, political leaders and the community encourages the development of a comprehensive post-disaster response (Haigh and Amaratunga 2010; Deeming 2019). Nevertheless, a significant challenge in socio-economically and politically unstable countries lies in the tendency for pluralistic approaches to be exploited for opportunistic purposes, particularly in consolidating political authoritarianism and bureaucracy. Some parties may prioritize individual interests, such as the popularity and moneymaking of local politicians and officials, rather than the greater collective objectives. This results in limited implementation of comprehensive post-disaster resettlement strategies that leverage expert and community knowledge (Turner 1976; Hamdi 2010; Lizarralde 2015; Sanderson and Bruce 2020).

9.2.2 Community Modifications to Improve Resettlement Outcomes

Due to the exploitation of pluralistic approaches, post-disaster resettlement programs, especially in socio-politically and economically challenging countries, are often inadequate in meeting the needs of disaster-affected communities. Community modifications to government-facilitated resettlement outcomes are one way of addressing this gap in meeting the needs. Although it is difficult for community modifications to overcome all of the shortcomings of provided resettlements due to a lack of resources (Lizarralde 2015), community-built or modified housing can be socio-culturally specific, economically and time efficient (Lizarralde 2015; Robertson 2018). The alterations can also fly below the radar of government policy making allowing for contextual solutions and bypassing bureaucratic red tape. Furthermore, community built, or modified housing can be:

"A reflection of user and social values defined not by economic and political qualification of what it [is], but by far more complex, nuanced, and qualitative political understanding of what it [does] for people and their communities" (Bower 2016, pp. 85–86).

In such a context, identifying the numerous ways in which disaster-affected communities contribute to modifying resettlement outcomes can be just as valuable as official post-disaster resettlement initiatives. They can also assist in revealing the limitations of official post-disaster reconstruction initiatives and identifying opportunities for community inclusion in future post-disaster resettlement program design.

9.3 Methodology

9.3.1 An Explanatory Critique Approach

An explanatory critique built around ontological knowledge that inquires into being or existence (Lacey 2002, 2007; Bhaskar and Collier 2013) is used in this research. This approach was chosen as it allows uncovering reasons underlying the resettlement using limitations, taking Rwandeniya and Ambalangodallawatta resettlements as case studies. An explanatory critiquing of the existing limitations also navigates to discuss an appropriate alternative mechanism for a resettlement delivery process tailored to the well-being of disaster-affected communities, considering community engagement as a critical aspect.

Ruwandeniya and Ambalangodallawatta resulted following the 2016 Aranayaka landslide. These two projects were chosen as case studies due to their divergent implementation processes, representing Sri Lanka's existing resettlement strategies, and the subsequent significant community modifications, representing the true well-being needs of the landslide victims. Both projects were delivered under the auspices of the Sri Lankan government. However, the Chinese government executed the Ruwandeniya as a doner-led project, while the Ambalangodallawatta was conceived as the Sri Lankan government's design proposal for an owner-led resettlement project.

Data is predominantly derived from the built environment and the inhabitants of the two resettlements. The data was collected by tracing the plan configuration and structure of 20 houses from each resettlement and interviewing inhabitants. Table 9.1 below presents the number of samples against the total number of houses in each resettlement. To assist in the analysis, each household was assigned a sample number from 1 to 20, and interviewees were de-identified through a code. A code contains the sample number along with the first letter of the representative resettlement's name (e.g. Interviewee-1R).

In addition, government perspectives are added to the data set, through five interviews conducted with the resettlement program facilitators representing the government (as shown in Table 9.2). The program facilitators are also de-identified through a code (e.g. Interviewee-PF-1); however, their positions or representative agencies have not been identified in the discussion to ensure anonymity. Data collection was conducted at these two sites from September to November 2022. At that time, Covid was still a concern, and public demonstrations were taking place in various parts of Sri Lanka following the country's economic and political decline. These conditions made data collection challenging, particularly in relation to engagement with communities and access to housing. However, strategic relationship buildings, and necessary health precautions allowed data collection to persist.

To analyze, architectural trace data and data from the two interview types are juxtaposed, providing a range of perspectives on government-facilitated resettlement processes, limitations to using resettlements and user-modifications to the resettlements. Ultimately, data leads to the discussion of an alternative approach to government-facilitated resettlements in Sri Lanka.

Table 9.1 Sample of interviewees from the community (Table by authors)

Resettlement site	No. of houses	Samples for the household studies and the interviews
Ruwandeniya	60	20
Ambalangodallawatta	52	20

Table 9.2 Sample of interviewees representing the service providers (Table by authors)

Role of the interviewee	No. of interviewees
Building industry professionals	2
Project administrators	3

9.3.2 *Sri Lanka's Government-Facilitated Resettlement Sites: The Case Studies*

See Fig. 9.2.

The Kegalle Landslide Resettlement Program (KLRP) was devised to resettle the 2016 Aranayaka landslide victims. Although specifically developed in response to the Aranayaka landslide the program was informed by Sri Lanka's long history of disaster susceptibility. According to Interviewee-PF-1, a program facilitator, the KLRP resulted from lessons learned from previous ad-hoc resettlement approaches following landslides and other disaster events such as floods and civil war in Sri Lanka. The KLRP also drew on several international recovery frameworks, including the Sendai Framework for Disaster Risk Reduction and the Hyogo Plan of Action, however, these were largely one-size-fits-all approaches and not tailored to the specific context of Sri Lanka.

Against that governance background, the KLRP has two approaches to resettle landslide victims: (1) individuals in private properties and (2) Communities in Government Resettlement Sites (GRS). Although the individuals in private properties approach presents unique issues of disconnection, this research focuses on the GRS approach. GRS involved in providing houses and infrastructure for the landslide victims in eight township developments (Divisional Secretariat, Aranayaka 2018; Fernando et al. 2021) delivered by 2022. The resettlements were programmed with a high level of government influence, motivated by the perceived need to get the townships built quickly and with efficiency of scale. However, the extent of power and autonomy afforded to resettlers in each resettlement project was demonstrably variable. The case studies discussed in this chapter, Ruwandeniya and Ambalangodallawatta (Fig. 9.3), are both representatives of the GRS projects.

Ruwandniya's housing comprises two-story units, where one unit consists of four houses, which are made up of two upper-floor houses and two ground-floor houses. There are 60 such houses in 15 housing units, and the ancillary facility buildings in the premises include a community center and a place of worship. According to Interviewee-PF-3, a program facilitator, the Ruwandeniya project was delivered as a donation that was designed to conform to a generic brief. The construction phase was completed by skilled laborers. Community involvement in planning and construction in Ruwandeniya was therefore limited, however, there are observable instances that the users have later attempted to modify the new dwellings to meet their well-being needs.

In contrast, Ambalangodallawatta consists of 52 individual houses and some ancillary buildings such as a nursery. Its land was divided into 500 square meter plots and allocated separately to disaster-affected families. According to Interviewee-PF-3, a program facilitator, Ambalangodallawatta was initiated by the Sri Lankan government prioritizing the needs of disaster victims. Accordingly, decisions such as house size, floor size and finishing materials were in the hands of the user community, while government representatives assisted in the building process. As Interviewee-PF-3 further explained, the KLRP also aimed to produce economically viable as well

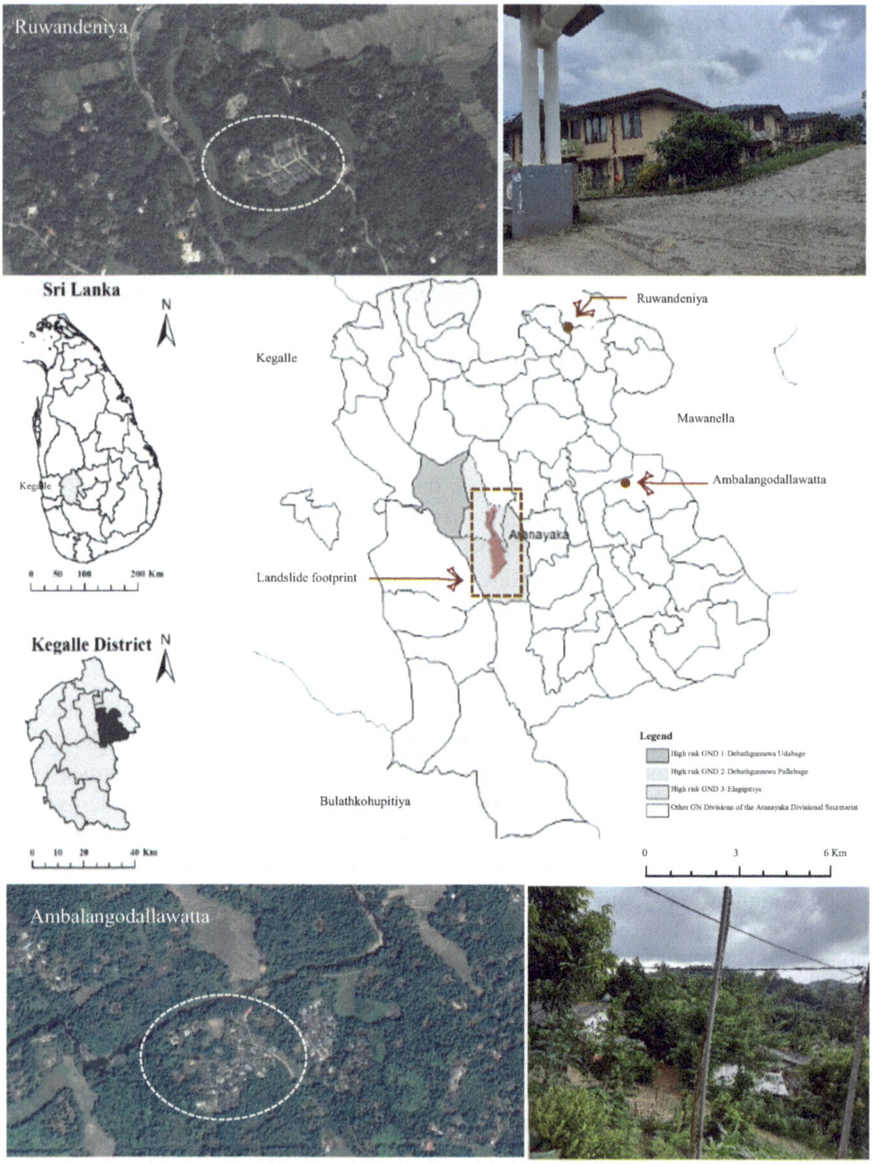

Fig. 9.2 Case study context and locational distribution of the two sites. Ruwandeniya (top) and Ambalangodallawatta (bottom) (*Source* Authors based on open sources)

Fig. 9.3 Two enlarged and decorated living rooms (*Source* Authors 2022)

as agriculturally and industrially competitive solutions. The KLRP policy also aimed to understand the diversity of needs affecting the disaster-affected communities. To that end, each plot was assigned a government-designed prototype house based on the land's topography, with community input to change the design, if needed, within the guidelines and available funding. Interviewee-2A, a 53-year-old woman, reported that many families greatly modified the prototype designs proposed for Ambalangodallawatta to meet their diverse needs. As reflected in the trace analysis, 20% of the Ambalangodallawatta families were constrained by the government's funding and design, while the remaining 80% constructed a bigger house with more features to meet their needs and aspirations.

Despite the two different delivery approaches, the analyses reveal common limitations in the use of resettlement outcomes that challenge the resettled communities meeting their well-being. The limitations are two types: Fundamental and deep-rooted. Fundamental limitations pertain to the built environment necessitating modifications, while the deep-rooted limitations emerge as consequences of community-involvements made to address the fundamental limitations.

9.4 Discussion: Limitations to the Use of Resettlement Outcomes

9.4.1 Fundamental Limitations

9.4.1.1 Contrasting Spatial Adequacy

The two resettlements revealed contrasting spatial adequacy for residents' existing and growing needs, limiting space utilization to meet the primary, but diverse, needs of 90% of the interviewees from the two resettlements. Additional space for daily activities such as gatherings, cooking and laundry were the primary spatial needs that were met. The minority of 10% have underutilized spaces in their houses. The different experiences of spatial consumption in parallel with the needs presented by two Ruwandeniya interviewees show that flexibility in space is important. Interviewee-8R, a 44-year-old man, reported that the government-provided spaces in his house are inadequate for accommodating daily activities. According to him, "the living room is too small. [The family] can't even sit together to watch TV" (Interviewee-8R, 2022).

From a contrasting viewpoint, Interviewee-5R, a 50-year-old woman, expressed satisfaction with the house layout since she lived alone. Regardless, she reported that she "wish [she] had a smaller house with a bigger garden" (Interviewee-5R, 2022).

To address these contrasting needs, Interviewee-8R had to resort to demolishing a permanent wall and enlarging the living room, while interviewee-5R opted to simply add ornaments and indoor plants (Fig. 9.3). Overall, the two interviews highlight the importance of spatial flexibility and provisions for customization to meet diverse and evolving user needs.

9.4.1.2 Lack of Socio-Cultural Connectedness

The built environment lacks the capacity to accommodate peoples' socio-cultural practices. This limitation arises from a resettlement strategic plan focusing on using land, funds, and technical assistance to construct high-density-to-land ratio shelters with the aim of housing the greatest number of disaster victims in each settlement. While the provision of permanent houses in large quantities may be well-intentioned by the Sri Lankan government, this densification approach has resulted in limited opportunities for socio-cultural connectedness.

For example, the building facades of Ruwandeniya limit the internal and external spatial relationships discouraging inhabitants' socio-cultural practices of interacting, sharing and caring. Addressing this limitation, approximately, 60% of the Ruwandeniya residents have added temporary front verandas to their houses. Explaining this, Interviewee-4R, a 69-year-old woman, reported that socio-cultural activities are primarily carried out in the streets of the new settlement, out of the doors, as no semi-public elements are added to the house design. However, "the situation has been

Fig. 9.4 Later addition of front verandah (*Source* Authors 2022)

somewhat improved by the later addition of the front verandah" (Interviewee-4R, 2022) (Fig. 9.4).

Even in Ambalangodellawatta, designed as a village of owner-driven scattered houses, a significant portion of the land is reserved for houses and roads, limiting spaces for socio-cultural activities. Interviewee-1A, a 55-year-old man, reports that he has observed a decrease in interaction among individuals since resettling in the Ambalangodallawatta, and he assumes this decline may be attributed to inadequate built environment support for such activities. He also stated that lack of socio-cultural connectedness may also lead to further difficulties, such as post-traumatic stress.

9.4.1.3 Livelihood Limitations

Income generation is limited within the resettlements due to housing priorities. The community interviewees are predominantly farmers or self-employed individuals, so the loss of their agricultural lands and self-employment market due to the landslide led to a pressing need for new income generation opportunities. However, the lack of spatial allocations and a job market tailored to their skills posed a significant limitation in the two resettlements. Interviewee-8R, a 44-year-old man, reported that "[he] was motivated to resettle in Ruwandeniya because the resettlement proposal promised spaces for establishing new businesses" (Interviewee-8R, 2022). However, that proposal had not materialized, limiting his livelihood opportunities.

Interviewee-15A, a 60-year-old woman, reported that the limited availability of ancillary buildings within the resettlement is also a part of the same problem. According to her, such buildings would have been good platforms for self-employment activities. Although there are community centers in the resettlements acting as main amenity buildings accessible for this purpose, she states that the positive contribution such a single building could make is minimal.

9.4.1.4 Limited Infrastructure

In both resettlements, the existing infrastructure is inadequate in meeting diverse community well-being. For example, the road network of the Ambalangodellawatta has been constructed with steep slopes, challenging pedestrians and drivers to traverse easily. According to Interviewee-4A, a 65-year-old man, "this resulted in increased time and energy spending on travel and transportation" (Interviewee-4A, 2022).

In another example, Interviewee-8R, a 44-year-old man, reported that the septic tanks and drainage systems in the Ruwandeniya were insufficient in capacity, leading the community to adopt temporary measures. He reported that in the absence of a donor's intervention to fix the matter, a group of residents have diverted their septic tank waste into open drains. This has caused air pollution and unsanitary living conditions within the resettlement, "but, there wasn't any other option" (Interviewee-8R, 2022).

Overall, these scenarios underscore that community-initiated solutions for resettlement limitations are often impromptu but not viable and long-term.

9.4.1.5 Absence of Collective Accountability

Ambiguous or unresolved tenancies cause situations where residents do not take accountability for places. For example, the property ownership of the Ruwandeniya was still not resolved during the data collection period in November 2022. The trace analysis reveals that unresolved tenancies in Ruwandeniya have resulted in spaces being left without maintenance or cleaning. Interviewee-1R, a 40-year-old woman, reported that Ruwandeniya has features such as stairs, gardens, ducts, rainwater pipes and drains for shared use, but no one legally owns them and does not care about maintenance.

Similarly, in Ambalangodallawatta, the residents in individual properties had not been assigned ownership rights through legal deeds. Interviewee-10A, a 69-year-old woman, expressed her unease about using the houses due to the lack of ownership rights, which has led to skepticism about participating in public works in the village and becoming a member of local societies.

These situations imply that not granting the right to own can create problematic conditions in resettlements; especially since it will not serve the purpose of disaster victims owning a new home after losing one.

9.4.2 Deep-Rooted Limitations

9.4.2.1 Isolation

Isolation in terms of socioeconomic and emotional factors is reflected in the analysis of interview data. Interviewee-8R, the 44-year-old man, reported that isolation in the

Fig. 9.5 The abandoned bridge structure (*Source* Authors 2022)

Ruwandeniya emerges because access to essential services such as schools, hospitals, and town centers is a significant challenge, and that is caused by their remote location and limited availability of transportation. He further reported that the Ruwandeniya residents have to cross a river to access facilities in the nearest township. An attempt by the residents to build a closer bridge to connect the resettlement to the township was subject to structural problems and abandoned when the government intervened (Fig. 9.5). So, the isolation issue thus remains a persistent concern.

Isolation also emerges because people don't get the opportunity to interact with other communities with diverse sociocultural and economic capabilities (Gilbert 2021). Ruwandeniya and Ambalangodallawatta interviewees agreed that interaction, once highly valued in their lives, is diminishing in the new settlements. Interviewee-4A, a 65-year-old man, expressed his reluctance of accepting the new place and interacting with the neighboring communities. According to him "[Ambalangodallawatta] doesn't feel like home" (Interviewee-4A, 2022).

These scenarios imply that it is important to design resettlements that foster connections and facilitate the socio-economical and emotional sense of resettled communities.

9.4.2.2 Identity Loss or Damage

Reflecting on the Interview data analysis, the identity of the two resettled communities is enforced by the challenging living conditions of the new places. According to Interviewee-5A, a 63-year-old woman, her family is identified as "extreme-poor and treated as if [they] recently received houses since being homeless [their] entire life" (Interviewee-5A, 2022). But in reality, the family has only been evicted from its original habitats and relocated to new homes.

Interviewee-PF-5, a project facilitator, reported that even six years after the landslide event, these groups of people are ascribed by society as re-settlers, and respect based on individuals' values is rarely received outside resettlement. This implies that the deep-rooted impact on identity and suggests that the sense of identity is not individually but collectively and continuously shaped.

From the community perspective, the livelihood changes from farmers or self-employed individuals to predominantly construction industry workers or the outright unemployed has also led to identity damage. Interviewee-1A, a 55-year-old man, reported that his family owned relatively large tea plantations, which were a significant source of income. After his inherited land was made unusable by the landslide, the new village was not provided with facilities for cultivation activities, leaving him jobless. He expresses his frustration at not being able to retain his occupational identity over the years, being newly identified as a jobless man.

Overall, the damage caused by disasters had resulted in limited community capacity to maintain their pre-disaster sociocultural and/or occupational identities. As new resettlements made little or no effort to replace these identities, sociocultural and economic identities of disaster-affected communities have been largely compromised.

9.4.2.3 Governance Problems

Over time, the power of authorities and community leaders have declined, and ceased initiating resettlement improvements. According to Interviewee-8R, the 44-year-old man, authorities or the community leaders did not have the will or capacity to address fundamental resettlement limitations, especially in a situation where financial resources are limited. On the other hand, Interviewee-13A, a 48-year-old man also expressed that a lack of interest in working towards common goals that benefit the community has caused the local government to lose interest.

The Ruwandeniya residents' attempt to build an unauthorized bridge after not getting a satisfactory response from the Sri Lankan government to their transport problems best exemplifies the lack of government interest in investing in the resettled communities' needs. Despite potential consequences that may arise from building an unauthorized bridge, Interviewee-8R argued that the residents should not be held responsible for that. The trace analysis revealed that the government had not yet intervened to rectify this problem, leading to an escalation of power conflict. This situation highlights the importance of effective communication and collaboration between the government and resettled-community members to avoid conflicts and ensure sustainable solutions to deep-rooted limitations.

9.4.3 Putting Forth an Approach to Future Post-Disaster Resettlement

The extent to which the built environment of the two resettlements and their modifications have affected resettled communities was discussed above. This section extends to proposing an alternative strategy for appropriate resettlement outcomes tailored to Sri Lanka's disaster-affected communities' well-being.

Based on the case study analyses and existing debates in the post-disaster literature, the primary argument of this chapter is that resettlement strategies appropriate to communities' well-being are crucial at the outset to mitigate the limitations of using resettlements. To further explain what the argument suggests, resettlement processes with meticulous planning with a significant pluralistic approach (Dai, Lee and Ngai 2002; Haigh and Amaratunga 2010; Sanderson 2010) and community-sensitive strategies (Lizarralde 2015) are likely to yield more favorable outcomes. However, this is rather idealistic in terms of practicality. From a practical standpoint, pluralistic approaches with significant community sensitivity may encounter limitations mainly for two reasons.

Firstly, the uncertainty surrounding a disaster (Rodríguez et al. 2007) and the urgency (Quarantelli 2000) of resettlement often lead to unclear communication between service providers and disaster victims. Although the literature suggests that disasters are opportunities for development (Quarantelli 1998; Collins 2018), this uncertain and urgent nature restricts opportunities for service providers to fully grasp the well-being needs of communities in the face of disasters. Secondly, the traumatic experiences associated with disasters diminish the community members' ability to effectively external communications (Moorthy, Benny and Gill 2018) of peoples' true recovery needs and aspirations. Consequently, essential aspects that should be considered in the early stages of resettlement processes can be overlooked, making resettled communities modify the built environments while occupying them.

Considering the challenges identified by analyzing the two Sri Lankan case studies that involved community modifications, we suggest two propositions for resettlements for their effective usage: collaboration and flexibility. This is particular for, but not limited to, Sri Lanka as its expertise and resources are not restricted to professionals or the government sector. Instead, the communities have distinct voices in this case to determine their well-being needs and to contribute through various channels. Therefore, disaster risk reduction and building construction industry experts, bureaucrats, political leaders, donors and the community need to collaborate. However, it is important to recognize that unlimited flexibility is not feasible within a collaborative approach. Instead, a balanced and inclusive strategy with all possible options foreseen is needed.

Therefore, we propose an approach that applies 'regulated flexibility', emphasizing significant flexibility for community autonomy with appropriate government influence. Accordingly, the flexibility in a resettlement program should be accompanied by specific guidelines and principles that draw upon the knowledge and collective wisdom of all stakeholders involved. Thus, striking a balance between flexibility and regulation allows individuals to modify the built environment to suit their living conditions but is controlled to the extent that the country's underdevelopment is no longer acute.

9.5 Conclusions

Limitations of government-facilitated resettlement outcomes are primarily due to a lack of sensitivity to community well-being that can be provided through the built environment. This lack of sensitivity is primarily due to inefficient and unclear communication due to the uncertainty of disaster events and the urgent need for shelter. Such communication barriers, especially in low-income countries, miss the opportunity to properly deploy the limited available resources in resettlement processes, making resettlements less likely to be effective for those who use them (Rodríguez et al. 2007; Alexander 2017; Moorthy et al. 2018).

Considering the Sri Lankan case study, although the country's government-facilitated resettlement programs are well-intentioned, limitations to use caused by the lack of sensitivity are prevalent. Some of these limitations are fundamental while some are deeply rooted. The study reflects that the government's unfit resettlement strategies have resulted in fundamental limitations, while deep-rooted limitations are the consequences of the communities alone attempting to address fundamental limitations.

As the analysis reveals, in the Sri Lankan context, the fundamental limitations are the results of fewer community well-being priorities during the resettlement delivery process. Notable limitations include contrastive spatial adequacy, lack of socio-cultural connectedness, livelihood limitations, limited infrastructure and absence of collective accountability. Isolation, identity loss or damage, and governance problems are the deep-rooted limitations created by user modifications to unfitting built environments. Both types of limitations hinder the overall sustainability of the resettlements and have created the need for an alternative post-disaster resettlement approach for Sri Lanka. Accordingly, these limitations should be factored in at the outset of resettlement planning, namely through the inclusion of community perspectives through active engagement and allowances for regulated flexibility in resettlement proposals.

Regulated flexibility focuses on creating a balance between flexibility and regulation, allowing the community to make modifications at each stage of development. The concept also upholds a follow-up mechanism founded on a set of tailored design and construction principles. This concept supports building a community's short- and long-term resilience and sustainability and reduces its vulnerability to future disasters.

References

Adger W (2000) Social and ecological resilience: are they related? Prog Hum Geogr 24:347–364. Available at: https://doi.org/10.1191/030913200701540465

Alexander D (2013) Planning for post-disaster reconstruction, p 12

Alexander DE (2017) How to write an emergency plan. Dunedin Academic Press Ltd.

Arbab P (2019) Global and globalizing cities from the global south: multiple realities and pathways to form a new order. Perspect Glob Dev & Technol 18(3):327–337. Available at: https://doi.org/10.1163/15691497-12341518

Bandara RMS, Jayasingha P (2018) Landslide disaster risk reduction strategies and present achievements in Sri Lanka. Geosci Res 3(3). Available at: https://doi.org/10.22606/gr.2018.33001

Bhaskar R, Collier A (2013) Introduction: explanatory critiques. In: Critical Realism. Routledge, pp 385–394

Bower R (2016) Who decides and who provides? the anarchistic housing practices of john turner as realizations of Henri Lefebvre's Autogestive Space. Altern: Glob, Local, Polit 41(2):83–97

Collins A (2018) Advancing the disaster and development paradigm. Int J Disaster Risk Sci 9. Available at: https://doi.org/10.1007/s13753-018-0206-5

Cruden D (1991) A simple definition of a landslide. Bull Int Assoc Eng-Ing Geol 43(1):27–29

Dai FC, Lee CF, Ngai YY (2002) Landslide risk assessment and management: an overview. Eng Geol 64(1):65–87. Available at: https://doi.org/10.1016/S0013-7952(01)00093-X

Daly P, Yenny R (2012) Cultural heritage and community recovery in Post-Tsunami Aceh. In: From the ground up: perspectives on post-tsunami and post-conflict Aceh, pp 57–98

Deeming H (2019) Professionalisation of emergency management: a chicken or egg conundrum? Cris Response J 14(4)

DeVotta N (2022) Sri Lanka's agony. J Democr 33(3):92–99. Available at: https://doi.org/10.1353/jod.2022.0042

Divisional Secretariat, Aranayaka (2018) Aranayaka arunella: rebuilding aranayaka following the 2016 landslide

Fernando N, et al. (2021) Life two years after relocation: status quo of natural hazard induced displacement and relocation in Kegalle, Sri Lanka. Project report. NBRO, Sri Lanka

George ASH, George AS, Baskar T (2022) Sri Lanka's economic crisis: a brief overview. PartnS Univers Int Res J 1(2):9–19. Available at: https://doi.org/10.5281/zenodo.6726553.

Gilbert P (2021) Creating a compassionate world: addressing the conflicts between sharing and caring versus controlling and holding evolved strategies. Front Psychol 11:582090. Available at: https://doi.org/10.3389/fpsyg.2020.582090

Haigh R, Amaratunga D (2010) An integrative review of the built environment discipline's role in the development of society's resilience to disasters. Int J Disaster Resil Built Environ 1:11–24. Available at: https://doi.org/10.1108/17595901011026454.

Hamdi N (2010) The placemaker's guide to building community. Taylor and Francis, Hoboken

Holling CS et al (1995) Biodiversity in the functioning of ecosystems: an ecological synthesis. Biodiversity Loss: Economic and Ecological Issues. Cambridge University Press, Cambridge, pp 44–83

IFRC (2017) Sri Lanka: severe floods and landslides—emergency appeal n° MDRLK005 Operation update n° 4 (One year consolidated report). Available at: https://reliefweb.int/report/sri-lanka/sri-lanka-severe-floods-and-landslides-emergency-appeal-n-mdrlk005-operation-update.

IFRC (2020) World disasters report 2020: come heat or high water—tackling the humanitarian impacts of the climate crisis together. N 978–2–9701289–5–3. Geneva: International Federation of Red Cross and Red Crescent Societies, Geneva. Available at: https://reliefweb.int/report/world/world-disasters-report-2020-come-heat-or-high-water-tackling-humanitarian-impacts

Konagai K, et al (2023) Early warning system against rainfall-induced landslide in Sri Lanka. In: K Sassa, et al (eds) Progress in Landslide Research and Technology, Volume 1 Issue 1, 2022.

Cham: Springer International Publishing (Progress in Landslide Research and Technology), pp 217–235. Available at: https://doi.org/10.1007/978-3-031-16898-7_16
Kruse S, et al (2017) Conceptualizing community resilience to natural hazards—the emBRACE framework. Nat Hazards Earth Syst Sci 17(12):2321–2333. Available at: https://doi.org/10.5194/nhess-17-2321-2017
Lacey H (2002) Explanatory critique and emancipatory movements. J CritAl Rism 1(1):7–31
Lacey H (2007) Explanatory critique. Dictionary of critical realism. 196–201'. Available at: https://works.swarthmore.edu/fac-philosophy/201
Lizarralde G (2015) The invisible houses: rethinking and designing low-cost housing in developing countries. Routledge
Malamud BD, et al (2004) Landslide inventories and their statistical properties. Earth Surf Process LandfS 29(6):687–711. Available at: https://doi.org/10.1002/esp.1064
March A, Kornakova M (2017) Chapter 15—planning for recovery: ideas and problematics. In: March A, Kornakova M (eds) Urban planning for disaster recovery. Butterworth-Heinemann, Boston, pp 231–248
MoDM, Sri Lanka, et al (eds) (2017) Sri Lanka rapid post disaster needs assessment: floods and landslides: May 2017. Colombo: ministry of disaster management & ministry of national policies and economic affairs, in collaboration with the United Nations, World Bank and European Union
Moorthy R, Benny G, Gill SS (2018) Disaster communication in managing vulnerabilities. J Komun: Malays J Commun 34(2):51–66
Perera ENC, et al (2018) Direct impacts of landslides on socio-economic systems: a case study from Aranayake, Sri Lanka. Geoenvironmental Disasters 5:11. Available at: https://doi.org/10.1186/s40677-018-0104-6
Quarantelli EL (1998) Major criteria for judging disaster planning and managing and their applicability in developing societies, p 43
Quarantelli EL (2000) Emergencies, disaster and catastrophes are different phenomena, p 6
Ratnayake U, Herath S (2005) Changing rainfall and its impact on landslides in Sri Lanka. J Mt Sci 2:218–224. Available at: https://doi.org/10.1007/BF02973195.
Robertson H (2018) Building in ways that meet the needs of Australia's remote regions. Available at: https://research.monash.edu/en/publications/building-in-ways-that-meet-the-needs-of-australias-remote-regions
Rodríguez H, et al (2007) Communicating risk and uncertainty: science, technology, and disasters at the crossroads. In: H Rodríguez, EL Quarantelli, RR Dynes (eds) Handbook of Disaster Research. New York, NY: Springer (Handbooks of Sociology and Social Research), pp 476–488. Available at: https://doi.org/10.1007/978-0-387-32353-4_29
Sanderson D, Bruce L (eds) (2020) Urbanisation at risk in the pacific and Asia: disasters, climate change and resilience in the built environment. Routledge, New York
Sanderson D (2010) Architects are often the last people needed in disaster reconstruction. The Guardian 3
Sangasumana P (2018) Post disaster relocation issues: a case study of Samasarakanda Landslide in Sri Lanka. Eur Sci J ESJ 14(32):1–1. Available at: https://doi.org/10.19044/esj.2018.v14n32p1
Turner JFC (1976) Housing by people: towards autonomy in building environments. Marion Boyars
UNDRR (2022) Terminology. Available at: https://www.undrr.org/terminology
Yin RK (2017) Case study research and applications: design and methods 6ed, 6th edn. SAGE Publications Inc., Los Angeles

Chapter 10
Community-University Efforts in High Vulnerability Contexts: Housing Reconstruction After the 2017 September Earthquakes in Mexico

Maria Isabel Perez Rodriguez and Chiho Ochiai

Abstract In September 2017, two strong earthquakes struck south-eastern Mexico. These events severely damaged housing in high-poverty areas, resulting in reconstruction efforts in communities affected by vulnerabilities. Given that being more vulnerable can reduce the capacity to confront disasters, it is imperative to study these cases in detail while considering the social and physical predispositions and their impacts during diverse recovery efforts. For this chapter, a case study of an affected community in Puebla, Mexico, and an ongoing University-led Housing Reconstruction Project (UHRP) are assessed to understand how recovery efforts work in highly vulnerable communities, given that multifactorial predispositions such as sociocultural and political/institutional factors influence and translate into material challenges during post-disaster housing reconstruction (PDHR). To study these multidimensional challenges, implications, and the dynamics of community-university cooperation, (1) official government data and a literature review of the 2017 earthquakes: policies, vulnerability, and geomatics; (2) observation and semi-structured interviews with beneficiaries, local people, university program leaders, and active participants in the recovery program; and (3) a housing survey of six houses in the UHRP were conducted. The results show that the UHRP demonstrates significant strengths during disaster recovery. The robust social networks, multidisciplinary knowledge, and long-term commitment of the university allowed for efficient mobilization, comprehensive disaster response, and enduring community impact. Additionally, cooperation based on trust and the implementation of livelihood-development activities showcased the potential for universities to work with communities against vulnerabilities.

M. I. Perez Rodriguez (✉) · C. Ochiai
Graduate School of Global Environmental Studies, Kyoto University, Kyoto, Japan
e-mail: misapr96@gmail.com

C. Ochiai
e-mail: ochiai.chiho.2x@kyoto-u.ac.jp

C. Ochiai et al. (eds.), *Disaster and Displacement*, Disaster Risk Reduction,
https://doi.org/10.1007/978-981-96-2583-3_10

10.1 Introduction

In September 2017, two strong earthquakes struck south-eastern Mexico. These events severely damaged housing in high-poverty areas, resulting in reconstruction efforts in communities facing pre-existing social vulnerabilities. Given that being more vulnerable can reduce the capacity to confront disasters (Johnson and Lizarralde 2012; Davidson et al. 2007; Rumbach and Makarewicz 2016), the consideration of these predispositions from a multidisciplinary approach, together with the understanding of the roles of multiple stakeholders is imperative for successful post-disaster recovery (Cardona 2004; UNISDR 2015). This is why a detailed study in an affected community in Puebla, Mexico, was selected due to two main characteristics: first, the high amount of damage, high levels of poverty, and low levels of coping capacity the inhabitants were facing; and second, the active involvement of a regional university through a *University-led Housing Reconstruction Program* (UHRP) as a response. The case study considers such vulnerabilities and their implications reflected in material and social challenges during post-disaster housing reconstruction (PDHR) efforts, and how effective collaborative efforts among civil society members (community-university-NGO) might play a relevant role when facing such complex pre-disaster conditions.

Hence, this chapter presents the conditions and reconstruction efforts in the community, highlighting key findings extracted from an analysis of the strengths and challenges of the university-led program in Alpanocan through a multidisciplinary approach based on qualitative analysis, to discuss the active role of universities as key actors in cooperative reconstruction programs.

10.1.1 Vulnerability and Post-Disaster Recovery

Contemporary disaster theory describes a disaster as more than an isolated hazardous event (United Nations Office for Disaster Risk Reduction UNDRR 2010; Economic Commission for Latin America and the Caribbean ECLAC 2014). The idea that "there is no such thing as a natural disaster" (Smith 2006) has led to studies that go beyond the norms by tackling the subject through the lens of social sciences, resulting in research with a multidimensional point of view that consequently acknowledges internal or structural risk factors like social, economic, political, or historical pre-existing conditions (also referred as vulnerabilities) and the lack of resilience (lack of coping capacity) as elements that could compromise the possibility for communities to confront hazardous events (e.g. Earthquakes, Floodings, Hurricanes) (Cardona 2004; Quarantelli 1998; Smith 2006).

This multidisciplinary study of disaster risk brings an understanding of external pressures affecting communities (Cardona 2004). Per modern studies, natural sciences can jointly work with applied sciences while considering perspectives from social sciences, leading to a deeper understanding of disasters and risk, considering "pre-disaster risk assessments" while working toward disaster risk reduction, recovery, and resilience for the most vulnerable (UNISDR 2015; Twigg 2006).

10.1.2 Post-Disaster Housing Reconstruction in High-Vulnerability Contexts

Low-income, marginalized, rural, and indigenous communities struggle severely when confronted with disasters, making post-disaster recovery (PDR) in these regions more challenging and complex (Johnson and Lizarralde 2012; Davidson et al. 2007; Rumbach and Makarewicz 2016; Ryder 2017). When disasters occur, they damage the living environment, affecting access to resources, health, education, and the economy. In general, disasters reflect both human and material losses (United Nations Development Program UNDP 2021). Among these losses, housing is one of the most visible ones and is strongly related to the victim's well-being, while being highly vulnerable to hazards at the same time (Felix et al. 2013). This situation makes post-disaster housing reconstruction (PDHR) a priority for PDR programs (UNDP 2021), due to the victims' immediate need for shelter after disasters and the necessity for livelihood restoration. Moreover, according to UN-Habitat (2019), around half of the total reconstruction costs in PDR are destined for housing repairs and reconstruction alone, which translates into several policies that involve a vast spectrum of stakeholders, including different government levels, NGOs (Non-government Organizations), volunteers, community members and CSO (Civil Society Organizations), academics and higher education institutions, as well as the private sector (Johnson and Lizarralde 2012).

Disasters affect housing in a multifactorial manner. Both physical and socioeconomic factors determine why houses suffer greatly from disasters. Physical factors such as lack of appropriate design, use of materials or non-engineered housing (UN-Habitat 2019) as well as social factors such as poverty, lack of supervision of policies, informal settlements, or uneven access to subsidies (UN-Habitat 2019; Johnson and Lizarralde 2012) make housing reconstruction extremely complex yet crucial for recovery.

10.1.3 Case Study and Methodology

The September 2017 Earthquakes in Mexico showcase the challenges of housing recovery in vulnerable regions due to (1) the fact that Mexico is considered at "High Risk" for disasters (Marin-Ferrer et al. 2017), (2) the alarming outcomes that the event caused mainly the poorer regions (Belisario Domínguez Institute 2017 and

CONEVAL 2020c), and (3) the scarce literature on this event from a multidisciplinary perspective.

Among the most affected communities, Alpanocan (a remote community in one of the poorest and highly marginalized regions, where the earthquake's housing damages were especially severe) (CONEVAL 2020c; Government of Mexico 2017) was selected as the case study due to the community's strong vulnerabilities, the severity of the housing damages, and the interesting dynamics between the community and a regional university to work on a housing reconstruction program.

The study followed a qualitative methodology through deductive hierarchical coding based on: firstly, a comprehensive literature review on PDR in Mexico and vulnerabilities; secondly, official data processing of the disaster and the community, newspaper survey on the local situation and local perception between 2017 and 2022; thirdly, semi-structured interviews with locals (6), beneficiaries (4), program volunteers (5), coordinators (2), and supervisors (1). Additionally, housing surveys on six homes from the UHRP and observation and participation in the ongoing recovery efforts during October and November 2022, to assess community-university cooperation in PDHR.

10.2 The September 2017 Earthquakes

10.2.1 Background

In September 2017, two earthquakes struck the southeastern region of Mexico (Fig. 10.1) on September 7th and 19th, affecting more than ten million people and causing physical damage to over 190,000 structures, mainly in the rural and remote regions (Belisario Dominguez Institute 2017).

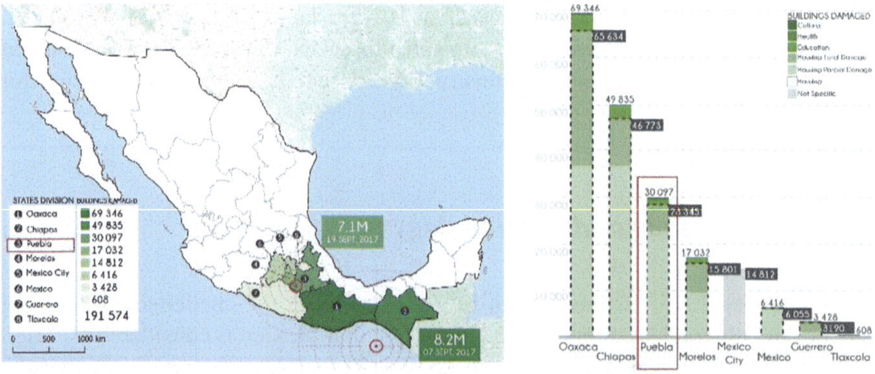

Fig. 10.1 **Left** Map of affected regions and epicenters. **Right** number and classification of damaged buildings (Author's elaboration with data from Belisario Dominguez Institute 2017)

The most affected states were (1) Oaxaca, (2) Chiapas, and (3) Puebla (Belisario Dominguez Institute 2017; Government of Mexico 2017), which are among the poorest in the country (with third, first, and fifth place, consecutively) (National Council for the Evaluation of Social Development Policies, CONEVAL 2020a pp. 23–28, 32–32, 39–41 2020b).

10.2.2 Government Policies and Programs for PDR

In response to the aftermath, the federal government implemented the national program "Fuerza Mexico" (Mexico's Strength, FM), focused on recovery and data collection. FM was applied for over a year, in charge of institutions such as the National Disaster Fund (FONDEN); Bank for the National Savings and Financial Services (BANSEFI); Secretary of Agriculture, Territorial, and Urban Development (SEDATU); and the Institute for the National Housing Fund for Workers (INFONAVIT). Victims of damaged or destroyed housing were supported either by subsidies (for informal or rural housing) or loans for social housing (for people with access to credit) (Fuerza Mexico 2019).

Due to elections that included multiple levels of government, in 2018, the country experienced a significant political shift when the Left-leaning party of MORENA (National Regeneration Movement) won the presidency of Mexico through his candidate Andres Manuel Lopez Obrador (Electoral Court of the Federal Law 2018). This political shift led to institutional modifications, including the previous reconstruction program FM, which was substituted by the "National Reconstruction Program" (PNR) in 2019 (Fuerza Mexico 2019; Government of Mexico 2020 pp. 12–22, 45–47, 65–72, 161–167).

The PNR showed significant differences compared with its predecessor. Firstly, areas with higher percentages of marginalization, violence index, and higher numbers of the indigenous population were prioritized, and technical assistance for construction, legal issues, and budgeting for victims who did not receive support from the first program was offered. Secondly, this assistance was in the form of temporary camps, where groups of volunteers led by the National Housing Commission (CONAVI) were temporarily placed in rural/indigenous communities (Government of Mexico 2020). However, help was not equally distributed, and logistic and organizational problems were reported (College of Mexico 2019, pp. 20, 21, 24–37, 88–91).

10.2.3 Support from NGOs

Despite government initiatives, the support from citizens, NGOs, and the private sector to help the victims was active throughout the different stages of recovery. International, national, and regional organizations were involved in reconstruction. While Red Cross Mexico, UNICEF, and over 13 foundations directly supported the national government initiatives through donations (Congress of Mexico City 2019, pp. 41–44), smaller NGOs built temporary and permanent homes in affected rural or vulnerable regions (TECHO 2017; Ayuda en Acción 2018).

10.2.4 Citizen Participation in Disaster Recovery

Mass mobilization from voluntary citizen groups during the emergency phase was observed in the affected regions in the form of rescue teams, debris cleaning, and essential item donations according to several newspaper articles that describe the national post-disaster landscape (Animal Politico 2017 2018; Sector 21 2017; Municipios Puebla 2022; Nodal 2022). However, responses from civil society are not an event isolated from this disaster, given that citizen groups tend to organize voluntarily to confront the situation, especially in vulnerable regions when the events exceed the government's capacity. Cases such as the Guatemala earthquake in 1972 (Gawronski and Olson 2013; Trudeau 1993) and the Haiti Earthquake in 2010 (Pyles et al. 2017; Green and Miles 2011) share patterns of high levels of citizen participation and low levels of government support. A similar case can be seen after the earthquake in Mexico City on September 19, 1985 (one of the biggest modern disasters in the history of the country), when both the government and academics highlighted the solidarity of civil society, which led the "Popular Housing Renovation Program" (RHP) to ultimately respond to victims' demands and needs (Government of Mexico 2017; Gamboa de Buen and Revah 1990; Reséndiz-Vázquez 2019).

10.2.5 University Support for Reconstruction and Recovery

A motivation to help, the high levels of necessity, and the fact that remote regions were the most affected during September 2017, led universities around the country and internationally (David Rockefeller Center for Latin American Studies -Harvard University 2019) to become actively involved in recovery (Table 10.1), and thus, making universities networking and technical hubs to directly support communities.

10.2.6 Social Service Practices in Mexico and its Influence on University Involvement in Disaster Recovery

The involvement of higher education institutions in social projects and professional activities in Mexico, also known as *Social Service* (Servicio Social), are professional activities (often in vulnerable regions) during the formative years of students as part of the university curriculum (Rodriguez et al. 1979 p. 20, Mungaray 2002 pp. 5, 29, 31).

Table 10.1 University and higher education institutions' involvement in Mexico as a response to the 2017 earthquakes

Universities/Higher education institutions			Support		Site	Timeline
No	University	Private or public	Support	Description	State	Phase of involvement
I-1	Iberoamerican University Puebla	Private	Multi-stakeholder initiative	Training program focused on Disaster Risk Reduction	Puebla	Recovery
			Post-disaster module	Post-disaster building prototype with local materials	Puebla	Recovery
I-2	Higher Education School of Jojutla	Public	Psychological support	Psychological support and aid for affected families in a vulnerable community	Morelos	Recovery
I-3	Autonomous University of the State of Morelos	Public	Emergency debris removal, aid	Bringing food and supplies to affected families, removing debris	Morelos	Emergency response
		Public	Permanent housing	Training and technical support for participatory design on permanent housing construction. Reconstruction of 47 housing units with vernacular architecture (collaboration with the National Institute of Anthropology and History [INAH] Morelos and civil society)		Recovery

(continued)

Table 10.1 (continued)

Universities/Higher education institutions			Support		Site	Timeline
No	University	Private or public	Support	Description	State	Phase of involvement
I-4	University of Colima	Public	Study on damaged housing units and buildings	Data compilation and research in affected community	Morelos	Recovery
I-5	UPAEP (Popular Autonomous University of the State of Puebla)	Private	Emergency debris removal, permanent housing design proposals	Emergency debris removal, course-related design proposals for affected families	Puebla	Emergency response and recovery
I-6	BUAP (Prestigious Autonomous University of Puebla)	Public	Emergency debris removal, aid	Bamboo temporary housing construction with local and donated materials in affected communities	Puebla	Emergency response
			Temporary shelters made of Bamboo		Puebla	Emergency response
I-7	Anahuac Mexico	Private	Emergency debris removal, housing reconstruction	Joint efforts with Anahuac Puebla for the University-led housing reconstruction program	Mexico City, Morelos, Puebla	Emergency response and recovery

These practices started in the 50s when UNAM (National Autonomous University of Mexico) created activities for students to link both the professional and academic aspects of learning (Rodriguez et al. 1979, p. 20) through voluntary social programs (Fig. 10.2).

This connection between community service and access to education, according to Mungaray (2002, p. 31,) comes from the "collective land rights and the obligation to offer free labor in favor of the community," which was common practice in pre-Columbian times done by the Aztecs. The social practices, Mungaray continues, are "one of the noblest institutions for social development" based on a "genuine feeling of solidarity that characterizes the Mexican idiosyncrasy" (Mungaray 2002 pp. 5–6).

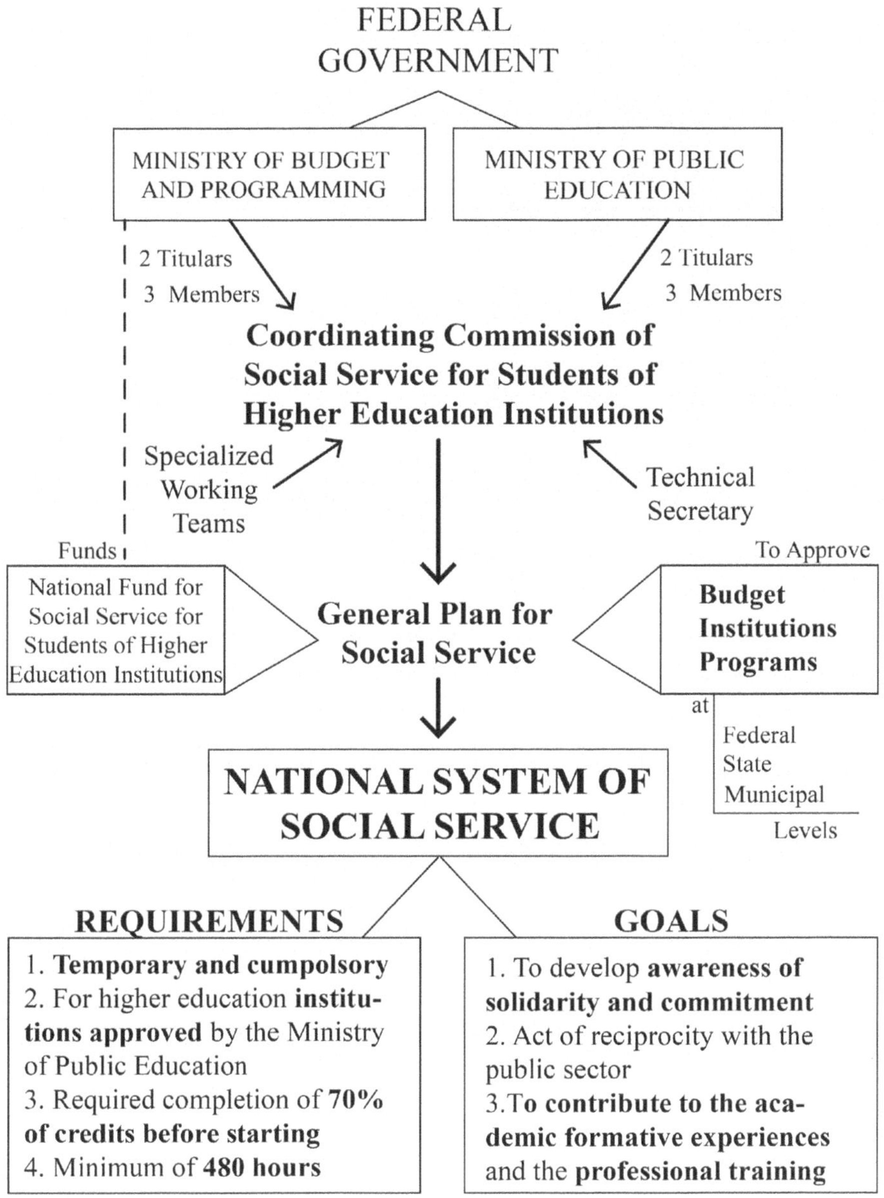

Fig. 10.2 A framework of the Social Service Practices in Mexico (Author's Elaboration Based on Government of Mexico 1982)

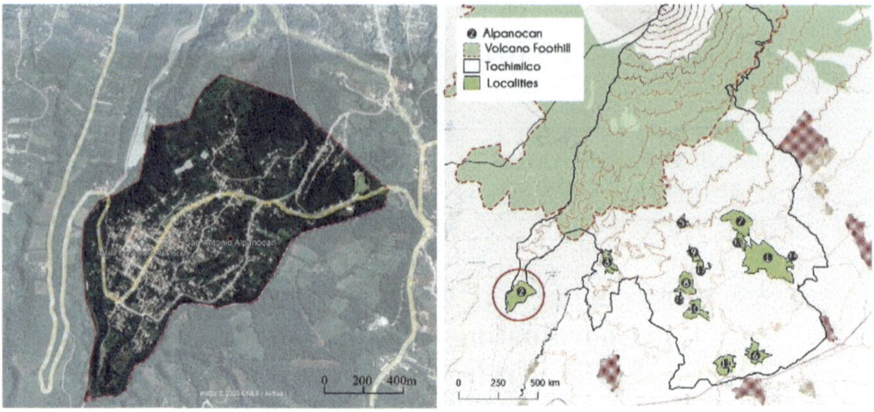

Fig. 10.3 **Left** Map of Alpanocan and Satellite View (Google Maps 2022). **Right** Map of Tochimilco (Author's Elaboration with data from INEGI 2021)

10.3 The Case of Alpanocan in Puebla

10.3.1 Puebla: Vulnerability, and Disaster

Puebla, located in central Mexico, is the fifth poorest state in the country, with 62.4% of its population living in poverty (CONEVAL 2020b). It was the third most affected state in 2017, with more than 30,000 buildings compromised by the earthquake (94% damage to housing alone) (Belisario Dominguez Institute 2017). The most impacted areas were the ones with higher poverty levels, and adobe housing, which is most common in rural regions in Mexico (Baquedano et al. 2021), was especially affected (Calvillo et al. 2021).

Among these highly vulnerable and affected communities, San Antonio Alpanocan (Fig. 10.3), a locality found in the periphery of the municipality of Tochimilco in Puebla, was already in a complex geographical situation because of its remoteness from the Capital City, where the State Government Office is located (approximately 48 km). This remoteness and difficult access predisposed the locality to neglect mobility, infrastructure, and communication issues that started even before 2017.

10.3.2 The Case of Alpanocan

The earthquake on September 19 was devastating for Alpanocan (Fig. 10.4), where approximately 346 houses were damaged (Government of Mexico 2017). Neither the local government nor the inhabitants were equipped to face the events due to poverty, health and infrastructure scarcity, marginalization, and a general lack of

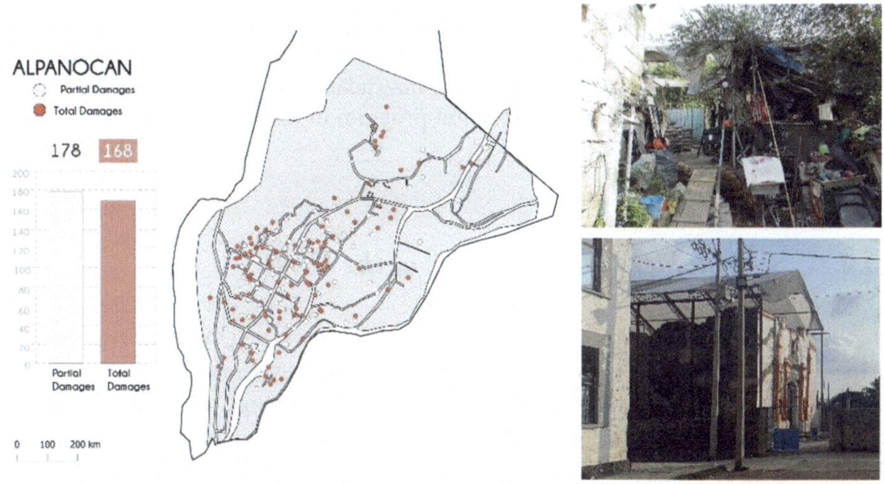

Fig. 10.4 **Left** Location of Housing Damages in Alpanocan (Author's Elaboration with data from Government of Mexico 2017). **Right** Photos of damaged houses and buildings from the Earthquake (Photos taken by the author 2022).

support from regional authorities, which affected the recovery phases. Even when national recovery programs arrived in the community (both FM and PNR), an equal distribution of said support was not achieved, resulting in additional programs from NGOs, such as TECHO, in the shape of temporary housing (Animal Politico 2017). Under these conditions, a UHRP, formed by joint efforts between people in the community, a regional university, and an NGO, stood out as a long-term, housing reconstruction proposal. The UHRP is held by ASUA (Social Action from Anahuac Puebla University) and it has been working in Alpanocan from 2018 to 2022 through permanent housing reconstruction and community development efforts.

10.4 University-Led Housing Reconstruction Program (UHRP) in Alpanocan

Social Action from Anahuac University (ASUA) oversees community programs by involving students and promoting their social responsibilities. ASUA has several initiatives to support vulnerable communities in Mexico. Their previous experience with similar construction programs led them to join key local actors in Alpanocan who were looking for support after the disaster. The university sought an alliance with an NGO called Construyendo (with technical knowledge of housing construction) to work directly and support the community through housing reconstruction.

ASUA's previous logistical experiences, the university's credibility (which brought donors and support from the private sector), and the interest and availability of hundreds of students to become volunteers, helped build more than 50 homes in Alpanocan between 2018 and 2022 through a new version of what was formerly known as “ASUA Construye,” and what was now adapted for housing reconstruction (called University Housing Reconstruction program or UHRP for reference in this text).

10.4.1 Stakeholders and Structure of the UHRP

The UHRP involves three main actors: the university (ASUA), the community (Alpanocan), and an NGO (Construyendo). The University works as a link between the NGO, the government, and donors, being mainly in charge of university volunteers, logistics, budget, and donations, whereas the NGO is mainly in charge of the technical process, materials, and design, while the community looks for possible beneficiaries and oversees the proper development, local logistics, and construction of the process (Fig. 10.5).

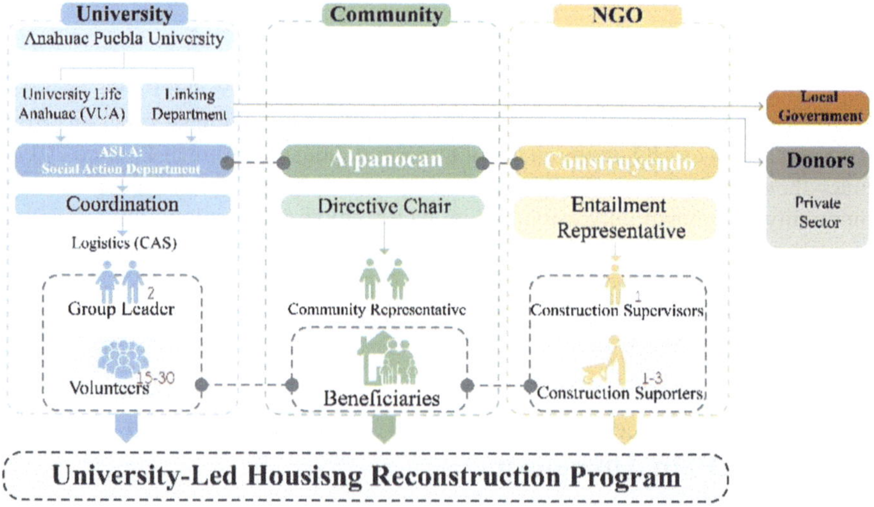

Fig. 10.5 Organigram of the actors and inner structures (Author’s Elaboration 2022)

10.5 Relationship Between University (Blue), Community (Green), and NGO (Yellow)

The UHRP works through six main phases: (1) Budgeting: the donors are contacted, funds are collected, and preparation begins; (2) Preliminary: the beneficiaries are selected, the community is organized, and the NGO checks the selected plots; (3) Preparation: the plot and the concrete firm are ready, and the logistics for the construction are prepared; (4) Housing construction: the main structure is built during a 3-day program; (5) Final Details: the NGO and beneficiaries finish the house; (6) Follow-up: community development programs addressing social, infrastructural, and health-related needs are held.

10.5.1 Community Participation

Community participation in PDR is essential for achieving natural resilience (Pyles et al. 2018; Rumbach and Makarewicz 2016; Twigg 2006). However, the type of participation can determine whether the project genuinely empowers the community or not (Varolgunes 2021; Davidson et al. 2007). Figure 10.6 shows each actor's role in UHRP during its main activities, concluding that the current program does include communities throughout the process; however, phases like design, do not include them at all.

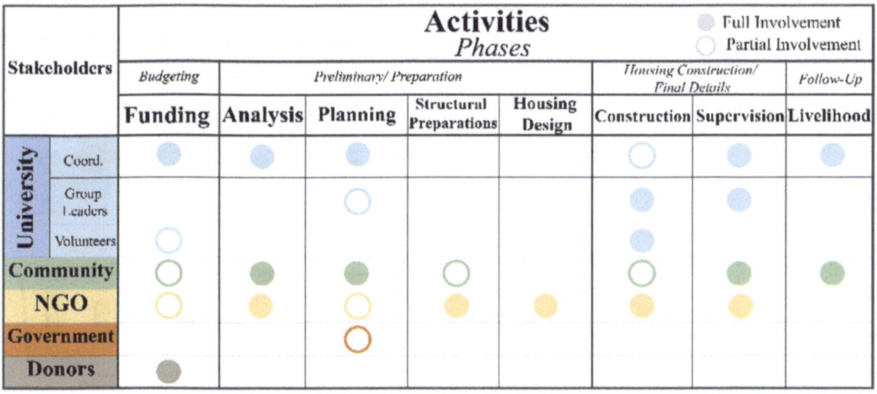

Fig. 10.6 Activities and actors involved in the UHRP (Author's Elaboration)

10.5.2 Construction Method

The UHRP permanent housing is built with prefabricated 3D sandwich panel modules, consisting of expanded polystyrene cores covered by a 3D electro-welded steel wire mesh (Qualy Panel, SA de CV 2011). The material's fast, durable, and easy construction and its lightweight characteristics make it a feasible solution for low-rise housing in seismic-prone areas (Refaei et al. 2015). The unit design consists of one-story homes of 38m^2 (previously 40m^2). The design is based on standard housing construction dimensions, designed by the NGO. The construction process is completed in a three-day program where around twenty volunteers per home, family members, and NGO technical assistants work together to finish the main structure. This is an ongoing program that currently occurs twice a year, and it is the main activity (Fig. 10.7).

Even if the main goal of the UHRP is based on housing reconstruction, the program has been working with the community since 2018, allowing for long-term support. This has led to the development of follow-up activities, or livelihood development activities to target local vulnerabilities like education, health care, and psychological support.

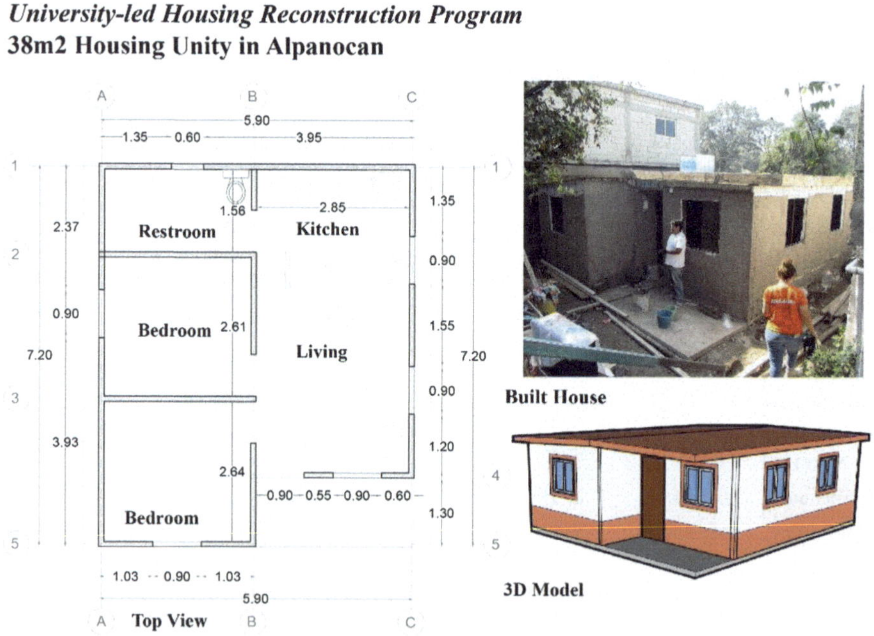

Fig. 10.7 Top View, 3D model, and one photo of the Standard 38m^2 Housing Model

10.6 Key Insights

UHRP efforts have received significant recognition from Alpanocan. However, the events' context, vulnerabilities, and complexity can lead to challenges and improvements for these or similar programs and the role of universities in PDR. The concluding section discusses key insights extracted from the program's efforts to understand the results and their impacts on the community.

10.6.1 Strengths of the Program

10.6.1.1 Trust Based on Long-Term Commitment

The response about how beneficiaries perceived the program was generally positive. Comments about appreciation towards the efforts from ASUA and their volunteers, and how they genuinely seem to care and bring efforts and support was highly valuable from them.

> "I honestly am glad they are doing it. I hope they keep supporting more people. For me, they will always be in my prayers. Why? Because they give everything to come all the way so a family can have a home, a roof, a place to live. And yeah, thanks to them, many families. You know, even after so many people lost everything in the earthquake, house, furniture, everything, it's good that they are still coming back to support them. Honestly, I have no words on how to thank them" (Female Beneficiary from 2018).

Comments about how, before the earthquake, people felt neglected and forgotten by the authorities, or how even during the national programs (both Fuerza Mexico and PNR), the lack of proper supervision led to some families receiving double support, while others did not at all were present. Moreover, the average response of Mexican citizens to their authorities tends to lean toward mistrust, generating uncertainty when disasters occur. This perception could be a factor in why, in Alpanocan, the commitment from the UHRP makes their efforts more trustworthy, so much so that, some beneficiaries mentioned the university as the main group they would approach in case of a similar disaster, even before the government or other institutions. These responses connect with the criticism that NGO programs are often short-term (which can lead to incomplete and lacking results) (Sapat 2016), whereas, conversely, when commitment is long-term, cooperation feels more stable and reliable. It seems that, for the beneficiaries, one of the most significant elements the program offers, besides housing reconstruction, is the commitment the university has shown to continue working with Alpanocan. Instead of offering temporary, short-term support, which could be perceived as neglect (like their experiences with regional authorities), the long-term commitment to community involvement and a relationship that has improved over time generates positive recognition and satisfaction from the community.

> "We try for the volunteers to talk, to interact with the families, for them to be at their service, to be aware of what the families need, because the head volunteer must delegate the

responsibilities of the emotional and human acts to the volunteers. The intention is for us to train them (referring to the volunteers), to help them to learn how to execute on their own, with the respect and appreciation we have for the family, and with the commitment we have as a university that says to have a human focus" (Male Program Coordinator)

10.6.1.2 The Importance of Livelihood Development Activities to Target Vulnerabilities

The link between disasters and vulnerabilities (Marin-Ferrer et al. 2017; Cardona 2004) shows that communities will not properly recover without considering social aspects, root causes, and complex unique challenges that made communities prone in the first place (Gilbert 1998; Economic Commission for Latin America and the Caribbean ECLAC 2014 pp. 18–23).

For Alpanocan, which has been deeply affected by vulnerabilities, the approach the UHRP brings based on an interest that goes beyond reconstruction and considers more than the immediate housing reconstruction needs, has been highly positively perceived by the community.

The reputation and trust the university holds amongst community members, together with human and material resources that the university has access to, make it possible for social programs to target livelihood issues. For instance, yearly medical brigades where the university provides free consultation and medicine are held, and legal and psychological consultations, educational programs for children, and aid collection during the first years of the COVID-19 pandemic are some of the most relevant activities organized and held directly by both the university and the community to target main issues (Table 10.2).

10.6.1.3 Active University Involvement in Housing Recovery and Reconstruction

As a response to the role universities have in the social and ethical development of Mexican women and men, the adaptation of social service practices for students proves to be part of engaging into the Mexican reality by bringing awareness and responsibility values with the hopes of finding better solutions against deep-rooted challenges (Mungaray 2002). Moreover, the correlation between Mexican higher education institutions and disaster response seems to be one of the key drivers for these practices, given that one of the first university-led social programs was the multidisciplinary team for emergency response after the Puebla Cyclone in 1958 (Rodriguez et al. 1979, p. 20), showing that there is a strong correlation between civil society disaster response and universities in Mexico, which not only brings more human and social resources to affected groups but makes young professionals and students aware of the realities and biggest necessities the country is currently facing.

Table 10.2 Main vulnerabilities affecting Alpanocan and the diverse programs led by civil society

Vulnerabilities	Program	Civil society organizations		Timeframe
		Anahuac Puebla university	NGOs	
		Departments	Involved	
Education	Interactive library renovation, book donations, children's day recreational activity	Department of social communication	n/a	Once a year
		Social action department	n/a	
Healthcare	Medical and psychological free consultations and medicine	School of medicine	n/a	Twice a year
		Social action department	n/a	
	Aid and sanitation kits	Social action department	n/a	Covid 19 Pandemic
Crime	n/a	n/a	n/a	n/a
Migration	n/a	n/a	n/a	n/a
Water	Rainwater collection and filter systems	n/a	OXFAM Mexico	One-time project
Poverty and Inequality	Free legal consultations	School of law	n/a	Once a year
		Social action department	n/a	n/a
Housing	Temporary housing reconstruction	n/a	TECHO	One-time project
	Permanent housing reconstruction	Social action department	Construyendo	4–5 times a year (2018–2020), 1–2 times a year (2021–2023)

The case of the UHRP in Alpanocan shows how, for students in universities, active formative programs force them to go beyond the classroom. For instance, one of the former coordinators explains the formative experience as follows:

> "The formative aspect of the university is that it helps you go to a community, to realize the different realities in which you live in, in Puebla city. In this case, you also put a lot of effort, you try very hard to put yourself into the service of others. So, I do feel that it is a change maker for many young people, which made them sensitized to certain circumstances that happen in so many areas in Mexico and the world. So, I feel that, in that sense, the project is amazing" (Male Program Coordinator)

10.6.2 Challenges of the Program

10.6.2.1 Unstable Funding

"Aid volatility and Aid fatigue" (Sapat 2016) are problems that can appear in long-term recovery programs held by NGOs and civil society. The UHRP faces similar challenges, given that, according to program coordinators and highly committed volunteers, there is less budget available each year compared to the program's first years. Comments about how there is little to no support from the local government or how the university budget for the program is slowly decreasing were mentioned. The current decrease in aid from former strong private sector donors who are reducing or completely cancelling their aid for future programs is especially worrying. The cause of this budget fatigue is a multifactorial issue that includes the natural decrease in the visibility of the disaster in traditional media when the event cools down. Additionally, the COVID-19 pandemic affected the consistent aid flow for the program, since the university and donors had to pause construction activities for more than a year for safety reasons.

With this into consideration, more stable funding through both private and public funding could be helpful given that funding reconstruction programs without public support seems to be unstable (Comerio 2014).

10.6.2.2 Need for Adaptable Housing

(1) Lack of Use of Local Knowledge and Resources

In post-disaster housing recovery, the consideration of local resources has been mentioned as a viable solution to improve economic and resource challenges (Twigg 2006; Chang et al. 2010). Especially for low-income settlements, when considering how non-supervised and non-engineered construction can become a risk to structural safety (Pribadi et al. 2014), accumulated construction knowledge from traditional techniques can become a safe alternative (Varolgunes 2021; Tucker et al. 2014). The literature in this regard continues to suggest that, because traditional techniques consider social, cultural, economic, and environmental conditions during the design, the inclusion of traditions can generate better satisfaction from the beneficiaries (Dikmen 2011).

Nevertheless, the promotion of traditional construction becomes a complex subject when considering the local and personal experiences of the affected communities. For instance, the fact that most of the damaged homes after the earthquakes in Guatemala in 1976, El Salvador in 2001, and Nicaragua in 1972 (Bommer et al. 2002 p. 18; Wright and Kramer 1973 pp. 19–20) were the traditional homes, makes it likely to change the perception of local communities regarding how safe their local construction is. In the case of the 2017 earthquakes in Mexico, and the case of Alpanocan, most of the affected homes were Adobe homes, making it common for beneficiaries to share their negative perception of the local techniques. However, younger interviewees responded more positively to the idea of preserving traditions and adobe construction for the sake of identity, which is one of their biggest concerns regarding social roots and cultural roots as an indigenous community.

Owing to these pre- and post-disaster conceptions about adobe construction, families disregard the traditional, which is also reflected in the positive level of acceptance of the UHRP construction material. UHRP's 3D sandwich panels are an easy-to-build technique that includes commonly found materials such as cement and gravel and uses simple tools. However, the panels themselves are not as economically viable for individual users as concrete blocks, and the lack of supervision and technical assistance for the panels, especially when modifications or additions (commonly found in informal housing) are not considered, could compromise the structure.

(2) Lack of Adaptable Design

Even though most beneficiaries positively accept the current housing design, a variety of designs and involvement of users in these phases of the process are highly recommended for PDR (Varolgunes 2021; Davidson et al. 2007). For Alpanocan, proper adaptability would respond to the challenges generated by slopes or narrow plots, as well as logistic-related challenges. For instance, in the case of families living in high-slope plots, the current model requires families to either excavate or fill the area, which could become an economic burden or increase the risk of landslides when unsupervised. In the case of narrow plots, the current housing design has a chance of not fitting the standardized dimensions, making the housing unit inaccessible for some due to the lack of an adaptable design.

These technical challenges show that although the current standardized homes generally work, the need for adaptable solutions in slopy, narrow, or irregular plots, or even if considering beneficiaries' unique needs, housing extension, or the inclusion of the community during the design process, could become beneficial for the program and families in the long term.

(3) Further Empowerment of Community Groups

Effective community involvement is necessary for establishing clear and transparent communication channels with community leaders (Davidson et al. 2007; Choguill 1996). Understanding local needs, preferences, and cultural nuances is paramount for creating appropriate disaster response strategies (Pyles et al. 2018; Twigg 2006). Hence, a more inclusive approach, involving the local community as active participants in decision-making processes, ensures the relevance and success of the initiatives.

Currently, the program has not yet taken a mentoring role in supporting this work. Hence, the study suggests that collaborating with local youth, Indigenous groups, women, and similar often marginalized collectives can become a strong and crucial strategy for ensuring resilience and long-term recovery. Moreover, by empowering local groups and providing them with a voice in decision-making processes creates the possibility of positioning the community as key actors in the development and implementation of disaster response initiatives to create self-sustainable communities.

10.7 Conclusions

Due to the complexity of Post-disaster Reconstruction in vulnerable regions, multi-disciplinary approaches can bring a clearer picture of the challenges such programs might face. Poor, indigenous, and marginalized communities tend to lack the infrastructure, communication channels, or even proper government support to face disasters. This can lead to unaccomplished or truncated recovery efforts, creating an environment of uncertainty. However, in times of hesitation, civil society members can join efforts toward recovery (Gawronski and Olson 2013; Reséndiz-Vázquez 2019). Moreover, the case of Alpanocan offers an interesting combination of several stakeholders from civil society (community-university-NGO), highlighting that, during disasters, higher education institutions have the potential to play more active roles with networking, academic, and physical support, to work as resourceful partners for PDHR. These community-university efforts have several advantages, such as the capacity of students and professors from multiple disciplines, the possibility of making long-term commitments for reconstruction programs due to an advantageous geographical location and human resources with a deeper understanding of the complex local vulnerabilities, which allow the programs to go beyond physical reconstruction by performing livelihood development activities focused on specific vulnerabilities. Nevertheless, challenges like the impacts of introducing new construction techniques and the lack of adaptability to housing designs must be considered when introducing projects in areas with unique traditions, contexts, and challenges. Finally, budget and resource stability must be guaranteed for long-term projects, and further empowerment of local leadership could lead to a more resilient long-term approach. Nevertheless, given that this is a highly specific case study, universities' roles, and similar cases of university-community cooperation for PDHR need to be further studied for a deeper understanding of such dynamics.

Acknowledgements I would like to express my deepest gratitude to Anahuac Puebla University, especially the members of ASUA, as well as the community of San Antonio Alpanocan and everyone who opened their doors and experiences, especially Damian and Eli for their trust and immense support during my fieldwork, and in general for their hard work in starting and maintaining the bond between the community and the university. I would like to extend my sincere thanks to the members of the youth group of Xalliquehuac for their time and support regarding the community, as well as a strong appreciation for their constant efforts to improve Alpanocan through culture and traditions.

References

Animal Politico (2017) In Alpanocan, Puebla, citizens rebuild their homes, they have not received nothing from the government [En Alpanocan, Puebla, Ciudadanos Reconstruyen sus Viviendas, del Gobierno no Han Recibido Nada]. Available at: https://www.animalpolitico.com/2017/10/alpanocan-puebla-sismo-gobierno

Animal Politico (2018) San Antonio Alpanocan, Puebla, waits for the promised federal and state resources for the reconstruction. [San Antonio Alpanocan, Puebla, Espera los Recursos Federales y Estatales prometidos para Reconstruirse]. Available at: https://animalpolitico.com/2018/03/san-antonio-alpanocan-espera-los-recursos-federales-y-estatales-prometidos-para-reconstruirse

Ayuda en Acción (2018) Reconstruction of communities in puebla [Reconstrucción de las Comunidades En Puebla]. https://ayudaenaccion.org.mx/actualidad/trabajamos-red-juntos-19s/. Accessed Nov 18 2022

Baquedano P, Eudave RR, Miranda FN, Graus S, Ferreira TM (2021) Traditional earth construction in Latin America: a review on the construction systems and reinforcement strategies. In: Masonry Construction in Active Seismic Regions, p 99–121. https://doi.org/10.1016/b978-0-12-821087-1.00011-9

Belisario Domínguez Institute, National Senate (2017) Assessing the damages, 7S and 19S: a month after the tragedy [Recuento de los daños 7S y 19S: a un mes de la tragedia]. Coordinación Ejecutiva de Investigación. No. 17. http://bibliodigitalibd.senado.gob.mx/bitstream/handle/123456789/3721/2017_16_NE_Recuento%20de%20da%C3%B1os_231017.pdf?sequence=1&isAllowed=y

Bommer JJ, Benito MB, Ciudad-Real M, Lemoine A, López-Menjívar M. A, Madariaga R, Rosa H (2002) The El Salvador earthquakes of January and February 2001: context, characteristics, and implications for seismic risk. Soil Dyn Earthq Eng 22(5):389–418

Calvillo AS, Guzmán EMA, Núñez MDCL (2021) Vulnerabilidad sísmica y la pérdida de la vivienda de adobe en Jojutla, Morelos, México, tras los sismos de 2017. Vivienda Y Comunidades Sustentables (10):9–29

Cardona O (2004) The need for rethinking the concepts of vulnerability and risk from a holistic perspective: a necessary review and criticism for effective risk management. In: Mapping vulnerability. Disasters, development, and people, p 37–51

Chang Y, Wilkinson S, Potangaroa R, Seville E (2010) Resourcing challenges for post-disaster housing reconstruction: a comparative analysis. Build Res & Inf 38(3):247–264

Choguill MBG (1996) A ladder of community participation for underdeveloped countries. Habitat Int 20(3):431–444

College of Mexico (2019) National reconstruction program PNR. An exploratory Analysis of the behavior of the program [Análisis Exploratorio del Diseño de los Programas Prioritarios de la SEDATU 2019–2020] No 3. Demographic, Urban, and environmental study center, p 11–15, 20, 21, 24–37, 88–91

Comerio MC (2014) Disaster recovery and community renewal: housing approaches. Cityscape 16(2):51–68

(CONEVAL) National Council for the Evaluation of Social Development Policies (2020a) Rural poverty in Mexico [Pobreza Rural] (pp 23–28, 32–32, 39–41). https://www.coneval.org.mx/Medicion/MP/Documents/PATP/Pobreza_rural.pdf

CONEVAL (2020b) Measuring poverty. Poverty in Mexico 2020. https://www.coneval.org.mx/Medicion/Paginas/PobrezaInicio.aspx. Accessed Feb 11 2023

CONEVAL (2020c) Poverty database 2008–2018. Accessed 10/13/2023: https://www.org.mx/Medicion/Paginas/Pobreza-municipio-2010-2020.aspx

Congress of Mexico City (2019) Earthquake of September 19th, 2017: statistic, financial and social aspects related to the reconstruction, recovery, and transformation of Mexico City [Sismo 19 de Septiembre 2017: Aspectos Estadísticos, Financieros y Sociales Relativos a la Reconstrucción, Recuperación y Transformación de la CDMX]. Government of Mexico. Institute of Legislative Research. https://congresocdmx.gob.mx/media/banners/d120319-2.pdf, p 41–44

David Rockefeller Center for Latin American Studies -Harvard University (2019) Beyond reconstruction. Post-earthquake recovery in Oaxaca, Mexico after the 2017 Earthquakes, p 17–39. https://issuu.com/mitarchitecture/docs/beyond_reconstruction_book_june2020_issuu. Accessed Feb 14 2023

Davidson CH, Johnson C, Lizarralde G, Dikmen N, Sliwinski A (2007) Truths and myths about community participation in post-disaster housing projects. Habitat Int 31(1):100–115

de Buen JG, Locouture JAR (1990) Mexican reconstruction and urban politics in Mexico City [Reconstrucción y política urbana en la ciudad de México]. Foro Int 30(4(120):677–694

Dikmen N (2011) Comparative analysis of permanent post-disaster houses constructed in Çankırı and Dinar. Disasters 35(2):404–416

Economic Commission for Latin America and the Caribbean (ECLAC) (2014) Handbook for disaster assessment. LC/L.3691 2013–817 UN. p 18–23c

Electoral Court of the Federal Law (2018) Federal court issues the dictum on the final count, declaration of validity of the electoral process, and declaration of the president-elect. [El TEPJF emite dictamen sobre cómputo final, declaración de validez del proceso electoral y declaratoria de presidente electo] Mexican Government Official Platform. Accessed Feb 24 2023. https://www.te.gob.mx/front3/bulletins/detail/3360/0#:~:text=Ciudad%20de%20M%C3%A9xico&text=Una%20vez%20efectuadas%20las%20operaciones,de%20la%20votaci%C3%B3n%20total%20emitida

Félix D, Feio A, Branco J M, Machado J S (2013) The role of spontaneous construction for post-disaster housing.

Fuerza Mexico [Mexico's Strength] (2019) Official historical platform database. Government of Mexico, 2017–2018. https://www.transparenciapresupuestaria.gob.mx/es/PTP/fuerzamexico. Accessed Feb 21 2023

Gawronski VT, Olson RS (2013) Disasters as crisis triggers for critical junctures? The 1976 Guatemala case. Lat Am Polit Soc 55(2):133–149

Gilbert C (1998) Studying disaster: changes in the main conceptual tools. In: Quarantelli EL (ed) What is a disaster? Psychology Press, Perspectives on the question, pp 3–11

Government of Mexico (1982) REGULATIONS for the provision of social service to students of higher education institutions in the Republic of Mexico. [REGLAMENTO para la prestación del servicio social de los estudiantes de las instituciones de educación superior en la República Mexicana]

Government of Mexico (2017) Official database. Government of Puebla. https://datos.gob.mx/busca/dataset/viviendas-danadas-por-el-sismo-del-19s. Accessed Feb 26 2023

Government of Mexico (2020) National reconstruction program. 1st Year Report 2019. [Programa Nacional De Reconstrucción. Informe de Primer Año]. p 12–22, 45–47, 65–72, 161–167

Green R, Miles S (2011) Social impacts of the 12 January 2010 Haiti earthquake. Earthq Spectra 27(1_suppl1):447–462

HOLCIM Mexico (2019) Holcim Mexico foundation gives 120 homes in areas affected by the earthquakes of September 2017 in the State of Puebla [Entrega Fundación Holcim México 120 Casas en Zonas Afectadas por los Sismos de septiembre 2017 en el Estado de Puebla]. https://www.holcim.com.mx/entrega-fundacion-holcim-mexico-120-casas-en-zonas-afectadas-por-los-sismos-de-septiembre-2017-en-el. Accessed Jan 30 2023

Johnson C, Lizarralde G (2012) Post-disaster housing and reconstruction. Int Encycl Hous Home 5:340–346

Marin-Ferrer M, Vernaccini L, Poljansek K (2017) INFORM index for risk management. Concept and methodology version 2017. In: JRC Science for Policy Report, European Commission, pp 10–12, 71–73, 82

Mungaray A (2002) Challenges and perspectives about university reciprocity through social service in Mexico. [Retos y perspectivas de la reciprocidad universitaria a través del servicio social en México]. Autonomous University of Baja California, pp 5–6, 29–31

Municipios Puebla (2022) Almost 5 years after the 19S earthquake, and there are still people without housing in Puebla [A casi 5 años del temblor del 19s aún hay personas sin vivienda en Puebla]. Available at: https://municipiospuebla.mx/nota/2022-05-13/puebla/casi-5-a%C3%B1os-del-temblor-del-19s-a%C3%BAn-hay-personas-sin-vivienda-en-puebla

National Center for Disaster Prevention Mexico (CENAPRED) (2019). Socioeconomic impact of the main disasters in Mexico [Impacto Socioeconomico de los Principales Desastres Ocurridos en la República Mexicana]. https://www.cenapred.unam.mx/es/Publicaciones/archivos/415-IMPACTO_SOCIOECONOMICO_2017.PDF

Nodal (2022) Five years after the tragedy, puebla remembers with pain and between unfinished work the 19S earthquake [A Cinco Años de la Tragedia, Puebla Recuerda con Dolor y Entre Pendientes el Sismo del 19S]. Available at: https://www.nodal.am/2022/09/mexico-a-cinco-anos-de-la-tragedia-puebla-recuerda-con-dolor-y-entre-pendientes-el-sismo-del-19s/

Pribadi KS, Kusumastuti D, Sagala SAH, Wimbardana R (2014) Post-disaster housing reconstruction in Indonesia: review and lessons from Aceh, Yogyakarta, West Java and West Sumatera Earthquakes. In: Shaw R (eds) Disaster recovery. disaster risk reduction. Springer, Tokyo. https://doi.org/10.1007/978-4-431-54255-1_11

Pyles L, Svistova J, Ahn S, Birkland T (2018) Citizen participation in disaster recovery projects and programmes in rural communities: a comparison of the Haiti earthquake and Hurricane Katrina. Disasters 42(3). https://doi.org/10.1111/disa.12260

QUALY PANEL SA DE CV (2011) Technical details Qualypanel Covintect

Refaie F, Elmihilmy M, Bahaa T (2015) Seismic behavior of sandwich panel walls. World Appl Sci J 33:1718–1731. https://doi.org/10.5829/idosi.wasj.2015.33.11.15622

Reséndiz-Vázquez, A. (2019). Urban resilience and post-disaster reconstruction. Evidences from Mexico and France. In: Urban resilience for risk and adaptation governance: Theory and practice, p 267–280

Rodríguez LY (1997) Roads and paths of the social service in Mexico: diagnosis of 10 national congresses for the university social service [Caminos y Rumbos del Servicio Social en Mexico: Diagnostico de 10 Congresos Nacionales de Servicio Social Universitario]. Universidad Iberoamericana, p 18–20

Rumbach A, Makarewicz C (2016) Affordable housing and disaster recovery: a case study of the 2013 Colorado floods. In: Coming home after disaster: multiple dimensions of housing recovery, p 99–112

Ryder SS (2017) A bridge to challenging environmental inequality: Intersectionality, environmental justice, and disaster vulnerability. SOCIAL THOUGHT & RESEARCH: a continuation of the mid-American review of sociology, p 85 115

Sapat A (2016) Civil society and recovery: nongovernmental organizations and post-disaster / housing. Ch. 17. In: Coming home after disaster: multiple dimensions of housing recovery. Routledge, p 263–276

Sector 21 (2017) San Antonio Alpanocan, between devastation and solidarity. [San Antonio Alpanocan, Entre Devastación y Solidaridad]. Available at: https://sector21site.wordpress.com/2017/09/25/san-antonio-alpanocan-entre-devastacion-y-solidaridad/

Smith N (2006) There is no such thing as a natural disaster. Social Science Research Council. https://items.ssrc.org/understanding-katrina/theres-no-such-thing-as-a-natural-disaster/. Accessed Jan 15 2023

TECHO (2017). TECHO [TECHO concluded its second construction phase of 100 emergency housing units together with communities from Oaxaca who were affected by the earthquake [TECHO concluyó segunda fase de construcción de 100 viviendas de emergencia junto a familias de comunidades de Oaxaca afectadas por el sismo] Available at: https://mexico.techo.org/techo-concluyo-segunda-fase-de-construccion-de-100-viviendas-de-emergencia-junto-a-familias-de-comunidades-de-oaxaca-afectadas-por-el-sismo/

Trudeau RH (1993) Guatemalan politics: The popular struggle for democracy. Lynne Rienner Publishers, pp 25–27

Tucker S, Gamage A, Wijeyesekera C (2014) Some design aspects of sustainable post-disaster housing. Int J Disaster Resil Built Environ 5(2):163–181

Twigg J (2006) Technology, post-disaster housing reconstruction, and livelihood security. BH Centre (ed), Disaster studies working paper no, 15

UN-Habitat/AXA (2019) Supporting safer housing reconstruction after disasters. Planning and Implementing Technical Assistance At large Scale, pp 22, 43

United Nations Development Programme (UNDP) (2021) Handbook on owner-driven housing reconstruction, pp 3, 30–36

United Nations Office for Disaster Risk Reduction (UNDRR) (2010). Terminology: Disaster. https://www.undrr.org/terminology/disaster. Accessed 20 Nov 2022

United Nations Office for Disaster Risk Reduction (UNISDR) (2015) Sendai Framework. Priorities for action, p 14 & 18–22

Varolgüneş FK (2021) Success factors for post-disaster permanent housing: example of Turkish Earthquakes. Turk Online J Des Art Commun 11(1):115–130

Wright RN, Kramer S (1973) Building performance in the 1972 Managua earthquake (vol 13). US Department of Commerce, National Bureau of Standards

Table References

I-1. Ayuda en Accion (2018) Reconstruction of the communities in Puebla [Reconstrucción de las Comunidades en Puebla]. Retrieved from: https://ayudaenaccion.org.mx/actualidad/trabajamos-red-juntos-19s/

I-2. Universidad Autonoma del Estado de Mexico. Retrieved from: https://www.uaem.mx/difusion-y-medios/publicaciones/boletines/atiende-la-uaem-a-poblacion-afectada-por-el-sismo-del-19-de-septiembre

I-3. Ortiz RM, Reséndiz RF, Ortiz CMJ, Martínez RM, Ortiz CM, Bazán CAG (2018) The earthquake of September 19. How have we faced the crisis in Morelos, Mexico? [El sismo del 19 de septiembre. ¿ Cómo enfrentamos la crisis en Morelos, México?]. Cadernos Metrópole, 20:325–345

I-4. University of Colima (2017) Viajan a Morelos Alumnos de Ingeniería para evaluar daños en edificios. Retrieved from: Universidad de Colima (ucol.mx)

I-5. U-Press (2017) Estudiante de Arquitectura Lanza Iniciativa para Apoyar a Comunidades Afectadas por el Sismo del 19s. Retrieved from: UPRESS - Estudiante de Arquitectura lanza iniciativa para apoyar a comunidades afectadas por el sismo del 19S (upaep.mx)

I-6. BUAP Mexico (2017) BUAP assumes its commitment to the victims of the earthquake [La BUAP asume su compromiso con los damnificados por el sismo]. Retrieved from: https://www.buap.mx/content/la-buap-asume-su-compromiso-con-los-damnificados-por-el-sismo

Chapter 11
Making New Community Relationships at Post-earthquake and Tsunami Relocation Sites in Tohoku, Japan

Chiho Ochiai and Kei Saito

Abstract Establishing a community organization, namely the Resident's Association (RA), at new relocation sites after a disaster is crucial because the living environment and social communication are highly dependent on it in Japan. This study aims to understand the people's participation in the resettlement plan and establishment of Ras. Interview surveys were conducted with residents, RA members, community groups, and local government officials in charge of residential planning and community establishment and support division. As a result, the council-based collective relocation project was able to select candidate relocation sites through consultation between residents and the local government. Many originally came from the same village and lived in the same temporary housing, where they had already established relationships and the community and the Community Development Council (CDC). Even if there were people who wanted to move in from a different area, they were able to smoothly integrate into the area based on the core group. On the other hand, in the city-guided collective relocation project of public housing, people who were gathered from different areas for many reasons had more difficulties in establishing a new RA. The study found that this group did not have a chance to participate in the decision-making stage of site selection or sections of building-type planning; they were simply accepted as planned. The establishment of community relationships and RA were greatly affected by whether prospective residents were included in the discussion at the initial stage of the housing reconstruction plan. It is important to demonstrate that the establishment of RAs leads to lively local activities and improves living environments.

C. Ochiai (✉)
Graduate School of Global Environmental Studies, Kyoto University, Kyoto, Japan
e-mail: ochiai.chiho.2x@kyoto-u.ac.jp

K. Saito
Graduate School of Engineering, Kyoto University, Kyoto, Japan

C. Ochiai et al. (eds.), *Disaster and Displacement*, Disaster Risk Reduction,
https://doi.org/10.1007/978-981-96-2583-3_11

11.1 Introduction

11.1.1 Research Background

Community building to establish social relationships is one of the most important parts of disaster recovery, as the lives of people depend greatly on their local communities (Murphy 2007; Aldrich 2011a). Several studies have shown that communities' formal and informal social ties help them recover after a disaster (Aldrich 2011a and 2011b). Among the stages of disaster risk reduction, the recovery stage is regarded as the most challenging, as it entails multidimensional perspectives, such as physical, socioeconomic, and environmental points of view (Davis 2007). The three main steps that people or communities decide to relocate, select newly resettled sites, and establish or re-establish the community, namely in the Japanese case, resident's associations (RAs), which are neighborhood associations traditionally established in Japan, greatly affect the local living environment after several years of settlement.

Considering the variety of current situations at different newly resettled sites, it is important to understand post-disaster reconstruction from the planning stage to the establishment of RAs in newly resettled sites from a long-term perspective.

The Great East Japan Earthquake (magnitude 9.0) that occurred on March 11, 2011, caused a tsunami with a height of approximately 10 m. A total of 120,000 houses were destroyed and many coastal areas were heavily damaged, forcing residents to relocate (Reconstruction Agency 2011). A new master plan was developed, suggesting that disaster-prone coastline areas should only have factories and commercial facilities. Therefore, many new residential lots and housing complexes were planned and constructed at newly resettled sites during the reconstruction process, including 18,000 units relocated to higher ground and 30,000 units of disaster public housing (Miyagi Prefecture 2011; Kesennuma City 2011).

To meet the high demand for housing on limited land, since the tsunami-affected area is a coastal area surrounded by mountains, residential lots and housing complexes have been constructed in various locations, forms, and under various living conditions. Therefore, many problems in the living environment have become evident in the five to six years since moving into the newly prepared plots and reconstructed houses.

Furthermore, the COVID-19 (coronavirus pandemic) that occurred between 2020 and 2022 has had a tremendous impact on community establishment and RA's activities. Therefore, this study aims to understand the processes, including the successes and difficulties that people and communities have experienced before and after settling in their newly resettled sites and homes.

This research on the lives of disaster-affected people and communities focuses mainly on two major discussions: (1) The establishment of people's lives and neighborhood relationships by their involvement in the planning and rebuilding stage of housing, which is the initial stage of reconstruction.

(2) The process of neighborhood relationship established, focusing on RAs after moving in several years.

This study examines two cases, namely the collective relocation for disaster prevention and disaster public housing. It consistently analyzes the process from the planning stage leading up to the establishment of RAs in chronological order, and discusses how those processes influenced the current living environment and community conditions.

11.1.2 Methodology

Based on the literature review, many studies have been conducted to target residential planning or community establishment; however, most of these studies were case studies of individual communities. On the other hand, targeting both residential planning and long-term community establishment was not the main focus of the studies since the target subjects were usually divided into engineering, architecture, and sociology. Therefore, a long-term perspective on the connection between the two has been overlooked until now. The objective of this study was to examine the establishment of RAs in newly settled residential sites, addressing the following questions: (1) When and how were communities involved in decision-making in the residential planning stage, and what was the process?; (2) How have community organizations like RAs been established and what is the key to binding people together?; and (3) How can RAs help improve newly resettled residential sites, both physically and socially?

To answer these objectives, the community organization and its involvement in the rebuilding process were confirmed based on the literature and city-government-related document review. Additionally, several interviews were conducted with city government officials to learn about the programs, efforts, and difficulties they encountered. Furthermore, interviews were conducted with the local residents of the two districts, including a head and core member of the RA and some local groups in each district. By conducting interview surveys, the process of their participation in residential planning and the establishment of RAs, management of the living environment, and issues and problems in each district were examined.

This study focused on Kesennuma City, Miyagi Prefecture, by reviewing the related literature and collecting written materials and information provided by city officials (Housing Division and Regional Development Promotion Division). Kesennuma City was selected as a study site because they had extensive damage in housing and they have different types of combinations of relocation and resettlement projects.

Based on these results, timelines were prepared in both the residential planning stage (before living in the new settlement) and community organization establishment stage (before and after living in the new settlement) to analyze the progress and differences of each case study site, as well as key times and points of establishing community relations, especially focusing on the establishment of RAs.

11.1.3 Study Site

Kesennuma City is located at the northeastern tip of Miyagi Prefecture, facing the Pacific Ocean to the east and bordering Iwate Prefecture to the north and west (Fig. 11.1). The mountains are close to the sea, and the terrain has little flat land, so there is little farmland (Kesennuma City General Affairs Department 1986); settlements have developed in those limited flat lands. Furthermore, many urban areas were formed in the Early Modern period by repeatedly reclaiming land from scarce plains (Kesennuma City 2021). This coastal region is known for its frequent occurrence of earthquakes, and tsunami attacks have been recorded many times since before the Edo period (1600-late 1800). In modern times, the 1896 Meiji Sanriku Tsunami, 1933 Showa Sanriku Tsunami, and 1960 Chile Earthquake Tsunami have caused enormous damage (Reconstruction Agency 2011). During the Great East Japan Earthquake, 1,246 people died and 16,459 houses were destroyed in Kesennuma City (Reconstruction Agency 2022; Kesennuma City 2022). The number of affected households was approximately 9,500, accounting for 35.7% of all households at the time.

Of the 56 post-disaster housing relocation sites, excluding large condominiums, three districts with the largest number of residential units, including detached and row houses, were selected as the case study sites. In District A, council-based collective relocation projects (CRP) (shown in red color), there are three housing types, namely city-guided collective relocation projects (green color), and disaster public housing projects (blue color) are mixed (Fig. 11.2-District A). There were 161 units in 123 buildings since some disaster public housing has three to four independent units. District B was originally planned as a combination of a city-guided collective

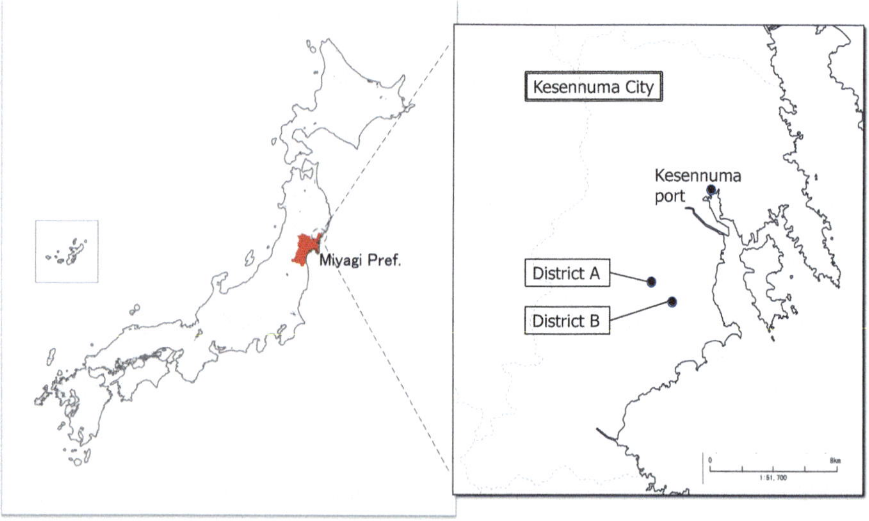

Fig. 11.1 Study site locations (Kesennuma city). *Source* Authors

relocation project and disaster public housing project (blue), but became a district for disaster public housing only (Fig. 11.2-District B). The area was originally covered with forests, but residential land was developed through logging to develop a residential area on high ground, which took longer than other site development. There are 244 units in 190 buildings, and it is the largest residential area in Kesennuma City, consisting only of detached and row houses. The area has been around for a long time and has close regional ties. Unlike newly developed residential areas such as District B, District A was developed on existing residential or agricultural land.

11.2 Housing Reconstruction in Tohoku

11.2.1 National Policy on Housing Reconstruction

Immediately following the earthquake, the national government set up the "Reconstruction Concept Council for the Great East Japan Earthquake" and formulated the "Seven Principles of the Reconstruction Concept" (hereafter Seven Principles), presenting basic ideas for future reconstruction. Item two of the Seven Principles states that "reconstruction should be centered on the region/community while taking into consideration the wide area and diversity of the disaster area." Thus, community participation has been considered important from the very beginning (Reconstruction Agency 2011). In Miyagi Prefecture, the "Miyagi Prefecture Earthquake Reconstruction Plan" was formulated. Because extensive damage had occurred throughout the prefecture, the period to achieving reconstruction was set at ten years; 2020 was set as the target for reconstruction. Moreover, the first item in the "reconstruction points" stipulates "relocation to higher ground" and "town development support" as specific initiatives (Miyagi Prefectural Government).

The "Kesennuma City Earthquake Recovery Plan" formulated by Kesennuma City stated that one of the seven pillars is a disaster prevention group relocation promotion project and a disaster public housing development project, and, simultaneously, the enhancement of the local community and promotion of consultation with citizens (Kesennuma City 2011).

11.2.2 Reconstruction and Community Rebuilding in Tohoku

A community participation council called the Community Development Council (CDC) was set up to discuss the selection of residential areas and the rebuilding of housing, where residents and the government held discussions. Research related to CDC by Takahashi et al. (2012) focused on the reconstruction process in the Yuriage district of Natori City, Miyagi Prefecture. After summarizing the changes in discussions held mainly by the CDC, they found that there was a difference of

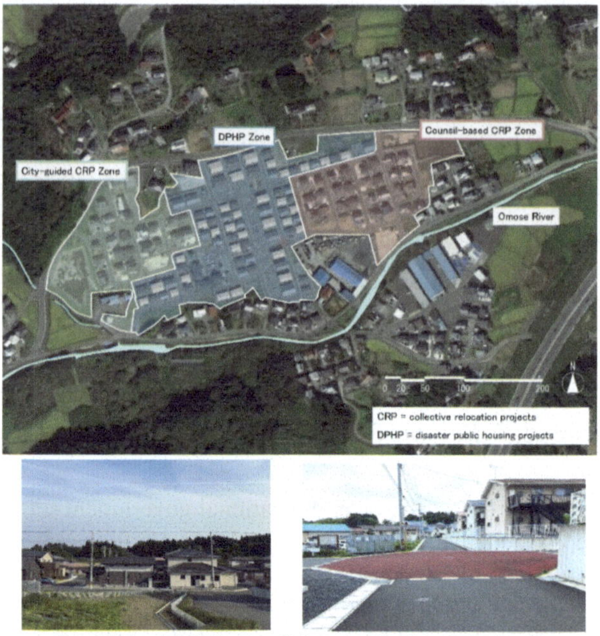

Council-based Collective Relocation Project (District A)

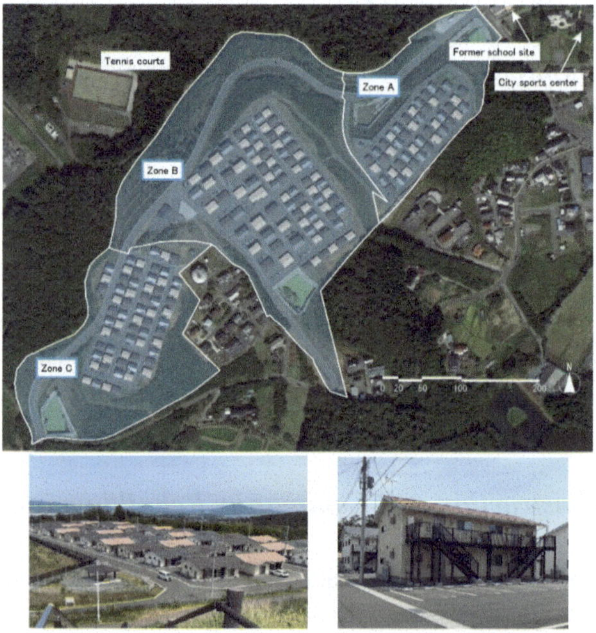

Council-based Collective Relocation Project (District B)

Fig. 11.2 Layout and housing types of Districts A and B. *Source* Authors

opinion between generations during consensus building. Furthermore, there was little experience in organizing and managing the council among residents, and organizational management was unstable. Focusing on collective disaster prevention relocation projects implemented in Kesennuma City and Miyako City, Mori et al. (2017) showed the composition and transition of the people involved in the process of formulating each plan and noted that the two cities have significantly different operational policies. In Kesennuma City, residents took the initiative to establish an RA to form a consensus for relocation. With the cooperation of external supporters such as consultants, it was possible to actively reflect their wishes and intentions in residential land plans. On the other hand, in Miyako City, the residents were not active at first, and the government took the lead in establishing and operating an RA for efficient consensus building; however, this caused residents to distrust the project, pointing out that one-way communication made it difficult to reflect their wishes and intentions. Tsukuda et al. (2017) analyzed changes in the results of a survey of people's intention to reconstruct public housing in Ishinomaki City, and discussed the degree and direction of change in intention for each household attribute. Here, households with more people and those who lived in their own homes before the disaster changed significantly. It was clear that many people were moving into disaster-prone public housing.

In research on the community formation process, Kato et al. (2021) discussed the establishment process of the local community and the actual activities of the newly established RA in the Yuriage district and analyzed the impact of the environmental transition process accompanying the relocation of the residence. He noted that when forming a new community, residents who have undergone a similar transition process after a disaster tend to clump together. In particular, the community that was established and its members were fixed during their time in temporary housing, making it difficult for other people to adapt to the neighborhood after moving into permanent housing. As shown in these studies, each site has different issues regarding the resident's involvement in the reconstruction plan and community establishment.

11.3 Community Organizations' Participation in the Planning Stage of the Residential Reconstruction Project

In this section, the processes of deciding and formulating a newly resettled site and the involvement of community organizations are discussed. There are two main approaches to formulating housing plans. Figure 11.2 shows the layout and housing types of two different residential reconstruction projects: the council-based collective relocation project and the city-guided collective relocation project. Figure 11.3 shows the decision-making process of two residential reconstruction projects.

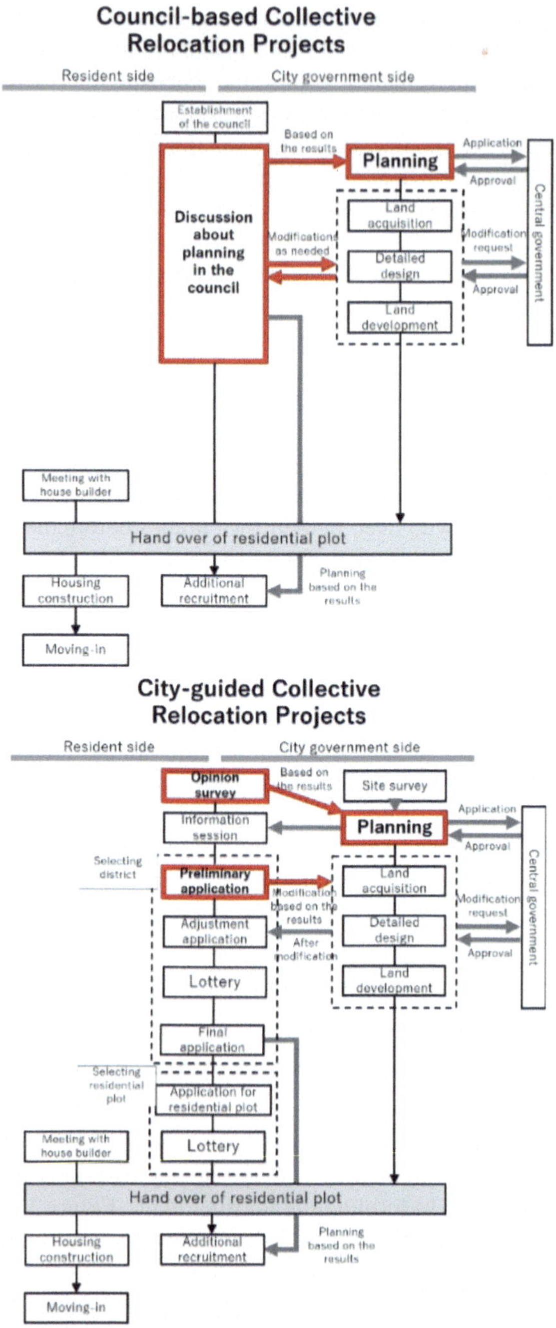

Fig. 11.3 Two different flows of residential reconstruction project planning

(1) **The council-based collective relocation project (District a)**

In the council-based collective relocation project, like District A, residents who lived in the same area before the disaster established the council (CDC) to make the collective relocation plan based on discussions among residents who wish to relocate collectively. At this point, the people in these districts had already established bonding since they had been living in the same area and established neighborhood relationships. After deciding where to relocate, they filed a petition with the city to develop the residential property. When selecting a residential site, locally elected city or prefectural council members in Districts A negotiated with land owners and the city department in charge of the residential plan. In response to this petition, the city formulated a housing development plan in consultation with the council. At this point, people in these districts dealt with the city's departments and landowners to ensure that their requests would proceed. While the relocation sites (newly resettled sites) were decided through discussions among residents, the selection of residential plots within the district was decided by lottery from the standpoint of fairness, so as not to affect the relationships between residents.

District A has public housing units in the same districts, as requested by residents from the same former district who have no willingness to rebuild new houses due to age or financial reasons. Therefore, the residential plan was formulated after the council determined the number of self-funded rebuilding and public housing units. Even though the lottery for public housing was held to be fair, for this reason, there was a limited number of residents from other districts to join the public housing units in District A.

(2) **The city-guided collective relocation project of public housing (District B)**

In the disaster public housing project in District B, the city's housing division first surveyed the disaster victims' intentions through a questionnaire survey targeting a whole city regarding their desired place of residence and building type (detached house, row house, or condominium). A housing plan was formulated based on the survey results. Then, the housing division held several briefings and consultation meetings for residents of each district to make the plan known to them. Residents had some time to decide which areas they wanted to move to and submitted their intentions to the city in the form of "provisional applications." The city aggregated the results and revised the plan in response to excesses or shortages of the desired types of housing and number of units. In this way, the people selected by lottery had no chance to discuss or form a council in advance, and people from different areas with different circumstances gathered in disaster public housing. Also, there were many reasons for the people to select this public housing site, such as losing lottery selections in other sites, preferably selecting a location far away from the seashore, preferring to live with pets, and people who were not able to respond in a timely manner. For this reason, there was no opportunity or organization for people to come together and establish a community relationship before they moved into public housing. Also, in this type, they lacked a chance to discuss with local government officials or other organizations to receive help or support.

(3) **The differences between the two types**

The major difference lies in the method of understanding the intentions of people who wished to live in newly resettled sites. In the former, residents and city departments were able to formulate and revise plans as needed through the council (CDC). The latter is characterized by the fact that the housing plan itself was created solely by the housing division, while only grasping people's intentions through city-wide questionnaire surveys and provisional applications. A lottery was held for this district where there was an excess of applicants even after the revision, and residents who were able to secure a place made a "final application" to determine their final intention to move in. If there were too many applicants even after the corrections were made, more careful adjustments were possible, such as making provisional applications again as a "re-provisional application" and holding a lottery if the excess was still not resolved.

Moreover, because District B was readjusted by logging and clearing forests owned by the government, it took longer than other districts to develop residential land, and it was one of the last districts planned in the city. Consequently, there were many cases in which vulnerable people moved in because they were not selected in the lottery at another relocation project site, or were unable to make a decision as quickly as others. Despite the fact that there were many requests for detached houses from the beginning, there were restrictions on land availability, so the housing division planned to build more row houses to compensate for the shortage; however, people's wishes did not change from detached houses to row houses. This has resulted in many vacant houses in District B.

11.4 Process of Establishing the Community Organization at Newly Resettled Sites

In this section, the establishment of a community organization (RAs), in rebuilding people's lives after moving into the post-disaster housing in newly resettled sites and its current situation are discussed. In the first stage, when residents moved into the newly constructed houses, some were self-built houses, and some were public houses, and discussions of the establishment of RAs began among residents. There are two main approaches to formulating RAs. Figure 11.4 shows two different processes of RA establishment, namely, the cases of Districts A and B.

(1) **District A**

In District A, the council was already established for the collective relocation projects since many of the residents were from the same district, and they were living in the same temporary housing, and communication proceeded smoothly. Also, residents of the city-guided collective relocation projects and disaster public housing projects moved to the same district sequentially. Among them, a person who later became the leader of the preparatory community group of RA moved into this district.

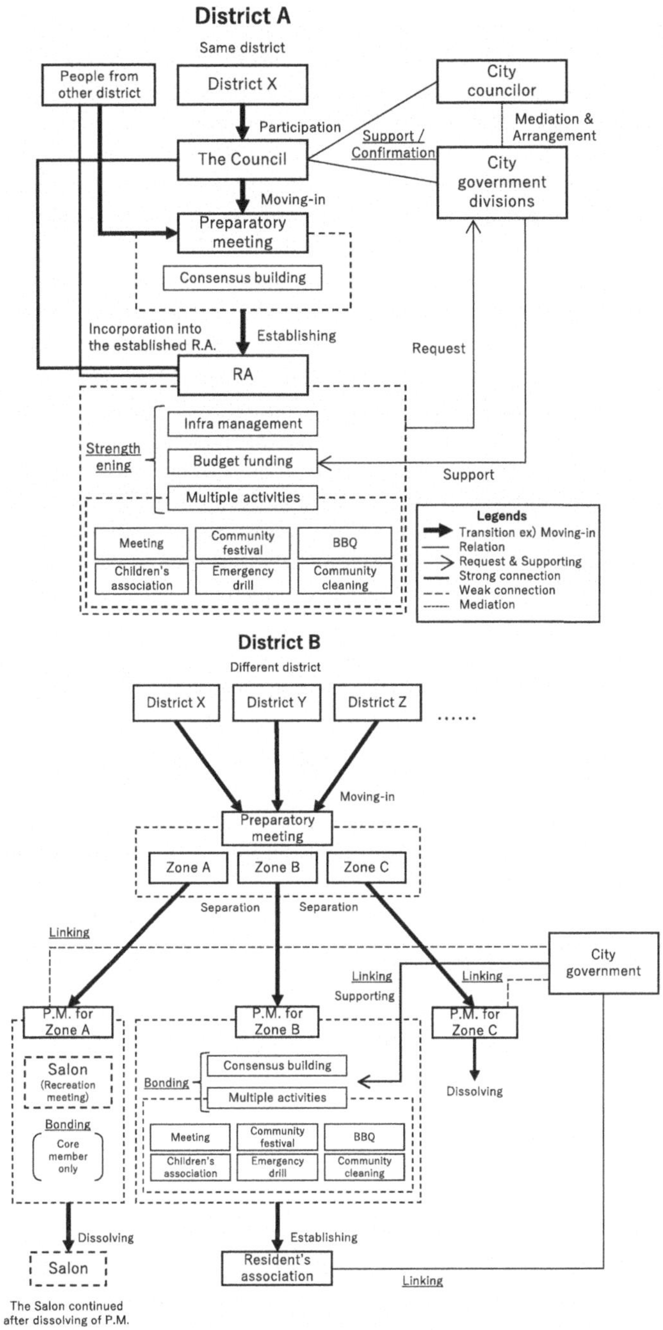

Fig. 11.4 Two different flows of the establishment of community organizations. *Source* Authors

After moving in, communication was promoted mainly by the people of the council-type collective relocation project, in which a relationship had already been built; thus, it seems that exchanges from different programs progressed smoothly. The leader, Mr. U, had long served as the head of the RA in the area in which he lived before the disaster and, therefore, had experience organizing the group and planning activities for the RA. Given this background, he visited a group of people in temporary housing to greet them and volunteered to be the head of the RA when they all moved into the new site. Immediately after people moved into District A in October 2016, he was actively engaged in activities within the district, such as distributing the city's public relations magazine to all households in the district twice a month as a free volunteer for a year until the RA was established. The RA aimed to understand the state of the new residents and simultaneously recognize Mr. U from the core group. The background to this was that the division in charge of community establishment at the city hall had told Mr. U that they wanted him to lead RA in the new site.

In this way, although bonding had already been formed in the core group, people from different areas and Mr. U himself gradually built up a friendly relationship with all the new residents. Therefore, compared to the original core group bonding relationship, another larger scale of bonding relationship was established.

In the initial stages, the city's community division hosted an event to solve future problems and deepen exchanges within the region. At that meeting, the community division proposed to establish a "preparatory committee for the formation of the RA"; following the suggestion, District A established a preparatory committee. With this establishment, several residents from each project would serve as preparatory committee members: four people from the disaster public housing development project, six from the council-type prevention project, and four persons from the city-guided prevention project, including the leader. Mr. U said, "The problem with establishing a new RA was that the participation rate of the residents might become low. To prevent this situation, it was necessary to get residents interested. To solve this problem, it was necessary to demonstrate the merits of the RA and start activities as soon as possible." On July 22, 2017, the first Preparatory Committee Board Meeting was held to elect the leaders, deputy leaders, and other officers of the RA. On September 16, the RA's founding general meeting was held, and the RA was officially established in District A. It was one of the earliest RA establishments in newly resettled sites.

After the RA was established, it actively requested that the city improve the living environment within the district. For example, asking the city hall to quickly build a community center, installing playground equipment in playgrounds, coordinating with the city hall and police to color marking intersections, installing curved mirrors, and negotiating with the post office to install post boxes. Various environmental improvements have been achieved under the initiative of the RA. Furthermore, the RA frequently holds events within the district and actively participates in local events to build relationships with neighboring communities. For example, ground golf practice sessions, tea time gatherings, and annual events such as New Year's parties and summer and autumn festivals have been held. Especially during summer festivals, the community center is decorated, exhibitions of residents' craft work are held, and

BBQs are held. Every year, a spring clean-up is held in June, and an autumn clean-up is held in October. As many as 100 people participated in these events.

In District A, they were forced to cancel major events due to the coronavirus pandemic, but by holding regular board meetings, they were able to constantly confirm the role of the RA, discuss issues, and continuously make efforts to organize small events in the district. Holding these events has the function of raising interest in improving the local environment and management of the RA. At the base, Mr. U provides strong leadership; simultaneously, the original core group members recognized his knowledge and social network and supported him in achieving goals and strengthening the RA and the local community.

(2) **District B**

As mentioned above, District B was built solely by a disaster public housing project, which is the only and largest in Kesennuma City. District B was divided into three zones; Zone A moved in December 2016, three months earlier than Zones B and C. In the beginning, people in District B were aiming to establish an RA jointly with Zones A, B, and C. However, most of the people in this district met for the first time; thus, it was decided between people who now became residents and the city's community division that it was necessary to start by getting to know each other first. Therefore, soon after the residents moved in, they held exchange meetings in each zone where they learned about each other's situations and elected preparatory committee members. Around the spring of 2017, a "joint preparatory meeting" was organized with the aim of establishing an RA in District B. In electing District B representatives, stagnation occurred due to a lack of consensus. Consequently, the aim was to establish an RA in each zone. Originally, they gathered in District B separately from different areas, so there was no basis for relationships, and it was difficult to elect the head of the RA. The results of the exchange meeting for the establishment of the RA did not appear easily, and District A decided to withdraw from the Joint Preparation Committee. In the background, there was a sense of distrust towards Zones B and C in conducting activities such as local events, and they wanted to independently apply for subsidies from the prefecture and continue their activities on their own. Districts B and C had been working together on the preparatory committee, but they also decided to split shortly thereafter, and all joint activities halted. Zone A formed a unique group called the Salon group and created a gathering for those who wished to gather. In Zone C, there are many young people raising children, so they are busy with their own lives, and it is difficult for them to participate in and agree to establish a community association. Under these circumstances, in Zone B, in cooperation with the community division, they continued the activities of the preparatory committee and finally established an RA in December 2022. In Zone B, to identify the residents living in the same area, they created their own resident list. They found that there were many elderly people aged 65–75 years, followed by households living alone, and there were a certain number of households with small children.

From these processes, it can be observed that Zone A was able to organize small-scale bonding, and Zone C did not build any social relations as a group during these years. However, Zone B maintained a good relationship and support from the community division; therefore, it was possible to establish the RA.

Differences in the implementation of local activities were confirmed for each zone. In Zone A, community activities were the first to be successful. From 2018 to 2019, local food parties, year-end parties, New Year's parties, and so forth were held, and 40–50 people from 44 households in Zone A participated. Moreover, the Salon group mentioned above continuously hosts activities, such as 20 meetings per year with subsidies.

In Zone B, information exchange meetings were held regularly, such as summer festivals, senior citizen parties, local food parties, Christmas parties, and year-end parties. The summer festival was held with a subsidy from the prefecture, and there were 97 participants, including some residents of Zone C who were still present immediately after the division and had a certain amount of interaction.

Community activities in District C operated smoothly from 2018 to 2019. Specifically, this includes posting in public relations papers, planning tea parties, setting up groups within the zone, and formulating rules for living, such as removing garbage. However, it was noted by the head of the Zone C preparatory committee that interest in the establishment of the RA among residents other than board members was low at that time. Problems in the cooperative relationship between the younger generation, who find it difficult to attend meetings frequently due to work, childcare, and so forth, and the older generation, who are highly motivated about community activities, led to the withdrawal of the younger generation. Furthermore, because the core residents who played a central role in local activities in the previous settlement often had a relatively solid living foundation, they left the district due to relocation. Such residents could find private housing land outside the district since disaster public housing is rental housing, and the price is adjusted based on their incomes.

While people were still struggling to establish the community organization and build relationships, in the spring of 2020, the new coronavirus pandemic began, and all these activities were forced to be canceled. In the early stages of forming a new community, being forced to take such a long middle step was a major obstacle to forming bonds.

11.5 Discussion

As a result of the study, the following three points will be discussed from the housing reconstruction plan and community formation perspective:

(1) Housing reconstruction planning and its impact on community formation

Depending on the stage of the housing reconstruction project that the community organization was involved in and the kind of social relationship and network the community organizations like RA possessed and used, it is expected to have a significant impact on the subsequent formation of community associations and local management. The "council" established by the residents was originally intended to conduct collective relocation on a regional basis as a "council-type collective relocation." However, the continuous repetition of consultations during the reconstruction process formed the basis for forming an RA and conducting its activities in the new district. For example, in District A, many residents were also active as council members when they lived in the same temporary housing. According to the current RA leader, many residents cooperated with local activities after moving into permanent housing because of their experience while in temporary housing. This means that the communication and formation of the group in the early stages of the reconstruction project can contribute to the subsequent formation of the local community. Furthermore, a person with a wealth of experience in RAs taking the lead in the community can strengthen social bonds but also use so-called "linking or bridging" social capital to benefit the community.

When selecting a newly resettled site, locally elected city or prefectural council members in Districts A with landlords and the city department in charge of the residential plan. In response to this petition, the city formulated a housing development plan in consultation with the council. People in these districts used so-called "linking" social capital to deal with cities and landowners to ensure that their requests would proceed.

On the other hand, even though the lottery of public housing was held to be fair, it was rare for many residents from other districts to join, as in District B, making it difficult to form a group and select a leader with the people's consensus. Moreover, they did not have any opportunity to be involved in the planning stage, which meant that they lacked the process of discussion and building social relationships. It was also not possible to obtain any link to the possible parties, as in other cases, because there was no established group that could serve as a basis for discussion among these people.

(2) Difficulties with community formation

In the case of District B, it became clear that a newly resettled site with people from different locations had a significant influence on the process of community formation. First, Kesennuma City's basic policy was to relocate while maintaining the community at its place of residence before the disaster. However, based on the survey of intentions and available land, it was decided to develop a disaster public housing site independently. In the disaster public housing development project, disaster victims from various locations in the city were eligible to move in; if the number of applicants exceeded availability, residents were selected by lottery. Owing to these factors, when the community reconstruction process began, there was no relationship between residents, and this caused the stagnation of the preparatory committee activities and a decrease in the motivation of residents, leading to the abandonment of the establishment of the RA. Furthermore, the progress of development and different numbers of residential plots in Zones A, B, and C hindered the development of a sense of unity among residents in the district. For example, moving in at different times gave some people a feeling of superiority over other zones, or some people felt that the largest zone should take leadership, although other zones had more experienced people in the RA. These small details made it difficult to build relationships during community formation. Therefore, at the planning stage, it is important to establish an opportunity to discuss the early stage of residential planning and consider the size of the zone and the timing of moving in.

However, in District A, in addition to the original core group bonding relationship, another larger-scale bonding relationship was successfully achieved by adding new people.

Through these activities, it was found that it is important to secure people interested in community activities and exchanges. Furthermore, it is important to demonstrate the benefits of RA activities and improve living environments. For example, events such as festivals and cleanups are frequently held in District A, and through these opportunities, exchanges between existing community residents and new residents have been created.

(3) The impact of COVID-19 and the importance of continuing community activities

The impact of COVID-19 was significant during the formation of new communities. In some cases, the relationships between residents that started to form were disrupted after three years of restrictions on community activities. Regarding the establishment of RAs, in districts where meetings and activities continued, even though they were small groups, activities after the end of COVID-19 could be carried out smoothly. However, many households no longer felt the need to form local communities such as RAs because they had already become accustomed to being without community groups, and it became almost impossible to establish RAs.

In District A, they were forced to cancel major events due to COVID-19, but by holding regular board meetings and so forth they were able to constantly confirm the role of the RA, discuss issues, and make continuous efforts to organize small events in the district. Holding these events has the function of raising interest in improving the local environment and the management function of the RA. Conversely, in District B, although the situation differed depending on the zone, the results were divided into districts where only some residents were able to hold meetings, districts where nothing was held, and districts where RAs were established.

11.6 Conclusion

The main contributions of this study are twofold. First, it was elucidated that the formation of community organizations, in this case, RAs was greatly affected by whether prospective residents were included in the discussion at the initial stage of the housing reconstruction plan. Many housing reconstruction projects tend to focus on the number of houses built and the speed of reconstruction, but it was found that having discussions with people at the early stage of a housing reconstruction project leads to the formation or strengthening of relationships among residents. Furthermore, the group of former community members has already established relationships and links with council members who can reach out to the city and landowners. Moreover, the initial plan to relocate residents of the same area to the same area was also effective in forming a local community. This is because they have a relationship in which they know each other face-to-face. Since they have already confirmed the skills of each individual, they can ascertain whether the person is capable of playing the role of a leader or board member of a community organization. However, residents who have already moved to planned residential land or houses by lottery cannot have discussions in advance, and it is necessary to get to know each other after moving in. This is very difficult and affects community formation.

Second, it is important to demonstrate that the establishment of RAs leads to lively local activities and improves living environments. In recent years, even in ordinary areas, community organizations such as RAs have faced the problem of low participation rates among the younger generation due to their many roles and time constraints. However, having seen a community organization like an RA improve the living environment, and experiencing activities in which families can participate, can provide opportunities for understanding and encouraging the participation of the young generation. Furthermore, the suspension of community activities owing to COVID-19 at newly resettled sites had a significant impact on community formation. In such cases, continuing activities, even if small, will lead to the smooth resumption of activities.

References

Aldrich DP (2011a) The externalities of strong social capital: post-tsunami recovery in southeast India. J Civ Soc 7(1):81–99

Aldrich DP (2011b) The power of people: social capital's role in recovery from the 1995 Kobe earthquake. Nat Hazards 56(3):595–611

Davis I (2007) Learning from disaster recovery; guidance for decision maker. International recovery platform (IRP). https://www.preventionweb.net/files/726_10110.pdf

Kato H et al (2021) The living environment and living style in disaster public housing in Yuriage district during the revitalization period after the great east Japan earthquake. J City Plan Inst Jpn 56(3):1501–1508 (in Japanese)

Kesennuma City General Affairs Department (1986) History of Kesennuma city nature edn 14 (in Japanese)

Kesennuma City (2011) Kesennuma city earthquake reconstruction plan (in Japanese). https://www.kesennuma.miyagi.jp/sec/s019/010/050/010/1389054300072.html. Accessed Jan 2023

Kesennuma City (2021) Kesennuma city great east Japan earthquake reconstruction record "Umi to ikiru" (Live with the sea) 18 (in Japanese)

Kesennuma City. https://www.kesennuma.miyagi.jp/index.html. Accessed Dec 2022

Mori S et al (2017) The administration-community relationship in plan formulation for projects for promoting collective relocation—Focusing on community groups' establishment and subsequent activities in Kesennuma and Miyako cities. J Arch Plan AIJ 82(734):929–939 (in Japanese)

Murphy BL (2007) Locating social capital in resilient community-level emergency management. Nat Hazards 41(2):297–315

Miyagi Prefecture (2011) Miyagi prefecture earthquake reconstruction plan (in Japanese). https://www.pref.miyagi.jp/site/ej-earthquake/list441-517.html. Accessed Jan 2023

Reconstruction Agency (2011) Reconstruction concept council for the great east Japan earthquake, (in Japanese). https://www.reconstruction.go.jp/topics/110811kaitei.pdf. Accessed Jan 2023

Reconstruction Agency. https://www.reconstruction.go.jp/topics/main-cat2/sub-cat2-6/index.html. Accessed Dec 2022

Takahashi Y, Ubaura M (2012) A research on the role of community organizations founded after the earthquake in the initial stage of disaster reconstruction planning—a case study of Yuriage area, Natori city, Report of the city planning institute of Japan (11):73–78 (in Japanese)

Tsukuda H et al (2017) A study on the change of housing recovery opinions based on public housing registration data. J Arch Plan AIJ 82(731):1–9 (in Japanese)

Chapter 12
Relocation and Rebuilding of a Religious Facility Destroyed in a Disaster in Japan: An Experience of Kumano Hongu Taisha After the 1889 Flood

Chiho Ochiai and Shinji Kajita

Abstract In the aftermath of a disaster, religious facilities such as churches, mosques, temples, and shrines have to be relocated and rebuilt, in addition to houses. These religious facilities are often the center of the community and can play symbolic and monumental functions following large-scale disasters. This study described the decision-making process, relocation, and rebuilding of the Kumano Hongu Taisha (Shrine) after the 1889 historical flood in Wakayama Prefecture, Japan. The study was conducted through a careful analysis of documents and receipts kept in the shrine storage. Kumano Hongu Taisha was originally located at present Ōyunohara, on a sandbank at the confluence of the Kumano and Otonashi Rivers. In 1889, the flood destroyed most of the shrine buildings, and the remaining three god halls were relocated to the hilltop site in 1891. Rebuilding work was carried out by carpenters from Wakayama City, managed by the Ministry of Interior and the Wakayama Prefectural Office. Furthermore, the parishioners as local people who were also affected by the disaster themselves, cooperated in various ways to rebuild the shrine, such as by removing driftwood, cleaning the timber of damaged buildings, and digging and saving stone lanterns. This study reveals the detailed relocation and rebuilding processes of Kumano Hongu Taisha.

C. Ochiai (✉)
Graduate School of Global Environmental Studies, Kyoto University, Kyoto, Japan
e-mail: ochiai.chiho.2x@kyoto-u.ac.jp

S. Kajita
Graduate School of Engineering, Kyoto University, Kyoto, Japan

C. Ochiai et al. (eds.), *Disaster and Displacement*, Disaster Risk Reduction,
https://doi.org/10.1007/978-981-96-2583-3_12

12.1 Introduction

12.1.1 Relocation and Resettlement of the Shrine Due to Flood

Post-disaster reconstruction includes not only the reconstruction of houses and infrastructure, which has been often discussed, but also religious facilities, such as temples, churches, and mosques, which are historically rooted in the local community, and serve as a base for people's hearts and minds, and have symbolic and monumental functions after large-scale disasters. However, their importance has been overlooked, and few studies have been conducted to understand how these facilities were rebuilt in the past.

Kumano Hongu Taisha, the main shrine of over 4,700 Kumano shrines nationwide, is one of the Kumano Sanzan shrines (three major shrines in Kumano) located in Hongu-cho, Tanabe City, Wakayama Prefecture, Japan (Fig. 12.1). Kumano Hongu Taisha was built in a place called "Oyunohara," located in the sandbank between the Kumano and Otonashi Rivers, for over 1,900 years, from its foundation until 1889. Looking back on its history, the shrine has suffered from disasters such as fires and floods many times but has been rebuilt repeatedly through the contributions of shrine workers and believers from all over the country (Hongu Town History Compilation Committee 2004).

However, the Great Totsukawa Flood that occurred on August 19, 1889 (Meiji 22 of the Japanese calendar) hit "Oyunohara," and many shrine buildings were washed

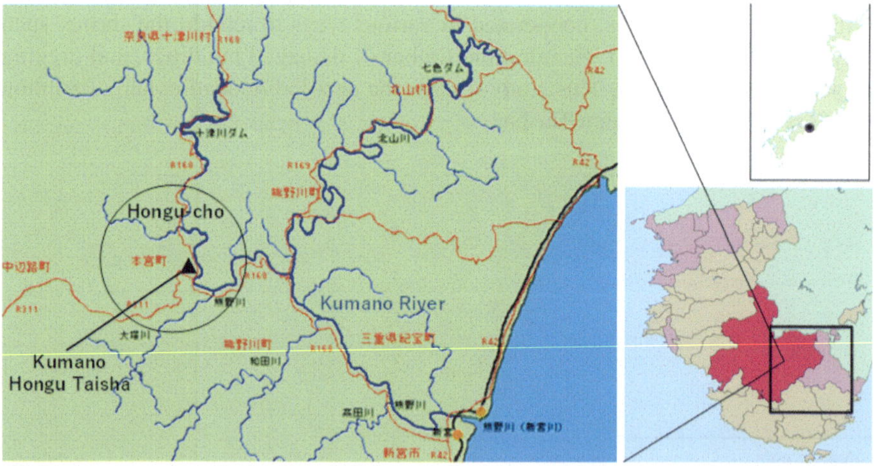

Right above map show the location of Wakayama Prefecture. Right bottom map in red shows the location of Tanabe City. The left map shows the location of Kumano Hongu Taisha and the river flow of Kumano River

Fig. 12.1 Location of the study site. *Source* Authors

away or heavily destroyed, which forced their relocation or rebuilding on a nearby hilltop. The relocation and rebuilding of the shrine were completed within only one year and seven months, which is considered a very short period. At present, all that is known about the relocation and rebuilding of this shrine is the fact that it has been relocated. Only some of the recorded materials related to the relocation and rebuilding process from flood damage have been deciphered, and details of its processes have not yet been researched.

The area that includes Kumano Hongu Taisha was registered as a "sacred site and pilgrimage route in the Kii mountain range" as a UNESCO World Heritage site in July 2004, making it a place visited by many tourists. Kumano Hongu Taisha is also known as a resurrection shrine. Therefore, it is important to understand the experiences of the shrine and the actual relocation and rebuilding processes in which the government and local residents were involved in 1889.

This study examined the relocation and rebuilding processes of Kumano Hongu Taisha, devastated by a flood in 1889. It considers the factors that make it possible to complete a project in a short period. Additionally, this study clarifies the details of Kumano Hongu Taisha's contributions, local residents, and local administrative agencies.

This study understands the relocation and rebuilding processes of Kumano Hongu Taisha, which was devastated by the flood in 1889. It considers the factors that made it possible to complete the project in a short period of time. In addition, the study clarifies the details of the contribution activities of Kumano Hongu Taisha, local residents, and local administrative agencies.

12.1.2 Methodology

This research begins by conducting a literature review based on previous research such as archival materials held by the Kumano Hongu Taisha Shrine. To organize the situation at the time of the flood in 1889 and the damage to the shrine buildings and attached buildings inside and outside the precincts, reviewed record documents in the possession of Kumano Hongu Taisha, like "Diary of August 19, 1889." Then, the diaries of the Kumano Hongu Taisha staff, written in 1889, and other recorded materials were deciphered.

Furthermore, we conducted an investigation into the storage of Kumano Hongu Taisha and interviewed shrine staff, Wakayama Prefectural World Heritage Center staff, Wakayama Prefectural Cultural Properties Center staff, and local residents (parishioners).

12.2 Rebuilding of the Religious Facilities

12.2.1 Rebuilding Shrines and Temples in Japan

Regarding the relocation and rebuilding of shrines and temples designated as important cultural properties of the country, shrines such as Ise Jingu and Kasuga Taisha, which are customarily relocated and rebuilt according to the Shikinen Rebuilding System, have been well studied. In addition, regarding the relocation and rebuilding of shrines and temples during the Meiji period, many studies have focused on the institutional background, such as the establishment of the modern shrine and temple construction system and the details of the regulations. There have been very few detailed studies on the relocation and rebuilding process of specific shrines and temples. Furthermore, most of these studies have focused on the relocation process and specifications of temples, and few studies have comprehensively analyzed the relocation and rebuilding process, including the institutional background (Ogashiwa 2020).

Many studies have been conducted on the relocation and rebuilding of shrines and temples regarding important cultural properties. Obayashi targeted buildings that had been relocated from among buildings designated as Important Cultural Properties (from shrines and temples to private houses and modern buildings) by 1998. He mentioned that the number of relocated shrines and temples decreased from early modern times to the Meiji and Taisho periods (late 1800–early 1900) and that there has been no significant change in the number of relocated buildings since then. The study also focused on two reasons why temple buildings were relocated after the Meiji era: disposal due to the ordinance separating Shinto and Buddhism and maintenance-related reasons such as location conditions (Obayashi 2005).

Regarding the relocation and rebuilding of shrines, much research has been conducted regarding the existence of the Shikinen Reconstruction System (periodic reconstruction of Shinto Shrines) for the former Kankoku Heisha Shrines and the diversion of old materials generated during the Shikinen Reconstruction of the Ise Jingu Shrine and Kasuga Taisha Shrine (Tani 1959; Tsunoda 2005; Yamano 2005). Kogashiwa and Miura's research is one of the few case studies on the relocation and rebuilding of shrines and temples during the Meiji period. Kogashiwa deciphered drawings, diaries, and cash flow reports from the Meiji period and explained the process of relocating and rebuilding Shigain beginning in 1877 (Ogashiwa 2020). Miura deciphered newly discovered historical materials concerning the preservation of Chikurin-ji Temple organized and analyzed the content of each historical material, and provided an outline of the current state of damage to Chikurin-ji Temple and the repair works before the relocation in the Meiji Era. Furthermore, details of the roof repair specifications were identified by comparing and verifying them using drawings and photographs (Miura 2017).

12.2.2 Kumano Hongu Taisha

Kumano Hongu Taisha, the subject of this research, is a shrine located in Hongu, Tanabe City, Wakayama Prefecture, and is one of the Kumano Sanzan (three temples). It is said that it was founded in the 65th year of Emperor Sujin (33 B.C.), but this is not certain. Furthermore, "Kumanoimasu-jinja (shrine)," which is classified as a Myojin Taisha in the Engishiki Jinmyocho, is the Kumano Hongu Taisha. The shrine was built on a sandbank between the Kumano and Otonashi rivers called "Oyunohara," about 700 m south of the current site. The Upper four shrines (first and second [together] Hall, third Hall, and fourth Hall), the middle four shrines (fifth Hall, sixth Hall, seventh Hall, and eighth Hall), the lower four shrines (ninth Hall, tenth Hall, eleventh Hall, and twelfth Hall), and 12 deities called the Kumano Junisho Gongen were enshrined there, and different deities were enshrined in the 13 subordinate shrines inside and outside the precincts (Fig. 12.2).

This "Oyunohara" began in ancient times and was the final destination of the Kumano pilgrimage, which was the act of praying for the Kumano faith that flourished in the Middle Ages. However, the Great Totsukawa Flood that occurred on August 19, 1899, hit Oyunohara, and many shrine buildings, such as the middle four major shrines, the lower four shrines, and the Setsumatsu Shrine in the precincts, were washed away (Fig. 12.3). Only the upper four shrines, which survived the erosion, were enshrined in the current shrine site, and the middle and the lower four shrines were enshrined together in one stone shrine in Oyunohara.

The former shrine site is located in the middle of the river, and the building surrounded by trees is Kumano Hongu Taisha.

Fig. 12.2 Former location of Kumano Hongu Taisha (Oyunohara). *Source* Kumano Hongu Taisha

Fig. 12.3 Photos of 1889 flood damages to the Shrine. Kumano Hongu Taisha

Regarding the rebuilding of the current main shrine, a measurement survey conducted by Mifune and Tamura confirmed the differences in common structures and detailed specifications among the three shrine halls (1st and 2nd deities in the 1st, 3rd, and 4th halls). All three shrines had cypress bark and cedar persimmon roofs. It is a lightweight structure that protects the enshrined deity from rain leaks due to repeated storms and flood damage and deterioration over time (Mifune and Tamura 2012). Sato reconstructed the scale of the Kumano Sanzan precincts and the layout of shrine buildings during the period when the Kumano faith was growing, based on travelogues and drawings written from the late Heian period to the early Middle Ages (8th to tenth centuries) and traced changes over time. Among them, it was stated that there were changes around the time of the fire in 1209 when the foundation of the main building was remodeled from the earthen floor to the current wooden platform, and railings were newly added to the main building (Sato 1975).

In addition, Mifune and Tamura who were in charge of the main shrine restoration work in the first of the two phases of the Kumano Hongu Taisha Shrine renovation project from 2010 to 2015, stated in the construction report that they investigated the state of damage and the building structure of the 12th, 3rd, and 4th halls of Shrine buildings, and indicated the necessary repairs in the foundations, wooden parts, roofs, decorative metal fittings, and tatami mats (Mifune and Tamura 2012). He explained the distinctive features of Kumano-zukuri, such as the construction that broke the left–right symmetry of the first and second halls, and the differences in the visible and hidden structures and details of the 3rd and 4th halls, which were built with the same shape and size.

12.2.3 Kumano Hongu Taisha, Disasters and Rebuilding

Kumano Hongu Taisha has experienced various disasters since its founding. The first recorded disaster was a fire that occurred in 1096, after which 13 fires occurred until 1899. Flood damage has increased since 1541, and 12 cases were reported by 1899. In addition, when classified by era, 12 fires and one flood occurred between ancient times and the Middle Ages (1573), and one fire and 11 floods occurred between the early modern period and Meiji 22 (1899). This shows that the content of disasters changed greatly as the times transitioned from the Middle Ages to early modern times. One of the reasons for the increase in the number of flood occurrences is the rise in the riverbed due to post-flood sedimentation.

The main and attached buildings, which were rebuilt during the Kyowa and Bunka periods, withstood the large-scale flood damage in 1848 and 1852, respectively.

With reference to the history of Hongu Town, there are two examples of shrine reconstruction in the early modern period: the Kyoho period (1721–1731) and the Kyowa/Bunka period (1779–1810) are explained. The Kyoho period has been reconstructed four times since the previous Keicho reconstruction: in 1619, 1630, 1680, and 1708. A plan was devised. After Tokugawa Yoshimune became the 8th shogun, the construction was approved in November 1721 and completed in 1731. A plan was devised for subsequent reconstruction during the Kyowa and Bunka periods, as most shrine buildings were destroyed in a fire in 1770. The project began in 1779, when the Shogunate approved it, and was completed in 1810.

Regarding funding, the presence or absence of financial support from the shogunate and the decrease in families of Shinto priests serving at shrines on a hereditary basis due to the disaster had a large impact on the funding period and funding amount. During the Kyoho period, there was support from the shogunate, so donations could be collected smoothly from senior government officials and feudal lords with large landowners throughout the country; however, there was no government support in the Kyowa and Bunka periods. Consequently, the shrine was unable to collect solicitation money and had no choice but to engage in entertainment, such as performances on the side of the road and begging for money from passersby. In addition, these donation activities were mainly carried out by shrine priests of the Kumano Hongu Taisha Shrine. However, during the Kyowa and Bunka periods, there was a great deal of staff breakup after the fire of 1873 of Meiwa. It is believed that the procurement period was prolonged because of a significant decrease in human resources.

12.3 Relocation and Rebuilding of the Kumano Hongu Taisha

12.3.1 Destruction and Damages

According to the shrine office diary of 1889, after the flood, books and other items stored in the office were taken to the hall of worship (*Reiden*) built in the middle of the former shrine grounds, and the shrine staff evacuated there. However, as the water approached the worship hall, they evacuated to the upper Kagura-sho (between the 12th and 3rd halls) using only the important items. Subsequently, the water continued to rise and approached the upper level, so the chief priests, *Negi and Shuten,* went to the 3rd, 4th, and 12th halls, respectively, to protect the sacred body. The water flooded

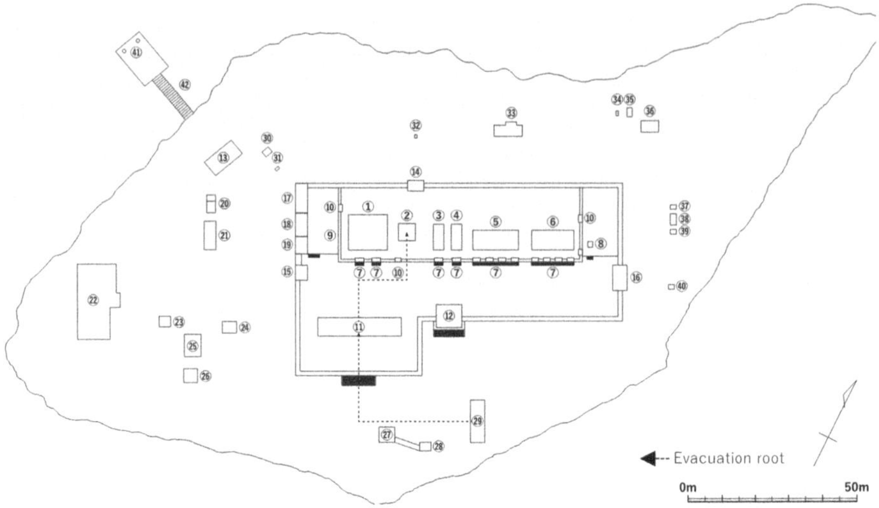

Fig. 12.4 Buildings and gates of Kumano Hongu Taisha in the former site. *Source* Authors

up to the top of the beams and was almost washed away, but the staff managed to survive by clinging to the eaves of the roof and the beams of worship.

As mentioned above, 12 deities (gods), called Kumano Junisho Gongen, were enshrined, and different deities were enshrined in the 13 subordinate shrines inside and outside the precincts. In addition, there were several gates, stages, places of worship, and offices in the 42 buildings in Oyunohara (the original location) (Figs. 12.3 and 12.4, and Table 12.1). The flood washed away most of them, and only the upper four shrines that escaped being washed away or totally destroyed were enshrined at the current hilltop site, while the remaining four middle shrines and four other shrines were enshrined together in a stone shrine in Oyunohara.

Flood damage extended beyond the main shrine building and attached buildings, with the greatest damage being caused by the many trees that covered the entire former shrine ground, which served as the guardian forest. From the newly discovered documentary materials (1890) during this research, it was confirmed that, as of June 8, 1899, there were 46 live trees, 71 fallen trees, and 348 standing dead trees.

12.3.2 Selection of the Relocation Site and Preparation of the New Site

The field survey conducted as the initial stage of the relocation work plan included (1) damage investigation conducted several times from September to October 1889 after the flood, (2) surveys to select buildings and areas for restoration in December of the same year, and (3) surveys to select a new relocation site in January of the

Table 12.1 List of buildings and gates and their damage level

No	Building	D/T	No	Building	D/T	No	Building	D/T
1	First/second god halls	D	15	West gate	T	29	Office	T
2	Ritual dance stage	T	16	East gate	T	30	Purification hall	T
3	3rd god hall	D	17	Poetry hall	T	31	Local shrine	T
4	4th god hall	D	18	Offering hall	T	32	Local shrine	T
5	Mid four gods halls	T	19	Music hall	T	33	Guardhouse	T
6	Low four Gods halls	T	20	Washroom	T	34	Local shrine	T
7	Gate A	T	21	Guardhouse	T	35	Local shrine	T
8	Gods halls	T	22	Meeting space	T	36	Resting room	T
9	Fence/wall	T	23	Kitchen	T	37	Worship hall	T
10	Small gate	T	24	Ritual dance stage	T	38	4 gods local shrines	T
11	Prayer hall	T	25	Treasure house	T	39	Worship hall	T
12	Middle gate	T	26	Library	T	40	Worship hall	T
13	Gate B	T	27	Stage	T	41	Torii gate	T
14	Gate C	T	28	Dance stage	T	42	Bridge	T

D: Damaged, T: Totally destroyed

following year 1890. Thereafter, specifications for the relocation and rebuilding work were prepared around May 1890, based on the survey results and drawings drawn before the flood.

(1) Damage investigations

The shrine office diary contains descriptions of the damage investigation in September and October 1889. Government officials of the Ministry of the Interior and Ministry of Finance Damage, prefectural officials such as the Wakayama Prefectural Governor and the Nara Prefectural Police Chief, and county and village officials such as the District clerks and Hongu Village Police inspectors joined the investigations to understand flood damage. In other words, the purpose of this damage investigation was to understand the actual flood damage situation and simultaneously collect information to discuss whether relocation work should be approved as a national project. In short, this is considered the basis for determining emergency maintenance expenses.

(2) Surveys to select buildings and areas for restoration and selection of new sites

Surveys to select buildings and areas for restoration were conducted by Wakayama Prefectural Secretary, one official, one civil engineering section member, and two technicians from December 8th to 14th, 1889. A survey to select a new relocation site was also conducted on the place known as the "Nakahirano," which was a candidate site for relocation. Then, the director of the Shrines Division of the Shrines and Temples Bureau of the Ministry of the Interior Director of the Wakayama Prefectural Office also visited the site of another candidate site called "Haraido" from January 19th to 21st, 1890.

Regarding the candidate sites where the territorial survey was conducted, based on the interviews, it was found that there was another candidate known as "Shimomukai." Takahiro is the land northwest of the former shrine site, and "Haraido" is the land north of the former shrine site, which is where the current shrine is located. "Shimomukai" is the land northeast of the former shrine site, which is currently used as residential and farmland. All these candidate sites are river terraces with an altitude of 105 m to 107 m, and they were planned on high ground where flood damage can be minimized, but in the end, the site was only approximately two-thirds the size of the former shrine site.

According to interviews, "Haraido" was selected because the land is along the Kumano pilgrimage route, and "Haraido Tenjin Shrine" is nearby. It is possible to visit Kumano Hongu Taisha along the same route. Also, it was thought that the former shrine site of "Oyunohara", can be seen to the south from the site, which was also a factor in the selection of the "Haraido".

(3) Purchase of the new relocation site

According to the shrine office diaries from April and May 1890, instructions were received from the person in charge of the civil engineering division of the prefectural office regarding the method of purchasing new shrine land. Subsequently, the purchase of land or requests from the Hongu villagers to donate land and money for the purchase were sent to Kumano Hongu Taisha.

(4) Civil works

Bidding for civil engineering work began in April 1890, and the bids were awarded in May of the following year. Several people submitted bids for civil engineering work at the new shrine site, and one of them, a resident of Hongu Village and a civil engineer won the bid. The Civil Engineering Division of the Wakayama Prefectural Office handled the bidding for the civil engineering construction. Plans and cross-sectional plans were drawn from the existing materials, and civil engineering work was conducted on land (approximately 9,000 m^2, 99 m wide × 90 m deep) consisting of the upper, middle, and lower tiers where the relocated structure was constructed.

From the above, it is clear that several candidate sites for a new shrine on high ground were considered and that the current site was selected based on local conditions, direction, and layout. This study clarifies the civil engineering work process by stating that a local contractor is selected through bidding and engaged in civil work.

12.3.3 Rebuilding the Shrine

Regarding the bidding for shrine-building relocation and rebuilding works, specifications were prepared in May 1890 by the Civil Engineering Division of the Wakayama Prefectural Office, and on-site explanations regarding the works were held on June 8th by four contractors from Tokyo, Osaka, and Wakayama. On June 23rd, a construction contractor from Tokyo won the bid.

First, a ceremony to temporarily relocate the deities (gods) was held on August 4, 1890, and the demolition of the main shrine began during the relocation of the shrine building. Simultaneously, as the location of the worship hall, treasury, sacrificial repository, and shrine office changed, the foundation changed from Kakezukuri (a structure fixed with some pillars to support the floor) to a stone wall foundation.

Figure 12.5 and Table 12.2 show the building arrangement of the newly relocated site after the completion of relocation and rebuilding. Figure 12.6 shows the implementation schedule for the relocation and rebuilding of Kumano Hongu Taisha. Figure 12.7 shows the current Kumano Hongu Taisha. Photographs of the remaining flood damage are shown in Fig. 12.8.

(1) Reuse of original shrine building material

The reuse of original building materials during the construction or rebuilding of shrines is common, including the Shikinen Sengu of Ise Jingu and the Kasuga Taisha Shrine. Based on the records of the "Contract Specifications" prepared at the planning stage of the relocation work and the "Construction Cost Settlement Book" made after the relocation work was completed, the reuse of the original building materials was already confirmed during the relocation and rebuilding of Kumano Hongu Taisha. The reuse of the original building materials varied depending on the extent of damage to the building.

For example, regarding timbers for the first and second deities' (god) halls (two combined as one building), the contract specifications specify that new cypress wood should be used for washed-away parts, and that the original timber should be used only when good-quality timber is available (with the exception of the two climbing railings). In addition, although the principle is to replace decayed parts with new materials, the directive also calls for original materials to be used for handrail coverings. Regarding the parts in which new materials were actually used, in the "Construction cost settlement book," it can be confirmed that 120 new Japanese cypress trees were used for the railing of hallways and the main deck, and 70 new cedar trees were used for under the roof materials. It can be seen that new materials, such as cypress and cedar, were used for other parts. Therefore, it can be said that many new materials were used for the washed away and decayed parts in the relocation and rebuilding of the first and second deities (god) halls.

Regarding stone materials, the contract specifications state that all stones attached to shrine buildings, such as gravel and curb stones, should be made from the original materials, and if severely damaged, they should be recycled and reused. Furthermore, in the "Construction cost settlement book," the cost of purchasing stones was not

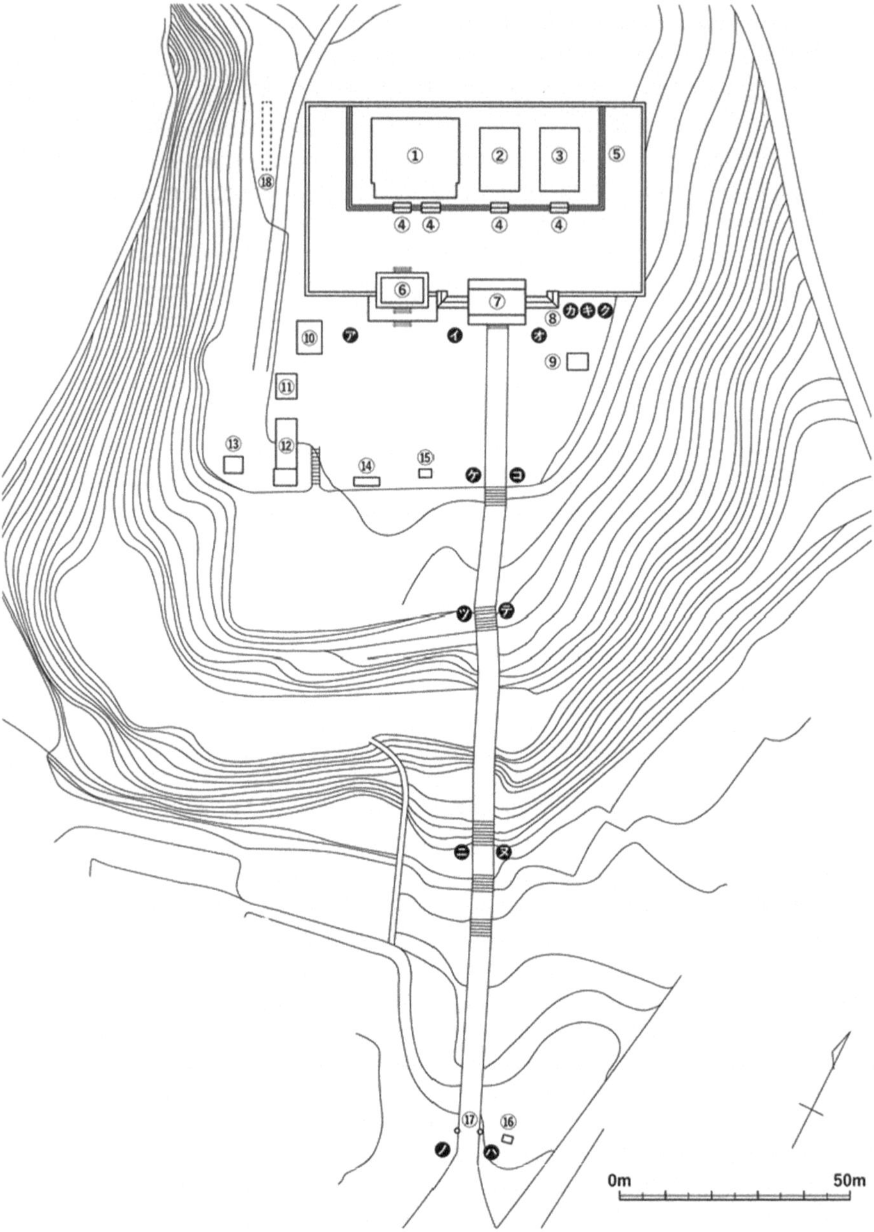

Fig. 12.5 Buildings and gates of Kumano Hongu Taisha in 1893. *Source* Authors

Table 12.2 List of buildings and gates in 1893

No	Building	No	Building
1	1st/2nd gods halls	10	Treasure house
2	3rd god hall	11	Storage of utensils for festivals
3	4th god hall	12	Office
4	Gate A	13	Storage house
5	Mid four gods halls	14	Place for visitors/prayers
6	Worship hall	15	Washing place
7	Gate B	16	Announcement board
8	Fence/wall	17	Torii gate
9	Divine food place	18	Place for ritual utensils
●	Stone lantern		

recorded, but only the cost of transporting the stones, so it can be said that all the stones used in the reconstruction of the first and second deities (god) halls were reused from original materials.

In addition, although some newly constructed gates were made entirely of new materials, the original materials were also used in many buildings, such as treasury houses, worship halls, and torii gates.

This survey also revealed that many of the people of Hongu-cho were engaged in a process called "Haiarai" to remove dirt from the wood using ash, as the original timbers of the flood-damaged shrine were covered with mud. This ash-washing process allowed for the reuse of many original timbers, and the contributions of the local people are also worth recording.

(2) Preparation of specifications and contractor selection

As mentioned above, a survey of materials revealed that a repair survey was conducted in December 1889, led by personnel from Wakayama Prefecture, and the specifications were prepared by May 1890. Regarding the shrine building relocation and rebuilding work, a site briefing session was held for four people from Tokyo, Osaka, and Wakayama, and on June 23, 1890, a construction contractor from Tokyo was selected for the work.

The contracting company responsible for shrine rebuilding work was subcontracted to "Miyadaiku," a shrine carpenter, to work with more than 20 workers from Wakayama City. Construction began on July 8, 1890. The work to relocate the shrine building was completed on March 22, 1891, one week before the Sengu Ceremony was held on March 29.

(3) Rebuilding

As shown in Fig. 12.5 and Table 12.2, together with the rebuilding of the 1st and the 2nd, the 3rd gods', and the 4th gods' hall, gate A, fence/wall, treasury house, tori gate, and other facilities, 18 buildings and facilities were newly built

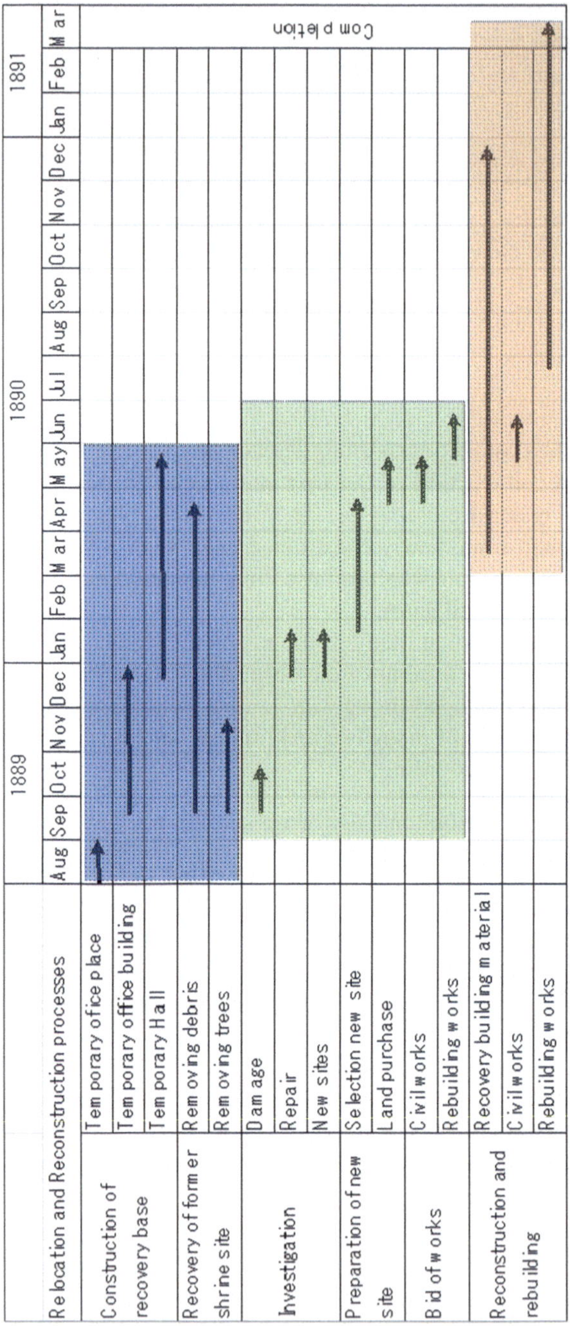

Fig. 12.6 Implementation schedule for relocation and rebuilding works. *Source* Authors

Fig. 12.7 Kumano Hongu Taisha. *Source* Kumano Hongu Taisha

Fig. 12.8 Flood damage remains on the building. *Source* Kumano Hongu Taisha

and reconstructed in total. The four main shrine buildings were demolished and rebuilt, while repairing parts of the building that had been washed away, so their planar structure hardly changed from what it was before the flood. Additionally, stone lanterns were unearthed and replaced at several sites.

It is considered impossible for a team of 20 carpenters and construction workers to complete such a large-scale relocation and rebuilding project in just eight months. By organizing the number of constructors for each "building recorded in the construction cost settlement book" by carpenter, woodturner, stone transporter/erector, timber transporter/carpenter's helper, ash washer, and other workers, the largest number of people involved in the enforcement of the first and second halls is 7,180. In addition, the numbers of people involved in the building of the torii gate, the former shrine shite's torii gate, and the third and fourth halls were almost the same, suggesting that carpenters, woodcutters, and various other workers were also responsible for the rebuilding or building each structure.

12.4 Reasons for the Quick Relocation and Rebuilding

Decision-making, financing, contractor selection, and the cooperation of many people are essential for the quick relocation and rebuilding of shrines. Here, we provide an overview of the procedure. Figure 12.9 presents an overview of the related parties and their relationships.

12.4.1 Financial Arrangement

The relocation and rebuilding of Kumano Hongu Taisha began with the submission of a reconstruction application to the Wakayama Prefectural Office on September 2, approximately two weeks after the flood damage. After several surveys were conducted and a reconstruction plan was prepared, on March 29, 1890, the Ministry of the Interior approved relocation and rebuilding and decided to provide the full amount of reconstruction funds; thus, relocation work began immediately. In other words, in the rebuilding of shrines in the early modern period, as explained in the previous section, the presence or absence of financial support from the shogunate and the degree to which shrine families were dispersed because of disasters had a significant influence on the period and the amount of funds raised for reconstruction. As the government provided all the funds for relocation and rebuilding, the Kumano Hongu Taisha Shrine was able to begin relocation work without wasting any time raising funds. This full funding for reconstruction was carried out in accordance

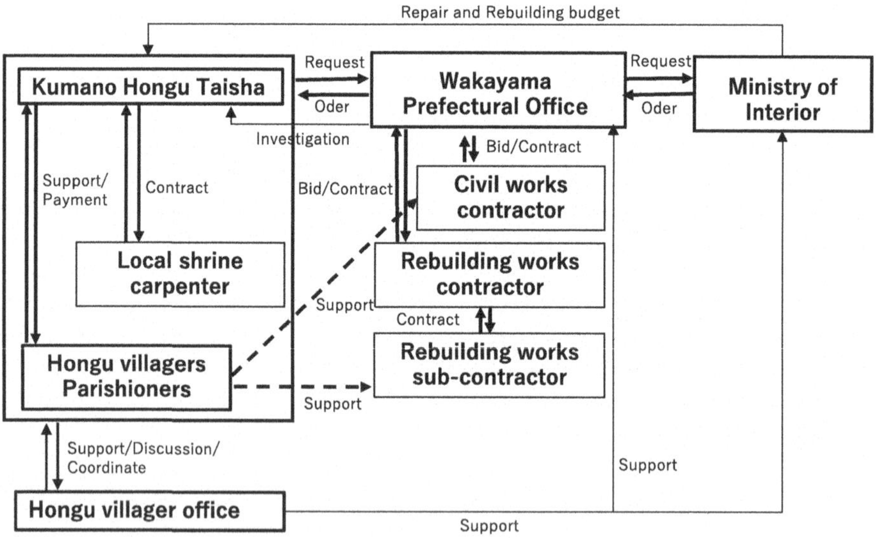

Fig. 12.9 Overview of related parties and their relationships

with the modern shrine maintenance system following the Separation of Shinto and Buddhism ordinance issued in March 1868 and is thought to have had a great influence on early relocation and rebuilding.

Since the modern era, each shrine in Japan has had its own rank, as determined by the modern shrine rank system. This system, promulgated in 1871, hierarchically classifies shrines. Under this system, Kumano Hongu Taisha was ranked as a "Kokuheichuusha," a national shrine enshrined by local government officials. The national treasury provides funds for annual festivals and government funds for shrine expenses.

From 1887, when this fund system began, until November 1890, when the ceiling limit for temporary maintenance expenses was stipulated, there was no upper limit on temporary maintenance expenses in the event of a disaster. Since Kumano Hongu Taisha received reconstruction permission from the Ministry of the Interior in March 1890, the shrine supplementary costs were paid by the National Treasury as temporary maintenance costs. Consequently, the relocation and rebuilding of Kumano Hongu Taisha was made possible as a national project and all funds could be raised immediately from the national treasury.

12.4.2 *Involvements of Different Offices and Contractors*

Among the administrative agencies, the Ministry of the Interior, as the decision-making body for relocation and rebuilding, conducts damage investigations and territorial surveys to approve the relocation work. The Wakayama Prefectural Office, as the construction supervisory authority for relocation and rebuilding, conducted all investigations and prepared specifications and bids for construction and rebuilding. The civil engineering division of the Wakayama Prefectural Office was also responsible for the planning and supervision of on-site construction during the construction and rebuilding stages.

Next, the contractors were divided into civil works, shrine relocation, and shrine rebuilding. A contractor from Hongu village was responsible for the civil works, and a contractor from Tokyo won the bid and subcontracted to a shrine carpenter with more than 20 workers in Wakayama City for shrine relocation work. The contractors from Wakayama City who were responsible for the on-site construction of the shrine building relocation and rebuilding work were made up of specialized workers such as carpenters, roofers, metal workers, plasterers, and civil workers, many of whom were skilled craftsmen and were involved in Wakayama castle restoration work.

12.4.3 *Participation of Local People*

The people of Hongu Village, who were victims of the flood as well as the shrine family of Kumano Hongu Taisha, were divided into Hongu Shrine carpenters and

Kumano Hongu Taisha parishioners. During the flood recovery stage, both participated in constructing temporary shrine buildings and removing the earth and sand from them.

It can be inferred that Kumano Hongu Taisha carpenters, Hongu parishioners, and other Hongu villagers were also counted as workers/laborers during shrine relocation and rebuilding and were engaged in the work. For the roles they played, they were paid wages by Kumano Hongu Taisha in accordance with reasonable working conditions, and it is easy to infer that these wages were a source of livelihood for the Hongu villagers after the flood. Furthermore, it can be said that mutual support existed between the people of Hongu Village and Kumano Hongu Taisha.

12.5 Conclusion

The decision to relocate and rebuild the Kumano Hongu Taisha, most buildings were destroyed and washed away by a major flood in 1889, to higher ground was a major project that involved not only Kumano Hongu Taisha and local people, but also the Ministry of the Interior and Wakayama Prefecture.

Until now, the details of relocation and rebuilding have been limited to the architectural details of the main shrine buildings. However, the details of the processes and decision-making have not been well investigated. This study contributed to clarifying various facts by carefully deciphering documents that had been stored in Kumano Hongu Taisha but had not been made public.

The factors that led to the early relocation and rebuilding of Kumano Hongu Taisha during the Meiji period can be broadly divided into institutional, architectural, and organizational factors.

First, regarding institutional factors, the national system to conserve shrines above a certain level, which is a modern shrine maintenance system, allowed the preparation of a budget for relocation and rebuilding after the disaster. This system helped shorten the period of gathering funding to realize the process.

As for architectural factors, the reuse of original construction materials (timber and stones) was a major factor. It was possible to procure construction materials in a short period of time while reducing the cost of procuring new materials.

The reuse of original construction materials (timber and stones) was a major architectural factor. Construction materials can be procured in a short period, while reducing the cost of procuring new materials.

Regarding organizational factors, the construction cooperation of the local people of Hongu Village and the organized contract system managed by the Wakayama Prefectural Office Civil Engineering Division made large contributions to efficient work. The local people of Hongu Village assist with the relocation work by donating or selling land for the new shrine site, procuring timber materials, and making advance preparations for on-site construction. The Wakayama Prefectural Office prepared the specification documents, contracted a major contractor in Tokyo, and supervised the work.

In addition, many other people involved, including parishioners, the Hongu Village hall, and donors, helped to quickly rebuild the shrine. Plans have also been proposed to rebuild other shrines that were not rebuilt in the area. Later, stone stairs were built to climb the current hill, and the Kumano Hongu Taisha assumed its current form.

Religious facilities have attracted people's faith and have been maintained and inherited since ancient times. However, there has not been much research on the reconstruction or relocation of religious facilities after disasters. Through the case study of the relocation and reconstruction of religious facilities, this paper has demonstrated the importance of cooperation and support not only with local residents but also with the national and local governments.

Hongu Taisha Shrine is a UNESCO World Heritage Site that attracts many worshipers and tourists. This study shows that the efforts and skillful work of many people involved made it possible to relocate and rebuild Kumano Hongu Taisha, but its value needs to be recognized along with its history of flood damage, relocation, and reconstruction.

Notes

(1) A shrine is a facility that enshrines the gods of Shinto, an ancient belief in Japan. Shinto is a so-called polytheistic religion, which is a religion of nature, and many gods called "Yaoyorozu no Kami", are enshrined at each shrine.
(2) The Shikinen Sengu (Shikinen rebuilding) is the largest ceremony at a shrine, where shrine buildings are periodically rebuilt and relocated, and the costumes and sacred treasures offered to the gods are also updated.
(3) The word "reconstruction" is used for the general translation of laws and regulations. In addition, when civil works are involved, "reconstruction" is used. Otherwise, this study uses the term "relocation and rebuilding."

References

Hongu Town history compilation committee (2004) Hongu town history, Hongu town

Mifune T, Tamura S (2012) Kumano Hongu Taisha first and second hall and other two building repair work report, center for cultural properties in Wakayama prefecture

Miura Y (2017) The restration of Chikurin-Ji main hall during the Meiji era. J Arch Plan 82(733):767–774. https://doi.org/10.3130/aija.82.767

Obayashi J (2005) Jyuyoubunkazai niokeru ichikutosonoriyuu (Relocation and its reasons for important cultural properties). Grant-in-Aid for scientific research (KAKENHI) (B) Year 2002–2004, pp 201–219

Ogashiwa N (2020) Characteristics of the reconstruction process of the Shiga-In Temple built in the Meiji period. J Arch Plan 85(773):1547–1555. https://doi.org/10.3130/aija.85.1547

Sato M (1975) On the arrangement of the three Kumano shrines from the latter half of the Heian period to the early middle ages. Report of Architecture Planning vol 236

Tani S (1959) Kyuukokuheisha niokeru shikinen zotaisei no chousa (Study of the Shikinenzotai system in the former Kankokuheisha). J Arch Hist Shokokusha 27:26–38

Tsunoda M (2005) Isejingu niokeru kozaitenyounoyousou (Aspects of reuse of old materials at Ise Jingu). Report (2) on Grant-in-Aid for scientific research (KAKENHI) (B) Year 2002–2004, 2005.03, pp 27–38

Yamano T (2005) Kasugataisha to Ichiku—Chuseiniokeru Kyuushadennnoyukisaki (Kasuga Taisha and its relocation – the destination of the old shrine building in the middle age. Report (2) on Grant-in-Aid for scientific research (KAKENHI) (B) Year 2002–2004, pp 137–140

Chapter 13
Indicators to Get You Home

Francis Hubbard, Regan Potangaroa, and Suzanne Wilkinson

Abstract This chapter delves into the challenges and significance of selecting reliable data and indicators in the context of disaster-induced displacements. When people are forced to leave their homes due to disastrous events, various entities such as aid organizations, local authorities, and national agencies come together to provide emergency response and support during the recovery phase. These organizations heavily depend on reliable data and established indicators provided by frameworks to guide their actions, assess the effectiveness of their interventions, and monitor progress. These frameworks promote the "Build Back Better" principle, which advocates for a comprehensive and holistic approach to resilience. This means practitioners must fully understand the intricate relationships and dependencies among the indicators within the framework. This understanding is crucial for making informed decisions regarding indicator selection. Unfortunately, the holistic approach has been identified as a significant weakness in the implementation process. The chapter utilizes a hierarchical decomposition algorithm proposed by Christopher Alexander to address this issue. It also draws upon a thematic analysis of 116 reports related to 40 interventions in disaster scenarios. These methodologies contribute to developing a rational process and workflow for constructing and selecting evidence-based indicators, considering the complex nature of displacement and the importance of a coordinated and holistic approach to resilience.

F. Hubbard (✉) · R. Potangaroa · S. Wilkinson
School of Built Environment, University of New Zealand, Wellington, New Zealand
e-mail: f.hubbard.nz@gmail.com

R. Potangaroa
e-mail: r.potangaroa@massey.ac.nz

S. Wilkinson
e-mail: suzanne.wilkinson@aut.ac.nz

C. Ochiai et al. (eds.), *Disaster and Displacement*, Disaster Risk Reduction,
https://doi.org/10.1007/978-981-96-2583-3_13

13.1 Introduction

When disasters cause displacement, humanitarian organizations, local authorities, and national agencies are mobilized to provide essential aid during the emergency response and subsequent recovery phases. These groups lean on their well-established organizational guidelines and frameworks to guide their actions and activities, assess the impact of their work, monitor ongoing developments, and gauge the success of their interventions through the framework's predefined indicators. The increasing occurrence and severity of disasters underscore the necessity for a unified and coordinated approach among all these entities to ensure an effective and efficient response. The goal of this response should be not only to help the affected community recover but also to strengthen its resilience, surpassing its pre-disaster state. This concept is encapsulated in the "Build Back Better" (BBB) principle, which advocates for a holistic approach to resilience, guiding the selection of appropriate indicators.

However, there have been noticeable shortcomings in the practical application of this resilience-building approach. This has led to calls for a more integrated and collaborative effort to enhance the effectiveness of disaster response (UNDRR 2015, 2022; UNDP 2020). This chapter delves into the vital role of indicators, highlighting the importance of reliable data in indicator selection, and introduces an innovative, evidenced-based methodology for this process. This is achieved through a hierarchical decomposition analysis of the 'Build Back Better Framework,' supported by a thematic analysis of 116 reports detailing 40 disaster interventions conducted by a leading international aid organization active in both immediate disaster relief and extended crisis scenarios.

13.2 Dimensions of Disaster Displacement

The consequence of disaster displacement is significant and far-reaching. Presenting numerous immediate and long-term challenges that can profoundly affect the lives of the persons of concern. In the immediate, displaced populations struggle to access necessities such as shelter, clean water, food, healthcare, and sanitation facilities. In the long term, the disruption of social networks, separation from family members, and the potential for trauma and psychological distress can further worsen their situation. Therefore, after any major disaster, there is a swift response. Taking necessary action to provide a platform for aid includes offering housing, psychosocial support, safe water, and sanitation, contact tracing, family reunification, access to healthcare, and the ability to store and prepare food (Noy and duPont 2016). To begin addressing these dimensions requires practitioners to engage with reliable data and navigate comprehensive frameworks to plan their actions and activities.

A framework acts as a standardized, reusable structure, guiding users toward a specific goal. It offers a strategic approach without dictating exact actions. Instead, defining key principles that highlight areas of focus. These principles, fundamental to the framework's logic, provide a solid foundation for decision-making. Frameworks typically follow a systematic structure, starting with broad objectives or "Priorities for Action," which are then dissected into specific themes or standards (Fig. 13.1). By addressing these elements, users can fulfill the overarching principles. Depending on the framework's complexity, additional layers may be added to further disseminate the themes. To facilitate effective implementation, frameworks include a set of indicators. They provide the greatest detail on how to plan interventions that will satisfy the framework's priorities (Bilau et al. 2018).

Indicators are vital, simplifying complex scenarios for better decision-making and spotlighting areas in need of attention. They also measure intervention success, providing insight into an intervention's strengths or weaknesses. Thus, choosing the right indicators is crucial, especially in the fast-paced, unpredictable environment of disasters. There are several types of indicators, such as benchmark indicators, which set baseline standards, and success indicators, which delineate what successful outcomes look like. In disaster displacements, aligning indicators with response and recovery goals is essential, ensuring they cover the necessary dimensions. Reliable data should underpin indicator selection, supporting successful interventions (Brown et al. 2008; Dwyer and Horney 2014).

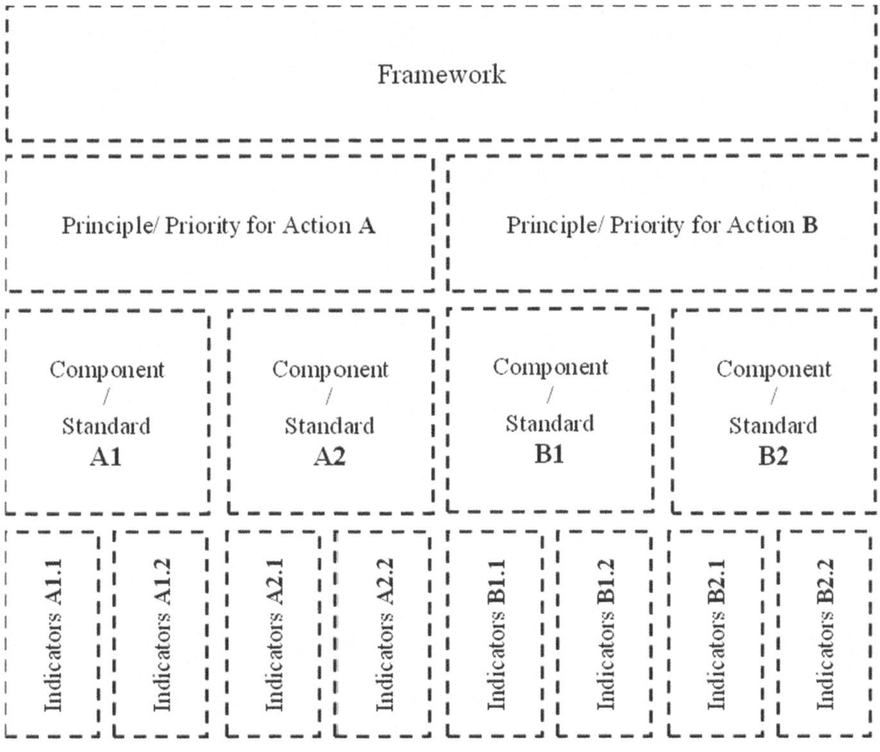

Fig. 13.1 Typical structure of a framework, by author

Reliable data is data that is considered dependable. Verified through field observations and accepted by reference groups such as cluster leads in the humanitarian action framework. Cluster leads are specific agencies appointed for their expertise in particular humanitarian sectors, and they play a vital role in coordinating various humanitarian actors, disseminating crucial information, and ensuring accountability in aid delivery. This cluster approach enhances collaboration, minimizes duplicated efforts, ensures clarity in leadership roles during emergencies, and emphasizes local involvement to ensure responses are contextually relevant and sustainable (IASC 2006).

Reliable data empowers practitioners to accurately evaluate the extent of displacement, discern the needs of the affected communities, and focus their response efforts effectively, providing insights into demographics, vulnerabilities, and specific needs. This enables the implementation of tailored interventions through carefully selecting appropriate indicators. Thus, accurate and reliable data forms the foundation for response efforts, guiding ongoing resource allocation, monitoring and evaluating recovery progress, measuring outcomes, identifying gaps, and justifying continuous interventions. However, collecting and analyzing reliable data in displacement scenarios is challenging due to factors such as access constraints, limited resources, and disrupted infrastructure–all characteristic of the rapidly changing environments in disaster situations. Inadequate or inaccurate data can lead to misguided response efforts and the misallocation of resources, exacerbating vulnerabilities and prolonging the suffering of displaced populations (Shalash et al. 2021).

13.3 Existing Frameworks and Methods of Indicator Selection

Aid organizations mobilize both nationally and internationally to disasters, aiding displaced populations. Working on the ground, they assess needs, coordinate with other actors, and ensure a swift and effective response. Aid organizations collaborate with local authorities and national agencies to coordinate their efforts and avoid duplication of services. In addition to their own organizational guidelines and frameworks, they adhere to coordinated frameworks such as the Sphere Handbook.

The Sphere Handbook was initially developed in 1997 to enhance the quality of humanitarian assistance and hold responding agencies accountable. This handbook has undergone several revisions, with updates in 2000, 2004, 2011, and 2018, each time drawing on the expertise of professionals, NGOs, governments, and United Nations agencies. The most recent edition of the Sphere Handbook, "Humanitarian Charter and Minimum Standards in Disaster Response," introduces four technical chapters, each divided into various components and standards. Every standard outlined in the handbook comes with key indicators, which are quantitative measures for minimum targets, as well as key actions, providing practical steps towards achieving these standards. These actions are akin to success indicators. In total, the framework encompasses 53 standards and 249 key actions (Sphere Association 2018).

The Sphere Association includes 47 member organizations, varying in size and reach. These members include large-scale entities like the International Federation of the Red Cross, which has 189 national societies globally, and Oxfam International, a confederation of 19 organizations operating in 90 countries. Additionally, smaller independent non-profits, such as the ACT alliance with its network of 152 churches, are also part of the Sphere Association (Sphere 2022).

In regard to indicator selection, the Sphere handbook acknowledges that it is not necessary to satisfy every key indicator across all standards. Stating that as humanitarian needs span across various sectors, a holistic approach is essential for an effective humanitarian response, requiring different sectors to work in tandem and collaborate extensively. The Handbook provides cross-references, and readers are encouraged to acquaint themselves with all its chapters when selecting indicators.

Other coordinated frameworks have been developed between aid agencies. One example is the "Characteristics of a Disaster Resilient Community: A Guidance Note" by Twigg. This framework, commissioned by six NGOs, was developed through extensive desk research and with contributions from prominent experts in the field. The primary objective of this document was to offer supplementary guidance for effectively implementing the Hyogo Framework for Action—the world's inaugural comprehensive strategy outlining the roles and responsibilities of various sectors and stakeholders in mitigating disaster losses (Twigg 2007; UNDRR 2007).

The guidance notes outline a vision of an ideal disaster-resilient community, specifying the critical elements necessary for its realization. Each of these elements is defined by distinct characteristics that depict resilience at the community level, complemented by features of a supportive environment that fosters and sustains these resilient traits. Though the characteristics are not explicitly labeled as indicators, they serve a similar function to success indicators (Twigg 2007). In response to practitioner feedback and insights gathered from case studies, the guidance note's second edition, published in 2009, introduced several updates and revisions. This edition proposed a comprehensive set of 28 components, encompassing a total of 167 unique characteristics, further refining the framework and enhancing its applicability in diverse contexts (Twigg 2009).

Twigg provides insights on the process of choosing appropriate indicators, noting that their placement within a single section of the framework might not be optimal, given resilience's holistic nature. He explains that the framework's structure is designed to help users navigate through its numerous parts, but functionally, indicators might be relevant to more than one thematic area or component of the framework. Twigg cautions that there is a risk inherent in any framework of isolating different elements too much and consequently ignoring the connections between them. It's crucial to remain aware of these interconnections across various themes and components to maintain a holistic understanding and application of resilience throughout the indicator selection process.

In 2015, the Sendai Framework was adopted at the Third UN World Conference on Disaster Risk Reduction. It introduced seven global targets, four priorities for action, and 13 guiding principles alongside 91 unique indicators to consider for implementation at local, national, regional, and global levels.

The Sendai Framework significantly changed from its predecessor, the Hyogo Framework for Action (HFA). Firstly, it shifted the focus from disaster management to risk management, emphasizing its application across all sectors. Secondly, it included considerations of man-made, technological, and environmental hazards alongside natural ones. Thirdly, it acknowledged the major gaps in establishing and selecting appropriate indicators at the local level within the HFA, posing significant challenges for decision-makers. As a response, the 'Words into Action' (WiA) series was established alongside the Sendai Frameworks.

The WiA guidelines provide practical and specific advice on implementing a people-centred approach to disaster risk reduction in line with the Sendai Framework. There are currently fourteen guides in the series, published between 2017 and 2022. The guidelines are designed to be flexible, allowing users to focus on relevant sections based on their interests and needs. Each guide discusses its relevance to specific targets or priorities within the framework and frequently references additional material. Some guides present clear process and target indicators, while others do not. Introducing an element of flexibility within the indicator selection process. However, this also adds a level of complexity to explicitly understanding the indicators of the Sendai Framework (UNDRR n.d).

Additional consideration must also be given to the Sustainable Development Goals (SDGs), established in 2015. Sustainable development and disaster risk reduction are closely related, as the impacts of disasters can hinder development. Therefore, there is significant collaboration between these two policy instruments. The Sendai Framework is intended to complement three of the 17 SDG goals comprising 11 additional specific indicators. However, the relationship, dependencies, and priority of these additional indicators within the Sendai Framework are not made obvious (UNDRR 2016).

When examining these frameworks, it becomes apparent that their scale is significant. Not only are there numerous organizations and agencies that adhere to these frameworks, but there is also a substantial number of indicators. This complexity is heightened by the fact that these frameworks promote a holistic approach to resilience. Implying that there are intricate relationships and dependencies between the indicators throughout the entire framework, and success depends on understanding and acknowledging these connections. In other words, each component cannot be considered in isolation; it must be viewed in relation to the others that make up the complete framework. As a result, the number of potential relationships to consider within a framework is proportional to the number of indicators squared, resulting in thousands or even tens of thousands of possibilities to be considered.

A notable limitation shared by all these frameworks is their lack of evidence-based relationships and dependencies between indicators, which would better inform a selection process. Currently, the selection process depends on the practitioner's judgment and their interpretation of the interconnectedness, dependencies, and priorities between indicators, applying that knowledge to the situation at hand. It requires sensitivity to holistic principles while maintaining relevance to their work by refining and narrowing down the volume of available indicators to a manageable level (Twigg 2009). This was identified as a fundamental weakness in the implementation of the HFA (UNDRR 2015). Furthermore, the consequences of poor indicator selection were underlined in 2020 when the 'Humanitarian Development Report' highlighted that engaging with the wrong indicator leads to a misdiagnosis of a situation and, consequently, the wrong actions (UNDP 2020). This concern has been echoed in the 2022 Global Assessment Report, suggesting that sub-optimal decision-making after a disaster remains a major obstacle to the success of the Sendai Framework's goals (UNDRR 2022).

13.4 Build Back Better and a Holistic Approach to Resilience

Holism, as an approach towards reliance, emerged in the early 2000s and was driven by the concept of 'Build Back Better' (BBB). BBB encapsulated a comprehensive goal: improving a community's physical, social, environmental, and economic conditions throughout the reconstruction phase, aiming for a resilient and sustainable comeback (Geis 2000; Monday 2002; Mannakkara and Wilkinson 2014; Khasalamwa 2009; Clinton 2006; Venable 2020). Jacquelyn Monday's paper was pivotal in this shift, presenting 'building back better' as a revolutionary change in how we mitigate future natural hazards. In her 2002 article, she advocated for a unified strategy, marrying vulnerability reduction, environmental sustainability, and social resilience within an overarching, holistic framework. Although these ideas weren't new, the BBB terminology made a profound impact, with its memorable alliteration adding emphasis and capturing attention.

The 2004 Indian Ocean Earthquake and Tsunami served as a critical point for BBB, necessitating an exceptional global response. Over 90 governments, 40 NGOs, and more than 500 independent organizations rallied together, a collective effort that likely propelled BBB into mainstream post-disaster recovery discussions (Kennedy et al. 2008; CARE International 2014). The concept gained further traction through William (Bill) Clinton's report as the Special Envoy of the United Nations Secretary-General. His publication, 'Lessons Learned from Tsunami Recovery: Key Propositions for Building Back Better,' provided early insights and recommendations on the global disaster response (Clinton 2006; Maly 2017; Fernandez and Ahmed 2019; Vahanvati and Rafliana 2019), establishing itself as a cornerstone text for BBB.

Growing in popularity, the concept of BBB was adopted by various organizations. Some welcomed its comprehensive themes in their guidelines, while others interpreted it narrowly, focusing solely on physical infrastructure restoration and often neglecting the social and economic aspects of recovery (Schilderman 2010;

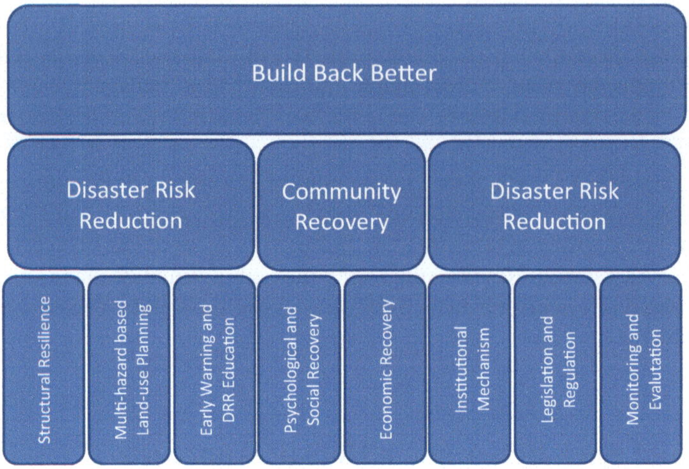

Fig. 13.2 'Build back better' framework (Mannakkara and Wilkinson 2016)

Mannakkara et al. 2018). This discrepancy has led to diverse messages being associated with the BBB concept that aim to be more specific in their focus. However, many alternatives, including the popularised 'Build Back Safer,' miss the significance of the core concepts related to BBB (Kennedy et al. 2008).

In response to this confusion around BBB principles, Researcher S. Mannakkara synthesized leading BBB frameworks in a PhD study at Auckland University in 2014. The result was the articulation of the core principles that contribute to an effective BBB response, producing a more unified and practical BBB framework. The shared core concepts from seven frameworks were analyzed. The findings were three broad categories: 'Disaster Risk Reduction,' 'Community Recovery,' and 'Effective Implementation.' An updated version of this framework in 2016 accompanied these categories (or principles) with eight components, as seen in Fig. 13.2 below. In 2018 this study was expanded upon further and a set of 87 indicators were published (Mannakkara and Wilkinson 2014; Mannakkara et. al 2018).

13.5 Selecting Appropriate Indicators

Developed by Christopher Alexander in 1962, the Hierarchical Decomposition System (HIDECs) is a methodology designed to understand the complex relationships and requirements within a community, using a small village in rural India as its initial application. Inspired by the Athens Charter of CIAM from 1929, which classified the city's functions into work, dwelling, recreation, and transportation, Alexander further defined the concept by treating a city as a living assembly with intricate component relationships. He highlighted the blurred lines between functions like dwelling and recreation, emphasizing the need for a nuanced understanding of component requirements and their interconnections to inform design and implementation (Alexander 1968, 2015).

In the original application, Alexander identified 141 village requirements through various sources, including villagers' expressed needs, national and regional standards, economic drivers, social purposes, and existing implicit relationships. He then mapped out the connections between these requirements, based on immediate perceptions of their interconnections. The process culminated in an algorithm that organized these relationships into a hierarchical structure, facilitating a comprehensive and holistic understanding of the community's needs (Alexander and Manheim 1962). This hierarchical analysis preserves the integrity of existing requirements while providing a rational basis for action and decision-making. The HIDECs methodology involves identifying and coding key variables, creating a matrix, identifying connections, and applying the algorithm to create a hierarchical structure.

This analysis is performed using the 'Build Back Better' framework, substituting its 87 indicators with the requisite variables or requirements. The ensuing matrix highlights the interrelationships among the indicators across various components. The HIDEC algorithm is subsequently applied to delineate relationships and dependencies among these indicators. This methodology ensures the preservation of the vital holistic aspect while simultaneously providing an evidence-based approach. This particular framework and its associated indicators were chosen as an example to demonstrate this methodology because they collectively embody a synthesized representation of fundamental components from various frameworks, encompassing perspectives from aid organizations, local entities, and national authorities, all addressing the issue of disaster displacement.

Figure 13.3 presents a sample portion of the matrix where the 87 indicators are assigned numerical codes ranging from 1 to 87. These indicators are categorically grouped according to the thematic classifications specified by the 'Build Back Better' framework. They fall under one of the nine core principles and are further categorized into three main components: "Disaster Risk Reduction," "Community Recovery," and "Effective Implementation." This list is replicated on the matrix's left side. Drawing from the list of indicators detailed in 'Resilient Post Disaster Recovery Through Building Back Better' (Mannakkara et al. 2018), the analysis seeks to identify explicit relationships between pairs of indicators, denoted by an 'x' marking their connection. Given the bidirectional nature of these connections, a diagonal symmetry is observed across the matrix, highlighted by the shaded blue section.

The numerical codes representing the indicators, along with their identified relationships, are then inputted into the HIDEC algorithm. Based on Christopher Alexander's theory, this algorithm generates a hierarchical, family tree-like structure, depicted in Fig. 13.4. Interpretation of this diagram begins at the lowest 'generational tier,' where each of the 22 subsets is clearly labelled. An evaluation of the indicators within these subsets takes place to ascertain a central theme, and subsets with similar themes may be grouped into clusters. This evaluative and clustering procedure is replicated across all generational tiers until distinct themes for each are

Category →				k Reduction (DR)												
Principle →				Multi-hazard-based land-use planning (2)									Early Warning and DRR Education (3)			
Component →				Risk-Based Zoning (d.)				Resettlement (e.)					Early Warnin g (f.)		DRR Educati on (g.)	
Indicator →				13	14	15	16	17	18	19	20	21	22	23	24	25
Disaster Risk Re	Multi-hazard-based land-use planning	Risk-Based Zoning	13		x	x	x		x			x	x		x	
			14	x		x	x	x	x	x	x	x	x	x		
			15	x	x		x	x	x	x	x	x			x	
			16	x	x	x			x			x		x		
		Resettlement	17		x	x			x	x	x	x				
			18	x	x	x	x	x		x	x	x		x	x	
			19		x	x		x	x		x	x				
			20		x	x		x	x	x		x				
			21	x	x	x	x	x	x	x	x				x	
	Early Warning and DRR Education	Early Warnin g	22	x	x					x				x	x	x
			23		x		x		x	x			x		x	x
		DRR Educati on	24	x		x			x	x		x	x	x		x
			25										x	x	x	

Fig. 13.3 Sample section of analysis matrix of connection between the 87 indicators of the 'build back better framework'

identified. Consequently, the initial 22 groups can be condensed into seven clusters, as illustrated in Fig. 13.4.

Figure 13.5 displays the post-analysis structure of the HIDEC, highlighting seven distinct themes that correspond to seven clusters at the lowest tier. Building upon this base are two additional tiers, culminating in the central interconnected factor: "engage with existing local organizations to support planning, implementation, and management of recovery." When navigating this diagram, practitioners should focus on their specific area of work. For instance, in the realm of risk reduction, a practitioner can refer to the box titled "understanding future risks and existing impacts." This allows them to recognize the connection and dependency between this theme's indicators and those addressing "legislative, governance, and regulation of recovery decisions." Additionally, this theme influences indicators related to "Educating on DRR & promoting lessons learned" and "Improving & expanding on knowledge," as seen in the tier below it. This approach provides practitioners with a clear understanding of

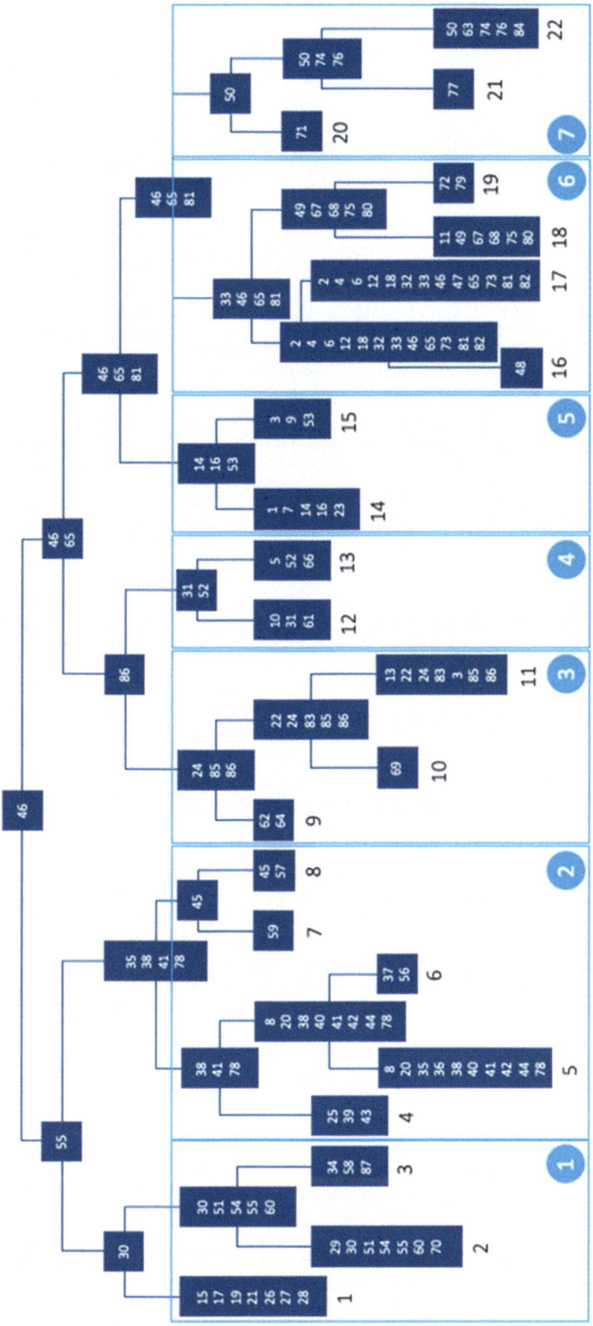

Fig. 13.4 Clusters identified in the HIDECs analysis of 87 indicators of the 'build back better framework'

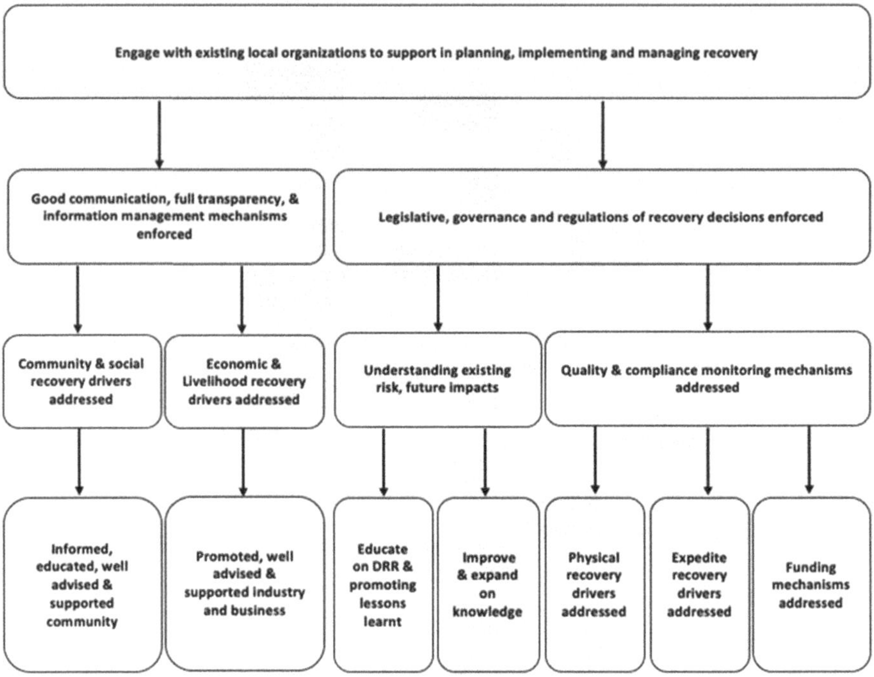

Fig. 13.5 Roadmap diagram for indicator selection

the most influential factors and their dependencies with others, all grounded in an evidence-based strategy.

The diagram acts as a valuable tool for indicator selection, clarifying the intricate relationships between indicators within this framework. It also deconstructs any pre-existing organizational structures applied to the variables, encouraging a comprehensive approach. This newfound awareness ensures that practitioners are well-informed, equipping them with the necessary tools to strategically plan actions and activities in the context of disaster displacement.

13.6 Humans in Humanitarian Deployments

It is vital that this roadmap diagram is not only based on the theoretical side but also rooted in practical application. This section details a thematic analysis of 116 reports, reflecting 40 interventions in disaster scenarios by a leading international aid organization. Their expertise spans a wide array of skills, encompassing disaster risk reduction (DRR), emergency preparedness, recovery, and stabilization. They collaborate with sixteen United Nations (UN) agencies or entities, frontline relief agencies, and local governments. During humanitarian crises, this organization, as a part of a global network of standby partners, extends additional support to the UN's response initiatives. Thereby, they will be operating around frameworks such as Sphere, The Sendai Framework, and the Sustainable Development Goals. Deployment spans from January 2018 to mid-September 2020 in stints between a month to a year. The extensive range of locations, stakeholders, and diverse contexts covered offers a unique and valuable perspective, as seen in Table 13.1.

The reports are divided into three documents. The main one is the End of Deployment Report (EoDR). EoDRs serve as a comprehensive written evaluation in the delegates' own words as they wrap up their assignments. Unlike conventional handover notes that predominantly focus on facts, EoDRs encompass personal reflections, in-depth analysis, and critical evaluations of each deployment. Their overarching aim is duel-faceted: to have a genuine, transparent perspective and to offer suggestions and insights to refine future deployment strategies. Structurally, these reports are segmented into distinct sections, including 'Administration and Deployment Support,' 'working context,' and 'results of deployment.' Each of these sections is further segmented by a series of prompts for the delegates to answer.

Two instrumental documents supplement the insight from EoDR: the Terms of Reference (ToR) and the Personal Evaluation Form (PEF). Drafted by the host organizations ahead of the deployment, the ToR outlines the delegates' anticipated role, the humanitarian response sector, the expected deployment duration, and the operational context. It also delineates responsibilities, thereby theoretically providing a roadmap of the delegate's mandate. Though ToR exhibits variations depending on the requesting entity, their commonality is in the intent to provide a clear focus on the actions and activities of the delegate and, thus, give guidance on the appropriate indicators for the intervention.

The second document, the PEF, provides a platform for the supervisors of the delegates' host organization to share feedback on the deployment. The PEF's structure closely mirrors that of the EoDR, prompting reflection on the success of essential activities, outputs, the impact and sustainability of the deployment, and the delegates' professional competencies. It also welcomes general comments and recommendations. The Significance of the PEF lies in its ability to present the alternative viewpoint, offering insights into successes and challenges from the host organization's perspective. A thematic analysis of the EoDR, ToR, and PEF recognizes the perspectives of both the host organization and the delegates, which is pivotal when discerning the realities of the indicator selection process. The two perspectives offer a more well-rounded understanding of the pattern and trends of the emerging themes from the thematic analysis.

Capacity building consistently emerges as a central theme throughout the reports. For instance, the Terms of Reference (ToR) consistently highlight the importance of capacity building in a delegate's roles and responsibilities. This underscores the critical role of capacity building in empowering partners, staff, and communities, equipping them to effectively address the challenges they face. This emphasis on capacity building is also evident at the End of Deployment Reports (EoDR), where it is recognized as the primary method through which delegates contribute to strengthening host organizations, enhancing national and local leadership, and ensuring the

Table 13.1 Agencies, sectors, roles, and countries encompassed in the reports

	Host Agency	Sector	Role	Country
01	UNHCR	WASH	WASH specialist	Lebanon
02	WFP	Shelter	Project manager/engineer	Myanmar
03	UNHCR	WASH	WASH officer	Zambia
04	UNHCR	WASH	WASH specialist	Iraq
05	UNHCR	Shelter	Shelter officer	Bangladesh
06	UNHCR	Shelter	Shelter officer	Bangladesh
07	FAO	DRM	Resilience expert	Philippines
08	NDMO	Shelter	Structural engineer	Samoa
09	UNHCR	Shelter	Site planner	Bangladesh
10	UNRWA	Shelter	Civil construction	Palestine
11	WFP	Shelter	Civil engineer	Uganda
12	FAO	Agriculture	Emergency programme officer	Ethiopia
13	UNHCR	WASH	WASH officer	Bangladesh
14	NEMO	DRM	DRM specialist	Tonga
15	FAO	Agriculture	Resilience review expert	Fiji
16	UNHCR	Shelter	Shelter officer	Bangladesh
17	UNHCR	Logistics	Coordination officer	Bangladesh
18	NDMO	DRM	DRM technical advisor	Vanuatu
19	UNHCR	Shelter	Associate site planning officer	Nepal
20	UNHCR	Education	Education officer	Uganda
21	UNHCR	MEAL	MEAL officer	Jordan
22	UNHCR	Logistics	Information manager	Uganda
23	UNHCR	Education	Assistant education officer	Jordan
24	UNHCR	Multi-Sector	Executive assistant	Bangladesh
25	UNHCR	DRM	Emergency coordinator	Bangladesh
26	UNHCR	Multi-Sector	Reporting officer	Bangladesh
27	UNHCR	MEAL	Associate MEAL officer	Lebanon
28	NDMO	DRM	Technical advisor	Vanuatu
29	PWD	DRM	Project management specialist	Vanuatu
30	UNRWA	Education	Education reporting officer	Lebanon
31	IOM	Shelter	Shelter manager	Bangladesh
32	NDMO	DRM	Resettlement coordinator	Vanuatu
33	UNICEF	DRM	Engineer adviser	Tonga
34	UNICEF	DRM	DRM specialist	Samoa
35	NDMO	DRM	DRM specialist	Solomon Islands
36	UNICEF	DRM	Emergency manager	Ethiopia
37	UNICEF	Shelter	Structural engineer	Malawi

(continued)

Table 13.1 (continued)

	Host Agency	Sector	Role	Country
38	PMO	DRM	Recovery coordination advisor	Vanuatu
39	PMO	DRM	Recovery coordination advisor	Vanuatu
40	OTP	DRM	DRM advisor	Kiribati

continuity of humanitarian efforts long after the delegates' commitments have ended. Even indirectly, there are consistent references throughout the reports to the delegates' commitment to sharing knowledge and strengthening institutional capacity through their specific technical skills.

Capacity development actions take various forms, including structured training, technical support, mentoring, and the improvement of both physical infrastructure and processes/systems. Furthermore, the Personal Evaluation Forms (PEF) underscore the perception of capacity building as a crucial element for success, particularly from the perspective of the host organizations when reflecting on the delegates' contributions. Taken together, the reports make it clear that capacity building is a multifaceted theme. The theme of capacity building underscores the importance of selecting appropriate indicators to measure and evaluate capacity development efforts comprehensively. The multifaceted nature of capacity building, as revealed in the reports, necessitates a nuanced approach to indicator selection to ensure that all aspects are adequately addressed.

Another prominent theme that emerges is flexibility or adaptability. The Terms of Reference (TOR) play a crucial role in clearly outlining objectives, outputs, and outcomes within a defined timeframe for the delegate to fulfill. Out of the forty interventions, 37 (92.5%) clearly stated their purpose, explaining why a delegate with specified skills or background was sent to assist in situations requiring external support. The strong emphasis on a clear purpose underscores the importance of selecting relevant indicators and understanding the key domains and dimensions to be addressed. It also recognizes that delegates must comprehend the connections to secondary or tertiary indicators to appropriately assess their impact on the overall framework. However, the End of Deployment Reports (EoDRs) did not address this aspect.

Considering the dynamic nature of humanitarian work, changes to the TOR are expected. In preparation for these changes, EoDRs request delegates to comment on any deviations from the TOR, defining a change as approximately 25% or more. Out of the 40 interventions, 15 provided comments on changes during their missions. Only one delegate explicitly mentioned a significant change to their terms of reference, citing a major breakdown in communication between agencies regarding objectives. Staffing issues and team changes were the most common changes reported, followed by the impact of COVID-19 and travel restrictions, which had implications for objectives and outcomes. These findings are significant because the selection of indicators relies on meeting expected outcomes. However, when circumstances change and prevent the achievement of these outcomes, there is a need to know what indicators ought to be considered to get back on track. Practitioners working directly in the field are often the first to observe changes in their area of expertise. It is crucial for them to demonstrate flexibility and readiness to modify their methods in response to these changes. This requirement for adaptability also applies to the indicators they use to inform their actions; these indicators must be updated and adjusted in line with changing data. Consequently, employing a solid, evidence-based strategy for selecting indicators is a vital element for successfully implementing frameworks at the field level. The theme of flexibility and adaptability emphasizes the need for indicators to be flexible enough to accommodate changes in the Terms of Reference (TOR). The dynamic nature of humanitarian work often leads to unexpected shifts in objectives and outcomes, and the indicator selection process must be adaptable to reflect these changes accurately.

The final major theme that emerged was the perception of success in terms of outputs, outcomes, and improvements. All delegates reported on the outputs of their missions in their End of Deployment Reports (EoDRs) to assess progress against target indicators and measure success. Many delegates went beyond their assigned duties, achieving additional outcomes when opportunities presented themselves. However, only one Terms of Reference (TOR) was amended to reflect the additional duties undertaken, aiming to better represent the delegates' success.

Recognizing improvements proved to be more challenging. Reports that provided insights into long-term improvements often came from multiple deployments in prolonged situations, such as refugee camps in Cox's Bazar, Bangladesh. In these cases, subsequent delegates commented on the situation upon their arrival and departure, allowing for the recognition of the impacts of previous missions. This suggests that identifying success indicators for shorter or one-off deployments has yet to be effectively accomplished without follow-up case studies. Understanding dependencies between success indicators could provide a roadmap for delegates to better recognize improvements in short-term or one-off deployments. This theme highlights the challenges in recognizing improvements, especially in shorter or one-off deployments. Understanding the dependencies between success indicators can provide valuable insights for delegates to better assess and recognize improvements in such scenarios.

Overall, these emergent themes underscore the complexity and importance of the indicator selection process in the context of humanitarian interventions, emphasizing the need for a robust, evidence-based strategy to guide this crucial aspect of implementing frameworks at the field level. The emergent themes highlight the intricate nature of indicator selection within humanitarian interventions, emphasizing the vital role of a strong, evidence-based strategy in guiding this process at the field level. In this context, the diagram becomes a valuable asset, shedding light on the complex interconnections among indicators within the framework. Moreover, it aids in dismantling any existing framework structures applied to these variables, fostering a more holistic approach. This heightened awareness equips practitioners with the ability to meticulously plan actions and activities in the challenging context of disaster displacement, ultimately contributing to more effective and informed disaster response efforts.

13.7 Conclusions

This chapter underscores the crucial role of dependable data and an effective methodology for choosing indicators in disaster displacement scenarios. It explores the challenges associated with obtaining reliable data and points out the deficiencies in the indicator selection processes of aid organizations, local agencies, and national authorities. An innovative method using HIDECs was introduced, advocating for a new approach to selecting indicators that emphasizes an evidence-based strategy while upholding a holistic approach. This was supported by a thematic analysis which shows indicator selections must consider capacity building at its core, remain flexible, and be clear in the dependencies between success indicators.

Moving forward, there is a need for additional research to refine the roadmap diagram introduced in this chapter, aiming for consistency and comparability across various frameworks and disaster contexts. Integrating the perspectives and experiences of displaced and disaster-affected populations is also vital, as it aids in selecting relevant indicators that genuinely resonate with their needs and aspirations.

Practitioners working in the field often contend with time limitations, a lack of readily available clear information, and rapidly evolving situations. These challenges can render the plethora of available indicators overwhelming. However, by adhering to a detailed guideline that considers the context, key domains, and reliable data, practitioners can attain a thorough understanding of the process involved in selecting evidenced-based indicators for disaster displacement interventions.

References

Alexander C (1968) HIDECS 2: a computer program for the hierarchical decomposition of a set which has an associated linear graph. M.I.T School of engineering, Cambridge

Alexander C, Manheim ML (1962) HIDECS 2: a computer program for the hierarchical decomposition of a set which has an associated linear graph, Cambridge, s.n.

Alexander C (2015) A city is not a tree: 50th anniversary edn. Sustasis Foundation

Bilau AA, Witt, E, Lill I (2018) Practice framework for the management of post-disaster housing reconstruction programmes. Disasters Cris Hazards Emergencies Sustain Dev

Brown D, Salto K, Spence R (2008) Indicators for measuring, monitoring and evaluating post-disaster recovery. Int Work Remote Sens Disaster Appl, Pavia

CARE International (2014) Learning from crisis: strengthening humanitarian response since the 2004 Indian ocean tsunami. https://reliefweb.int/sites/reliefweb.int/files/resources/Tsunami%20Report_2014_web.pdf

Clinton W (2006) Lessons learned from tsunami recovery, key propositions for building back better. Prevention Web. https://www.preventionweb.net/files/2054_VL108301.pdf. Accessed 29 Apr 2023

Dwyer C, Horney J (2014) Validating indicators of disaster recovery with qualitative research. https://www.ncbi.nlm.nih.gov/pmc/articles/PMC4323412/

Fernandez G, Ahmed I (2019) Build back better approach to disaster recovery: research trends since 2006, s.I.:s.n.

Geis D (2000) By design: the disaster resistant and quality-of-life community. Nat Hazard Rev 1(3):151–160

IASC (2006) Guidance note on using the cluster approach to strengthen humanitarian response. https://interagencystandingcommittee.org/system/files/2021-03/Guidance%20Note%20on%20Using%20the%20Cluster%20Approach%20to%20Strengthen%20Humanitarian%20Response.pd

Kennedy J, Ashmore J, Babister E, Kelman I (2008) The meaning of 'build back better': evidence from post-tsunami Aceh and Sri Lanka. J Contingencies Cris Manag 16(1):24–36

Khasalamwa S (2009) Is 'build back better' a response to vulnerability? Analysis of the post-tsunami humanitarian interventions in Sri Lanka. Nor J Geogr 83(1):77–88

Maly E (2017) Building back better with people centred housing recovery. Int J Disaster Risk Reduct 29:84–93

Mannakkara S, Wilkinson S, Potangaroa R (2018) Resilient post-disaster recovery through building back better. Routledge, s.l.

Mannakkara S, Wilkinson S (2014) Re-conceptualising "building back better" to improve post-disaster recovery. Int J Manag Proj Bus

Monday JL (2002) Building back better; creating a sustainable community after disaster. Nat Hazard Inf

Noy I, duPont IV W (2016). The long-term consequences of natural disasters—a summary of the literature. Working papers in economics and finance, school of economics and finance, victoria business school

Schilderman T (2010) Putting people at the centre of reconstruction. In: Lyons M, Schilderman T, Boano C (eds) Building back better; delivering people-centred housing reconstruction at scale. Practical Action Publishing, s.l., pp 7–39

Shalash A, Abu-Rmeileh NM, Kelly D, Elmusharaf K (2021) The need for standardised methods of data collection, sharing of data and agency coordination in humanitarian settings. https://gh.bmj.com/content/bmjgh/7/Suppl_8/e007249.full.pdf

Sphere (2022) Who are our memebers? https://spherestandards.org/wp-content/uploads/sphere-full-members-list-1.pdf

Sphere Association (2018) The sphere handbook: humanitarian charter and minimum standards in humanitarian response. Fourth edn ed. Geneva, s.n

Twigg J (2007) Characteristics of a disaster-resilient community: a guidance note. https://www.preventionweb.net/files/2310_Characteristicsdisasterhighres.pdf. Accessed 23 Mar 2023

Twigg J (2009) Characteristics of a disaster-resilient community. https://actionaid.org/sites/default/files/characteristics_of_a_disaster_resilient_community.pdf. Accessed 23 Mar 2023

UNDP (2020) Human development report 2020. New York, s.n

UNDRR (n.d.) Words into action. https://www.undrr.org/publications/words-intoaction#:~:text=Words%20into%20Action%20(WiA)&text=WiA%20are%20a%20series%20of,Risk%20Reduction%20(DRR)%20practitioners. Accessed 2023

UNDRR (2007) Hyogo framework for action 2005–2015: building resilience of nations and communities to disasters. https://www.preventionweb.net/files/1037_hyogoframeworkforactionenglish.pdf. Accessed 28 Mar 2023

UNDRR (2015) Sendai framework for disaster risk reduction 2015–2030. https://www.preventionweb.net/files/43291_sendaiframeworkfordrren.pdf. Accessed 12 Mar 2023

UNDRR (2016) Implementing the sendai framework to achieve the sustainable development goals. https://www.undrr.org/publication/implementing-sendai-framework-achieve-sustainable-development-goals

UNDRR (2022) Global assessment report on disaster risk reduction, our world at risk; transforming governance for a resilient future. Geneva, s.n. UNDRR (n.d) Words into action. https://www.preventionweb.net/sendai-framework/words-into-action. Accessed 27 Apr 2023

Vahanvati M, Rafliana I (2019) Reliability of build back better at enhancing resilience of communities. International J Disaster Resil Built Environ 10(4):208–221

Venable C, Javernick-Will A, Liel A (2020) Perceptions of post-disaster housing safety in future typhoons and earthquakes. Sustainability 12(9):3837

Part III
Reconstructing the Sense of Home and Community Agency in Resettlement Re-Covery

Chapter 14
Long-Distance Collective Relocation and Resettlement: The Case of Disaster Recovery and Domestic Migration in Japan

Miwa Abe, Chiho Ochiai, and Tadashi Uchiyama

Abstract This study deals with two government-led long distance collective relocation and resettlement (later refer to LDCRR). In 1889, a flood in Totsukawa Village, Nara Prefecture, Japan, caused landslides and the collapse of an earthen dam, damaging many villages and forcing many residents to migrate to Hokkaido, more than 1,200 km away. When Sakurajima in Kagoshima Prefecture erupted in 1914, residents of villages damaged by pyroclastic flows and volcanic ash were relocated to Kimjeong City on the Korean Peninsula, more than 600 km away from Kagoshima. Currently, the mainstream of mass relocation for the purpose of community maintenance and disaster recovery is to move within a short distance and among members of the same community. This study aims to discuss the possibility of LDCRR and resettlement based on the cases of the 1889 flood and 1914 volcanic eruption, the longest distance relocations and resettlement in Japan. Based on a literature review and interviews, the study clarified the factors that led to the long-distance mass relocation, which is discouraged in modern times, and the policy background of the time, as well as the administrative considerations that were implemented to ensure the continuity of life at LDCRR sites. The results show that in both cases, the government took the initiative in developing infrastructure such as roads, paddy fields and railways, and that relocation was carried out in a way that combined the securing of a place to live with livelihood support for disaster recovery purposes. In addition, the government tried to distribute land and housing as much as possible so that the relocated sites would be the same as the original settlements. On the other hand, many people had difficulty adapting to the new environment because of the completely different

M. Abe (✉) · T. Uchiyama
Undergraduate School of Humanities and Science Department of Community and Social Studies, Tokai University, Shibuya, Japan
e-mail: abe.miwa.k@tokai.ac.jp

T. Uchiyama
e-mail: uchiyama.tadashi.k@tokai.ac.jp

C. Ochiai
Graduate School of Global Environmental Studies, Kyoto University, Kyoto, Japan
e-mail: ochiai.chiho.2x@kyoto-u.ac.jp

C. Ochiai et al. (eds.), *Disaster and Displacement*, Disaster Risk Reduction,
https://doi.org/10.1007/978-981-96-2583-3_14

climate and soil characteristics, and over half of the first migrants left the area. As time went by, mixing with migrants from other regions showed the integration to tolerate cultural diversity. In Hokkaido, even now, more than 100 years later, there are still interactions between the original place and the destination of relocation. Also, in the case of Sakurajima at Kagoshima Pref., the human relationships that had been formed at the relocation site after the postwar evacuation continued even after the evacuation. We believe that long-distance migration, depending on the way migration policy is implemented, can also allow human relationships to continue even after a long period of time has passed.

14.1 Introduction

14.1.1 Research Background and Purpose

The concept of the phenomenon of climate change-induced migration and forced displacement already emerged in the late 1970s (Brown 1976). Since 2010, 21.5 million people have been displaced from their homes due to climate change-related disasters, according to the UNHCR, and food and water shortages have been caused not only by sudden disasters but also by climate change. It is predicted that there will continue to be a combination of causes, including the difficulty of accessing natural resources and the expansion of urban areas in developing countries (Newsweek 2016). However, among environmentally induced migrants, especially with regard to domestic migrants, the ambiguity in definition makes it difficult to ascertain the actual number, and a clear definition and comprehensive data collection are needed to implement effective resettlement measures.

Japan is no exception, with the number of disasters fluctuating and increasing, and in the last 50 years alone, more than 30,000 families have been massively relocated within the country as a result of disasters. If individual relocations are included, the number of environmentally induced domestic migrants is expected to further increase. On the other hand, despite the fact that forced evictions associated with disasters have been continuously identified in Japan in the past, there is very little recognition that these are environmentally induced domestic migrations, and domestic mass relocations associated with disaster-induced forced evictions are currently being implemented solemnly, integrated into the disaster recovery process.

LDCRR has been implemented around the world, and there are many examples of post-disaster relocation in Japan. In recent years, most mass relocations after disasters have been short-distance relocations in order to avoid major changes in the living environment of disaster victims. However, once a large-scale disaster such as the Nankai Trough, which is currently predicted to occur, occurs, such short-distance relocations may not be possible. Despite this current situation, studies focusing on LDCRR are limited, and although there are records on the adverse effects of LDCRR, there are few studies that actively review LDCRR. It is significant to examine LDCRR beyond the traditional living area in order to consider its operation as an option for housing reconstruction after a disaster.

This study focuses on "long-distance collective relocation and resettlement (later refer to LDCRR)," which is when people affected by the disaster are relocated to the outside of neighboring prefectures, it is not actively chosen for disaster recovery purposes nowadays. The purpose of this study is to summarize the process after relocation and to clarify what kind of support and consideration are necessary for "LDCRR" to become an option for post-disaster reconstruction, from the perspectives of policy background at the time, continuation of life at the relocation site, and the link between the relocation site and the original site.

Focusing on the "resettlement" process, which is not a temporary "moving of residence" after a disaster, but rather the need to rebuild a new life in a new place after relocation, the project will organize the long-term process of resettlement.

14.1.2 Methodology

In this study, we compare a case of LDCRR following the flood of 1889 from Totsukawa Village, Nara Prefecture to Shin-Totsukawa Town, Hokkaido. The case study of the 1914 volcanic eruption in Kagoshima City, Kagoshima Prefecture (formerly East Sakurajima Village and West Sakurajima Village) relocated to Kimjeong City, Jeollabuk-do, Korean peninsula. Because both cases involved mass relocations more than 100 years ago, we conducted a literature review based on administrative documents, newspapers of the time, and records of the relocated residents, as well as interviews with the descendants of those who migrated/emigrated. Outline of hearing subjects is shown in Table 14.1. In both areas, field surveys have been conducted and research results have already been reported (Abe et al. 2020, 2023). Based on these results, the present study adds a comparative analysis of past cases of LDCRR in terms of policy background, residents' movements, and long-term changes.

14.2 Recovery from Disasters and Collective Relocation

14.2.1 Post-Disaster Collective Relocation

Unlike development-induced relocations, such as those associated with dam development or road expansion, mass relocations after disasters are characterized by the fact that residents do not have sufficient time to individually consider where they will live next or to discuss alternative locations with other residents of the community. In the past, many victims of the Florence Island earthquake (1992), the Great Hanshin-Awaji Earthquake (1995), the Papua New Guinea tsunami (1998), and the Indian

Table 14.1 About the hearing subjects

Hearing subjects (age at the time of interview)	Outline of the subject	Survey year
Female in her 80 s	Descendants of victims who emigrated from Sakurajima to South Korea. Born in the relocation area, he lived there until the age of 15, and lived in Kagoshima after the postwar evacuation	2018
Male in his 60 s	Third generation of Koreans still residing in the relocated Korean	2018
Male in his 50 s	Third-generation immigrant from Totsukawa village	Nov. 12th, 2021
Female in her 50 s	First-generation immigrant from Totsukawa village (by marriage)	Nov. 13th, 2021
Male in his 50 s	Third-generation immigrant from Toyama prefecture	Nov. 12th, 2021
Male in his 50 s	Employee of Shin-Totsukawa town hall	Feb. 20th, 2023
Male in his 50 s	Totsukawa village office employee	Mar. 13th, 2022
Female in her 40 s	Totsukawa village office employee	Jan. 22nd, 2023

Ocean tsunami (2004) were forced to relocate from their homes. However, gathering multiple communities of different living environments, religions, etc., tends not to be generally chosen for post-disaster resettlement involving relocation, not only because it makes establishing the community difficult, but also because it is not sustainable, with some households leaving the new community as time goes by. It has also been pointed out that when environmental transitions occur as a result of moving one's living base, people are forced to reconstruct the relationship with the living environment that they have cultivated over the years. (Iwaisako et al. 1996). Under these circumstances, the purpose of "reconstruction of livelihood" through disaster recovery has been discussed in various fields, including sociology, cultural anthropology, political science, and architecture, regarding collective relocation after the disaster (Cernea 1988; Oliver 1991; Badri et al. 2006; Iuchi 2015). Other studies include an assessment of long-term occupancy conditions after relocation and interpretation from a social capital perspective (Abe et al. 2023).

Even for disaster recovery purposes, victims face involuntary or forced relocation or displacement for a variety of reasons, including physical loss of land, political influence, and financial problems. These involuntary or forced migrants are referred to as "refugees", "asylum seekers", "resettlers" or "internationally dis-placed people". The World Bank (2010) argues that involuntary migration occurs when the decision to migrate is made and forced by outsiders and there is no possibility of staying. It also states that involuntary migration can be divided into three types depending on the circumstances leading to migration: migration and resettlement due to development, migration due to disasters, and migration due to conflict (FMO 2010; WB 2010). On the other hand, some researchers note that the terms involuntary migration, forced migration, involuntary resettlement, and forced migration tend to be used interchangeably to refer to forced population movements that take place under impoverished conditions, often without the consent or choice of the affected population, and that there are few studies on disaster resettlement, as existing involuntary relocation is broadly divided into relocation due to conflict and development (Quetulio-Navarra et al. 2013).

One of the challenges of collective relocation after a disaster is the short time available to make the decision to relocate. There have been cases where people have given up on rebuilding at the original site due to the extent of the damage and decided to relocate early, only to find themselves wishing they had stayed, or regretting that they had stayed. Another issue is the lack of long-term evaluation after relocation. There have been few studies on the impact of social capital and other social factors, and no evaluation indicators have been developed.

It is said that post-disaster reconstruction does not proceed at the same pace in all areas of the city (Eugenie 1994; Kamel and Loukaitou-Sideris 2004; Aldrich and Crook 2008; Wood et al. 2010). It is necessary to understand the backgrounds that have forced people to move their residence collectively rather than individually, and to respond to various issues such as economic independence at the new location, securing education, and consideration for the relationship with the former location, based on an understanding of the regional characteristics. However, as mentioned above, the challenge in implementing collective relocation for the purpose of disaster recovery is that, despite the need for such careful consideration and support, there is not enough time for consensus building before the relocation takes place.

The Law for the Promotion of Collective Relocation for Disaster Prevention Purposes in Japan states that the target of the law is for 10 or more households to move together to other land. For this reason, this study focused on cases in which ten or more households were collectively relocated after a disaster to an area with a different living environment and culture across prefectural borders, prior to the enactment of the Collective Relocation Promotion Law in 1972.

14.2.2 Disasters and Collective Relocation in Japan

In Japan, where "restoration in its original form" had been the principle, the introduction of the concept of "improved restoration" was an epoch-making reform of the system, triggered by the torrential rain disaster in July 1972. This disaster triggered the establishment of the law caused damage in 32 prefectures from the Tohoku to Kyushu regions, leaving 447 people dead or missing (database of Japan Meteorological Agency), 2,977 houses completely destroyed and 10,204 houses half destroyed (database of Japan Meteorological Agency). The law was enacted in an extremely short period of time, and although problems were pointed out from the outset, priority was given first to rebuilding the lives of those affected in each region.

From 1972, until 2006, "collective relocation and promotion projects" based on this law were implemented in a total of 35 municipalities and 1,834 houses (Ministry of Land, Infrastructure, Transport and Tourism 2012). In addition, approximately 30,000 households affected by the Great East Japan Earthquake relocated their residences through this project.

The collective relocation and promotion project is designed to promote the collective relocation of dwellings from areas that are not suitable for habitation in areas where disasters have occurred or where there is a threat of disasters. The project provides partial subsidies to municipalities that develop housing complexes, relocate dwellings, purchase former relocation sites, etc. Residents must have five (ten in some areas) or more households to be relocated together.

In some cases, it is difficult to obtain a consensus among these five (or ten) cohesive households, especially in collective relocation for the purpose of disaster recovery in depopulated areas. In Japan, a survey showed that 90% of citizens intend to settle down (Ministry of Land, Infrastructure, Transport and Tourism 2010), suggesting that collective relocation should not be considered simply for disaster recovery or for depopulation issues.

In 1972, at the 69th session of the Diet's Special Committee on Disaster Countermeasure[1]s, a Diet member who checked the affected areas reported that the damage was beyond imagination and that the local government had suffered irreversible damage (Diet's Special Committee on Disaster Countermeasures 1972–1). Clearly, the magnitude of the damage could not be dealt with by the conventional disaster response system alone, and it was suggested that legislation for collective relocation should be considered. Later, in 1972, the 70th Extraordinary Diet session passed the "Law Concerning Special National Financial Measures for the Project for Promoting Collective Relocation for Disaster Prevention," allowing people to move their places of residence before and after disasters for disaster prevention purposes (Diet's Special Committee on Disaster Countermeasures 1972–2). However, it was anticipated that many residents would be reluctant to leave their longtime homes, even though they knew they were at risk from disasters, and the government (Ministry of Home Affairs) estimated that only about 40% of those who planned to relocate would do so. This shows that the choice of relocation destination and site was a difficult decision for people to make.

[1] Unlike the regular Diet, which meets on a regular basis, the Diet meets on a temporary basis during times of disaster to discuss disaster countermeasures.

14.2.3 Before the Disaster Prevention Collective Relocation Promotion Project Was Developed

Even before 1972, when the collective relocation and promotion project began, several villages were relocated collectively, including Inadani in Nagano Prefecture in 1961, Nishitani in Fukui Prefecture in 1965, and Kurokawa in Niigata Prefecture in 1967, all of which were triggered by flood damage.

In the flood disaster in Nagano Prefecture, where villages were buried by mudslides, it was decided that it would be more desirable to relocate the residents to safer locations than to carry out disaster recovery work on site. Therefore, instead of implementing the agriculture and forestry disaster recovery project as subsidized by the government, the government subsidy originally intended for this project was transferred from the then Ministry of Agriculture and Forestry to the Ministry of Home Affairs to be used for the collective relocation project subsidy (Abe 2015). This method of collective relocation is called the "Inadani method," and was used as a reference in all affected areas until the Disaster Prevention and Collective Relocation Promotion Project was established.

While prewar collective relocation in Japan enabled long-distance migration due to the development of local public infrastructure and support measures that covered even post-relocation livelihoods in line with national land development (Abe et al. 2020), once the current subsidized system was in place, the emphasis changed to housing reconstruction to rebuild livelihoods, including livelihoods (Abe et al. 2023). Therefore, the prevailing trend now is to keep relocation distances as short as possible and to prefer relocated members to be in the same region.

The case study of collective relocation from Totsukawa Village of Nara Prefecture to Hokkaido in 1889, and the case of Kagoshima Prefecture due to the Sakurajima eruption in 1914, were also conducted prior to this 1972 event. In particular, these two cases of long-distance relocation took place at a time when Japan was developing Hokkaido and its colonies, and unlike today's collective relocation and promotion projects, the relocation method involved not only housing reconstruction but also a mixture of projects with development purposes, such as livelihood support and the construction of railroads.

14.3 Case Comparison of Long-Distance Collective Relocation and Resettlement in Japan

Here, two cases of long-distance collective relocation discussed above in Japan are compared and discussed based on information obtained from reviewing the village/town history, interviews with the descendants of migrants.

14.3.1 From Totsukawa to Shin-Totsukawa (Collective Relocation Due to Flooding)

A major flood in the Kumano River basin in August 1889 caused devastating damage in six villages[2] called Totsukawa-go, Yoshino County, Nara Prefecture (present-day Totsukawa Village). In Totsukawa alone, 168 of the 12,862 residents were killed, and 426 houses were completely destroyed or washed away due to the record-breaking rainfall of over 1000 mm per day. Including half-destroyed houses, a quarter of all houses were damaged. Approximately 3,000 people lost their livelihoods as a result of the flooding, and 2,691 of these victims began to consider migrating to Hokkaido in October of the same year (Kabata and Kobayashi 2006; Shin-Totsukawa Town History Compilation Committee 1991).

Hokkaido, about 1,000 km away from Nara Prefecture, was about to adopt a colonization plan and begin prefecture-wide development in 1904.

Ultimately, 2,489 people from 600 households decided to migrate to Hokkaido in search of a new place to live, and in October 1889, they took three separate trips from Kobe to Otaru by boat (Shin-Totsukawa Town History Compilation Committee 1991). They overwintered in a place that had been developed near where they had settled, and in June of the following year, 2,230 people from 537 households crossed the Ishikari River and entered the Tokutomi River basin (present-day Shin Totsukawa Town), known as the Tokku Wilderness. It took about eight months from the time they left Totsukawa in Nara Prefecture to the time they arrived at their relocation site. Although it took eight months, what is noteworthy in this case is the short time it took to reach the decision to relocate. The people were affected by the disaster in August, and they had already begun moving to Hokkaido by October, two months later, indicating that the time until the relocation decision made was extremely short.

In those days, many disaster victims, not only from Totsukawa, migrated to Hokkaido in search of a new life. 55 families migrated from Gifu Prefecture in 1895 after the flood and earthquake, settlers from Aichi Prefecture after the Nobi Earthquake in 1891, and 400 families from Yamanashi Prefecture in 1908 and 1909 after the flood disaster in Yamanashi Prefecture. Many people continued to migrate after the Great Kanto Earthquake of 1923 and World War II (Shin-Totsukawa Town History Compilation Committee 1991). As these people mixed and mingled, a new town was formed, which is now Shin-Totsukawa Town.

---

[2] Totsukawa Township consists of Kita Totsukawa Village, Totsukawa Hanazono Village, Naka Totsukawa Village, Nishi Totsukawa Village, Minami Totsukawa Village, and Higashi-Totsukawa Village, which became the current Totsukawa Village after a merger on June 18, 1890.

14.3.2 Sakurajima to Kagoshima, Kirishima, and Korean Peninsula (Collective Relocation Due to Eruption)

When Sakurajima in Kagoshima Prefecture erupted in 1914, residents of villages damaged by pyroclastic flows and volcanic ash relocated to the city of Kimjeong on the Korean Peninsula, more than 600 km away in a straight line from Kagoshima. As for nearby relocation destinations, various locations in Kagoshima and Miyazaki prefectures were designated for relocation, and people moved to one of these places at the discretion of their families (Sakurajima Town History Compilation Committee 2006). At the time, there were two villages in Sakurajima, Higashi Sakurajima Village (population 8,463) and Nishi Sakurajima Village (population 12,939), and the number of casualties was 30 people. The two villages together had 97% of their population engaged in agriculture, so even if they did not suffer physical damage, those whose cultivated land was considerably damaged in terms of rebuilding their livelihoods.

In the selection of candidate sites for relocation, the envoys were dispatched to the prefecture or neighboring prefectures to begin the selection process. In the end, there were 1,002 households and 6,245 people who wanted to move to the sites designated by the government, and 2,066 households and 14,587 people who moved on a voluntary basis (Sakurajima Town History Compilation Committee 2006). The Korean Peninsula[3] was the designated site for relocation by the government. It is known that the number of migrants to the Korean Peninsula was higher than those relocated to designated domestic relocation sites, although the amount of aid and supplies received was only slightly higher (Abe 2016).

On April 8, 1914, 53 people from 10 affected households traveled north on the Kagoshima Main Line by train, passing through Moji and taking a ferry from Shimonoseki to Busan. In Jeollabuk-do, which was under colonial rule at the time, the Toyo Takushoku Corporation (hereafter, Oriental Development Corporation), which was in charge of colonial development, was commissioned to handle the relocation.

Figure 14.1 shows two case study sites of long-distance relocation. Figure 14.2 shows the current condition of Shin-Totsukawa Town, which was taken in 2020. Figure 14.3 is the current condition of the Korean peninsula which was taken in 2019.

[3] Japan annexed Korea in August 1910 and ruled Korea for a long period of time until September 1945.

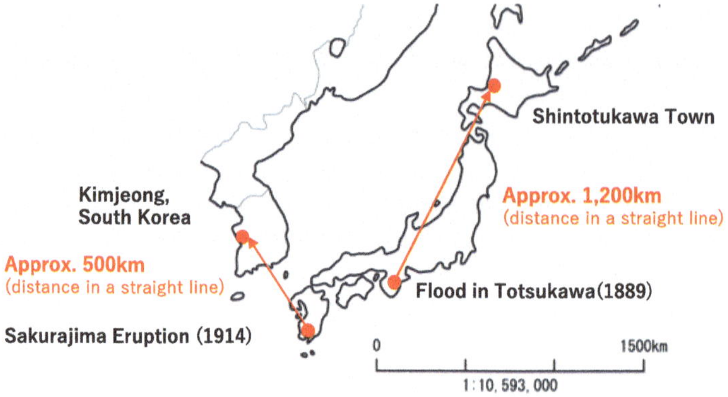

Fig. 14.1 Locations of long-distance relocation. *Source* Authors

Fig. 14.2 Current situation in Shin-Totsukawa, Hokkaido (A view of rice fields developed by migrants). *Source* Authors

Fig. 14.3 Current situation in Korea in 2018. *Source* Authors

14.3.3 Factors that Led to Long-Distance Collective Relocation and Resettlement, and the Policy Context of the Time

In both areas, we know that the decision to relocate was heavily influenced by people from the local area.

In the case of the 1889 flood migration, it began when bureaucrats from Totsukawa who were living in Tokyo and were concerned about the plight of their hometown. When they received detailed information about the damage, they realized that their hometowns could not support the existing population and decided that they should be relocated to other areas. At that time, there was still much-undeveloped land in Hokkaido, and the government was encouraging people to move to Hokkaido. Knowing this, bureaucrats from Totsukawa investigated Hokkaido and requested the cooperation of the then Governor of Nara Prefecture and the Commissioner of Hokkaido. As a result, they obtained promises of full cooperation from both Nara Prefecture and Hokkaido.

Because of the cold winter and the distant location of Hokkaido, the disaster victims were very anxious, so the town office asked Hokkaido officials to explain the current situation when they were in Tokyo. About two months after the disaster, after a meeting with the village mayors, they decided to start moving to Hokkaido.

At the time, Hokkaido was in the process of creating a system for northern security and development of the country's land, and the colonization plan was adopted in 1886, with the development of the entire area about to begin. In particular, the development of the Ishikari Plain, where the disaster victims were to be relocated due to Karafuto management and defensive measures against Russia's southward expansion, was an urgent issue, and permission was granted for their relocation. Therefore, when the Hokkaido Government was in the process of pioneering the area and the migrants from Totsukawa started their lives in Hokkaido, land division and road construction projects were carried out in parallel. Today, the roads from those days are still in use, and the plots of houses and fields remain as they were then.

Similarly, in the case of relocation due to the 1914 eruption, the implementation of long-distance relocation was triggered by a network of people from the village. At the time of the disaster, people from Nishi-Sakurajima Village were working for the Korean Governor-General's Office and made a request to the then Governor of Kagoshima Prefecture regarding the emigration of the disaster victims, attaching even the emigration regulations. The proposal included the following: "In order to provide the islanders, who are forced to leave their families, with a foundation for eternal peace of mind and a chance at life, we would like to suggest that they emigrate as a part of a program of the Oriental Development Corporation as a measure for reconstruction. Due to the agricultural season, the emigration must take place during the spring season from March to April, and since the emigrant recruitment period is limited, if they miss this recruitment, they will have to wait a year for the next one. We hope that the prefectural government will negotiate with the Oriental Development Corporation to open the door for emigration recruitment".

The governor of the prefecture at the time made a petition that included the free transfer of government-owned land and subsidies for relocation expenses, and eventually a plan was announced to relocate people from Kagoshima Prefecture, mainly to government-owned wilderness areas.

We believe that one of the factors behind the acceptance of such long-distance relocation by the disaster victims was not only the generous support by the government, but also the pre-existing infrastructure such as railroads and roads, as well as the consideration for their livelihood in the new land (Mori 1984).

The departure of those who wished to relocate to the Korean Peninsula was on April 8, 1914. Since they were affected by the eruption in January of the same year and left for the Korean Peninsula in April, it is clear that they had only about three months to leave, similar to the case in Nara Prefecture in 1889. Unlike the cases in Japan where the migrants were relocated to designated resettlement areas, the migrants who obtained previously cultivated land were able to engage in agriculture immediately after their relocation. In addition, the road leading to the station for the migrants was also in place at the time of their relocation. Therefore, in the case of relocation to the Korean Peninsula, the reconstruction of livelihood was faster than in the case of domestic migration, although the reconstruction was done by relocation caused by the same disaster.

14.3.4 Migrants' Efforts to Continue Living in the Relocated Site

In the 1889 case, upon relocation, the six village heads of the then Totsukawa Township held a meeting to discuss the disposition of common property in the village, and 17 resolutions were made, ranging from the division of property among the migrants and the bond between the original village and the new site (destination of migration). Based on the premise that even if they are separated by relocation, "they will continue to hold on to their causal ties and history as belonging to the same Totsukawa," it is promised that the common property that cannot be divided will be well managed by the original village. These decisions were put into writing and widely disseminated to the population. The village leaders, led by the village mayor, discussed and committed to a long-term relationship between the original village and the new site, and also decided on and announced the method of property division. This was not only to allay the fears of the villagers about relocation, but also to discuss "what the village should be like for generations to come. This action was taken to alleviate concerns about the original land and common property, and although the decision to relocate was made at an early stage, it is clear that the relocation was based on the consensus of the migrants.

The 600 families who decided to relocate headed for Hokkaido under the slogan, We will surely build a second hometown. The emigrants drafted an "Migrant Pledge

(Table 14.2)" to ensure the success of the settlement and to establish rules for migration, and not only the village chiefs but also the residents themselves set common goals for the implementation of it and for rebuilding their lives at the destination. The pledge not only states that the migrants will help each other through the uncertainty of migration, but also includes a plan for what to do if they do not abide by the rules. The existence of such written pledges suggests that common goals for village formation were set and made available for all to see.

This kind of written pledge among migrants or between migrants and the people of the original village is not common in other cases of relocation, and differs from other collective relocations case, especially having such a written pledge.

In addition, the distribution of land among the settlers was carried out by lot drawing. Therefore, each family's land was called "winning land. It was also found that, depending on the allocation of land, people from the same village were clustered together to some extent. It can be inferred that the lottery was conducted to determine the plots after consideration was given to the proximity of the settlements from which the settlers migrated when conducting the lottery. It is also recorded that once the winning lots were decided, in some cases, after discussed among themselves and exchanged the winning lots (Totsukawa Village 2012). In terms of occupation, the main occupation before migration was forestry, but in the villages where the migrants settled, they switched to rice cultivation. However, due to the lack of rice farming skills, some migrants left Shin-Totsukawa Town because they could not secure sufficient harvests.

The aspect of relocation after the 1914 eruption disaster is different from the Totsukawa case. Even though it was colonization, the emigration to another country's territory was managed by the Oriental Development Corporation, which took care of

Table 14.2 Immigration pledge overview

1	Since we have separated from our blood relatives and migrated, we are more united than ever before to help each other and to avoid creating conflicts
2	No other work is allowed until the land is cultivated, except with permission from the government or during the off-season
3	The common property inherited from the original village is the basic property of the new village and should not be divided and consumed by each individual under any circumstances
4	Each one of us should make efforts to be thrifty (Do not make any unnecessary purchases, keep your house simple and neat, refrain from feasting and drinking except on anniversaries and holidays, do not open restaurants in the village, and wear cotton clothes as much as possible)
5	Establish schools, promote education, and enroll children in school
6	Respect the rules of etiquette, and will not disrupt public morals and be laughed at by the public
7	Those who disobey the above will be warned and admonished by the village leaders. If the situation is not improved, the village leaders will cut off their friends from the village

the emigration, including the emigration plan and rules regarding life in the emigration destination. Since the emigration of the disaster victims was incorporated into the Japanese colonization program for the Korean peninsula, the aspect of colonial settlement was more significant than the aspect of support for the disaster victims. At that time, eleven emigration missions had been dispatched to the Korean Peninsula by October 1921, and the victims of Sakurajima were included in the fourth emigration mission. The Sakurajima disaster victims were part of the fourth group of emigrants, some of whom had come for the purpose of recovery from the disaster, and some of whom had emigrated for the purpose of cultivating farmland on the distant Korean Peninsula. At the time, 570 households, including those from Sakurajima, had settled in Kimjeong-gun, Jeollabuk-do. Although the houses and fields were also clustered by place of origin, the source of relocation was not taken into consideration as in the case of Totsukawa. They were mixed with emigrants from other areas and their housing locations were scattered. The emigrants had to adapt to the cold weather of the Korean peninsula and had to change their occupation from fruit cultivation on Sakurajima to rice cultivation in their new land. Those who lacked rice-firming skills relied on Korean people who had been engaged in rice cultivation for some time, some started a business selling local products to China, and others cultivated fruit trees. Compared to the case of Totsukawa, the adaptation to the environment and industry was more flexible.

14.3.5 Relationships Between Original Villages and Migrants

In the example of relocation from Totsukawa, 120 years after migration, the program continues with exchange programs for children and administrative officers, who travel to each other's villages on school trips and for training. In addition, every year a ceremony is held in Shin-Totsukawa Town to commemorate the establishment of the town, and the village mayor and other officials from Totsukawa Village of Nara attend the ceremony. This ongoing personal or institutional exchange has been useful in times of emergency, such as in the event of a disaster. For example, the disaster caused by Typhoon No. 12 that hit the Kii Peninsula in late August and September 2011 was known as the Kii Peninsula Flood, and in Totsukawa Village in Nara Prefecture, a series of mountain landslides occurred, collapsing approximately 260 hectares in the entire village. In Totsukawa Village, seven people were killed and six have been missing. At the time of the flooding, Shin-Totsukawa Town dispatched three support staff to Totsukawa Village to work on disaster recovery for two months, and the residents of Shin-Totsukawa Town collected donations and contributions (Mori and Mori 1984). At the time of their immigration in 1889, the village leaders pledged to "maintain the ties and history of belonging to the same Totsukawa," and as promised, over the past 120 years, the village and the town have maintained their ties and inherited its history for generations to come.

As mentioned earlier, the transition from forestry to rice cultivation was very difficult for the migrants from Totsukawa. After nearly 10 years of clearing forests for settlement, setting up houses, and finally getting to the point of growing rice, they were unable to succeed due to a lack of knowledge about irrigation facilities and pest infestation. Over time, the migrants left and then began to mix with migrants from other prefectures. When migrants from Toyama Prefecture, a rice-producing region, entered the area, rice cultivation in Shintotsukawa Town developed in a stable manner. In the process, Totsukawa's migrants allowed other cultures to enter their daily lives, such as the lion dance of Toyama Prefecture and the portable shrines of Tokushima Prefecture in the local festivals, and flexibly adapted their own cultures and lifestyles. The current festivals in Shin-Totsukawa Town are a mixture of cultures brought in by migrants from various regions.

On the other hand, in the case of those who migrated from Sakurajima, it is known that they did not actively try to bring their culture of original village. Upon emigrating, the families of the victims from Sakurajima sold much of their property, including their housing lots, to their relatives who remained in Sakurajima. From the stories of second-generation emigrants who grew up on the Korean Peninsula at the time, it is clear that they became aware of "Kagoshima" only through the language used by their families and neighbors, and through the local cuisine of kara imo mochi (sweet potato cakes) and dango jiru (dumpling soup). They did not actively communicate these cultures to their children either, but understood them through the language and food used by the people when they first arrived in Kagoshima as a result of the postwar evacuation. In the case of emigration from Sakurajima, while those who migrated to neighboring prefectures such as Kagoshima and Miyazaki actively utilized the name or image of "Sakura" (a part of Sakurajima) such as new town's name or school seal to remind them of their original village, and left traces such as inscriptions on stone monuments about emigration.

Unlike other cases in Japan, there was no village head or other leader representing the community, and when there were disputes, local elders would come out to calm them down, and the police would maintain security and public order. The people interacted through an elementary school's sports day, where local children and adults gathered to learn about the newcomers' faces, businesses, families, and other aspects of the community. The elementary school's sports day was an opportunity for Japanese immigrants to get together and build a network of immigrants.

14.4 Conclusion: Implication of Long-Distance Relocation as an Option for Rebuilding People's Life

Why was "long-distance" collective relocation, which is not actively chosen in recent years, possible at that time? This section summarizes what we have learned through a comparison of past cases.

- In both relocation cases, the government led the development of roads and other infrastructure, and relocations were carried out in a manner that combined the securing of residential areas for disaster recovery purposes with livelihood support, and sites with relatively good conditions were selected, such as railroad development that was in progress close to the relocation site.

- Although the time to build consensus among the migrants was very short (two to three months), the leaders led the consensus-building process, and they understood the concerns of the disaster victims and held explanatory meetings.
- In Totsukawa, the same considerations as today, attentions were given to keeping the similar living environment and community conditions should be established from the original village to relocated sites. On the other hand, in the Sakurajima case, we confirmed that setting up the new community and common goals among migrants from different regions other than the Sakurajima community's disaster recovery. Therefore, migrants from Sakurajima did not actively try to inherit their culture at the newly relocated sites but rather built a new life.
- In both cases, the migration was to a place with a completely different climate and soil conditions, which made it difficult to adapt to the environment, and many people left the area. Over time, the process of mixing with migrants from other regions showed the flexibility to tolerate cultural diversity.
- Regarding the relationship with the original villages, there was a case where measures were actively taken from a long-term perspective in Totsukawa, and a case where people chose to cut off the relationship in order to seek a more affluent life in the new site in the Sakurajima case. However, in the Sakurajima case, the outbreak of World War II made it not possible to evaluate life after emigration from a long-term perspective.

Based on the two cases of LDCRR and resettlement, we believe that collective relocation projects should be considered for the reconstruction and continuation of "livelihood" rather than simply as a means of housing reconstruction, even for the purpose of disaster recovery. It is not enough just to have houses and roads, in the long run, new people will begin to mix with people from other regions in the relocated areas, and cultures will become more diverse. We need time to take a perspective of what will happen in the long run. It should not be limited to the restoration of the housing that is in front of us now but should be carefully conducted by asking what will happen in the future and what we want to do.

Above all, time is short before the decision to relocate residences after a disaster. The question will be how much effort will be made to alleviate the concerns of the disaster victims. In order for collective relocation to become a part of housing reconstruction choices, we need to be confident that careful work will be done to address these recurring questions. To this end, it is necessary to have leaders or external human resources who can link the process to necessary policies, and organize issues, and set goals not only for disaster recovery but also for the continuation of long-term livelihoods.

Acknowledgements This research was supported by JSPS KAKENHI Grant Number JP (17K17954). We would like to thank the people of Totsukawa Village, Nara Prefecture, and Shin-Totsukawa Town, Hokkaido, for their cooperation in the survey.

References

Abe M, Onishi Y, Hasegawa A, Honma R, Shimoda S (2020) Selection of relocation site for the victims considering their quality of life—a case study on the group migration for disaster recovery from south Kyushu to South Korea after the eruption of Sakurajima in 1914, Jyusoken research papers and practical research report No. 46, pp 167–178

Abe M, Uchiyama T, Ochiai C (2023) Social capital and its transition that enabled long-distance collective relocation after disasters—the case of 1889 flood reconstruction and domestic migration in Japan. Int J Disaster Risk Reduct 96:103945

Abe M (2016) Development of legal systems of collective resettlement after flood in Amakusa, the process and present life at resettlement site. Kumamoto University Policy Res 7:75–86

Aldrich DP, Crook K (2008) Strong civil society as a double-edged sword: siting trailers in Post Katrina New Orleans. Polit Res Q 61(3):378–389

Badri SA et al (2006) Post-disaster resettlement, development and change: a case study of the 1990 Manjil earthquake in Iran. Disasters 30(4):451–468

Brown LR et al (1976) Twenty-two dimensions of the population problem, world-watch paper 5, Worldwatch Institute

Cernea MM (1988) Involuntary resettlement in development projects: policy guide-lines in world bank financed projects. The World bank, Washington DC

Diet's Special Committee on Disaster Countermeasures (1972) Proceedings of the 69th diet session special committee on disaster countermeasures no.4, diet proceedings search system. https://kokkai.ndl.go.jp/#/detailPDF?minId=106914339X00419721011&page=1&spkNum=0&current=1. Accessed 27 Jul 2015

Diet's Special Committee on Disaster Countermeasures (1972) Proceedings of the 70th diet session, special committee on disaster countermeasures, subcommittee on basic issues of disaster countermeasures no.1, diet proceedings search system. https://kokkai.ndl.go.jp/#/detailPDF?minId=107004345X00119721107&page=1&spkNum=0&current=-1. Accessed 27 Jul 2015

Eugenie R (1994) The social geography of disaster recovery: differential community responses to the north coast earthquakes. Assoc Pac Coast Geogr Yearb 56:49–74

Forced Migration Online (FMO) What is forced migration? http://www.forcedmigration.org/whatisfm.htm. Accessed 10 Sept 2010

Hiroshi I, Miura K, Maki N, Kobayashi M (1996) Relocation to restoration housing due to the south-west off Hokkaido earthquake: natural disaster and environmental transition no. 2, academic abstracts. E-2, architectural planning ii, dwellings and residential areas, Rural planning, education, pp 137–138

Iuchi K (2015) Planning resettlement after disasters. J Am Plann Assoc 80:413–425

Japan Meteorological Agency. https://www.data.jma.go.jp/stats/data/bosai/report/1972/19720703/19720703.html. Accessed 27 Jul 2023

Kabata F, Kobayashi Y (2006) Totsukawa flood and emigration to Hokkaido (Totsukawa Suigai to Hokkaido Ijyu in Japanese), Kokonshoin

Kamel N, Loukaitou-Sideris A (2004) Residential assistance and recovery following the northridge earthquake, Urban Stud 41(3):533–562

Ministry of Land, Infrastructure, Transport and Tourism (2010) Summary of the "Survey on Awareness of Residential Areas" Report. https://www.mlit.go.jp/common/001208168.pdf

Ministry of Land, Infrastructure, Transport and Tourism. Status of Implementation of the Disaster Prevention Group Relocation Promotion Project. https://www.mlit.go.jp/toshi/content/001414763.pdf. Accessed 30 Oct 2012

Mori H, Mori I (eds) (1984) Kaikyuroku Totsukawa Migration, Shinjuku Shobo

Newsweek (Japan edn) (2016) The world's ticking time bomb of environmental migration. https://www.newsweekjapan.jp/picture/185372.php. Accessed 20 Aug 2022

Oliver SA (1991) Successes and failures in post-disaster resettlement. Disasters 15(1):12–23

Quetulio-Navarra M, Niehof A, van der Vaart W (2013) Social Capital in voluntary displacement and resettlement. Int J Soc Sci (2)

Sakurajima History Compilation Committee Sakurajima-cho Local History (1988) Asahi Printing Co. (3)
Shin-Totsukawa Town History Compilation Committee (1991) Published by Shin-Totsukawa Town, the Shin-Totsukawa Centennial History
Summary of the Survey on Awareness of Residential Areas Report, Ministry of Land, Infrastructure, Transport and Tourism. https://www.mlit.go.jp/common/001208168.pdf (2010)
Totsukawa village general affairs division (2012) Village Newsletter Totsukawa (612)
Wood N, Burton C, Cutter S (2010) Community variations in social vulnerability to cascadia-related tsunamis in The U.S. Pacific northwest. Nat Hazards 52(2):369–389
World Bank (2010) Involuntary resettlement. http://web.worldbank.org/WBSITE/EXTERNAL/TOPICS/EXTSOCIALDEVELOPMENT/EXTINVRES/0,,menuPK:410241~pagePK:149018~piPK:149093~theSitePK:410235,00.html. Accessed 11 Oct 2019

Chapter 15
Local Community Agency in Post-disaster Displacement

Penny Allan, Martin Bryant, and Andrew Toland

Abstract This chapter addresses the community agency fostered by local knowledge of unique biophysical contexts in post-disaster decision-making around displacement and its alternatives and contrasts this with government-driven managed retreat strategies. Having investigated the aftermath of the February–March 2022 floods in the Northern Rivers Region of New South Wales, Australia, and its global comparators, the research draws upon a discussion of a community's landscape knowledge that comes from living on a floodplain, the issues associated with temporary and permanent displacement, and the institutional language of "natural disaster" and its reductive understanding of landscape processes, as well as the language of "displacement" itself. We find that the tactics of community-led, localized, bottom-up initiatives provide an important counterbalance to top-down generic policy strategies which can sometimes constrain adaptive planning and innovation. We suggest place-specific design and policy directions that might enhance a community's resilience and foster greater knowledge, respect, and care for the contingencies of catchment-scale landscapes. In this light, we see permanent displacement as a last resort.

15.1 Introduction

The river was once healthy, and the locals knew it well. The truth is that people are both the problem and the solution, not the river. The only way to learn is from the river, like our old people did when they sat on its high banks of red and green. (Costello 2022)

P. Allan (✉) · M. Bryant · A. Toland
School of Architecture, University of Technology Sydney, Ultimo, Australia
e-mail: penny.allan@uts.edu.au

M. Bryant
e-mail: martin.bryant@uts.edu.au

A. Toland
e-mail: andrew.toland@uts.edu.au

C. Ochiai et al. (eds.), *Disaster and Displacement*, Disaster Risk Reduction,
https://doi.org/10.1007/978-981-96-2583-3_15

This chapter highlights the importance of place-specific community agencies in policymaking for post-disaster permanent displacement and its alternatives. We look at the ways in which disasters often operate as a highly specific set of interactions between unique biophysical contexts, human activities, and infrastructures. We draw attention to how these interactions, together with the language of displacement in the context of disasters, shape the current framework of responses, including social displacement. While existing landscape architecture scholarship addresses the place-specific landscape systems that are implicated in natural disaster situations, as well as place-specific landscape risk mitigation and community resilience (Claghorn et al. 2015; Mazereeuw et al. 2017; Lunghino et al. 2020), there is little scholarship addressing issues of post-disaster displacement from a landscape perspective (an exception is Yarina et al. 2019). On the other hand, in disciplines focused on humanitarian crises, the literature on post-disaster displacement and its impact on communities (Adams 2009; Rhodes et al. 2022; Mosneaga 2023) tends not to address the specificities of landscape and place in relation to questions of community preparedness and response.

Our research offers a reframing of the concept of disaster-related displacement centering on place-specific landscape processes and their interaction with human activities, communities, and their communications.

We explore this in relation to the impact of the February–March 2022 floods in the Northern Rivers region of New South Wales, Australia. The floods caused substantial damage to physical infrastructure, particularly in the main regional center of Lismore, and left tens of thousands of people temporarily displaced, with many facing permanent displacement. We consider the impact of these floods on communities in Lismore and elucidate the very place-specific relationship between landscape and settlement, using examples of this community's response to understand the capacity to self-organize recovery in the wake of disaster. Considering these findings, we examine global precedents of flood-affected settlements that highlight a spectrum of top-down to community-led adaptive responses where landscape-focused community agency is, to varying degrees, a driver for either staying in place or displacement.

Finally, we offer a set of design and policy directions to enhance community agency and resilience in ways that are more in line with the contingencies of specific environmental contexts. This will generate strategies to enhance effective reconstruction and consider relocation, especially in the face of intensifying climate change effects.

15.2 Community Displacement from a Landscape Perspective

Urban resilience relies on communities understanding and adaptively responding to the range of place-specific risks. Awareness of risk enhances community cohesion and resilience, but relatively recent developments in infrastructural technology, together with risk aversion, have encouraged governments to attempt to neutralize hazards by constructing defensive infrastructure like levees and dams to protect flood-prone areas or sea walls to defend against events like flooding, storm surges, and rising sea levels. Paradoxically, each layer of protection diminishes a community's resilience (McBride 2013; Bryant et al. 2013; Nurandini 2019). Without the cultural memory of past events, and with the assumption that governments have removed risk, people forget and become complacent. Then, if defensive infrastructure fails and emergency services are slow to respond, as happened during the 2022–2023 summer floods in Lismore, the diminished capacity to recognize and respond to local disaster indicators and timings (Šakić Trogrlić et al. 2022) can be disastrous.

While knowledge of landscape can empower communities in the face of events, the way knowledge is communicated can shape perceptions, influencing whether events are seen in vastly different ways: maybe as disasters; or maybe as part of a regenerative environmental cycle. Da Cunha (2019) highlights how colonial mapping shaped knowledge of rivers and floods on the Indian subcontinent, leading to defensive responses like levees, dams, canals, and other engineered infrastructure. A broader view of the hydrologic cycle might see water as part of the land's processes, recognizing their interconnectedness and the vital role of the monsoon in sustaining human and non-human life on the Gangetic plain for millennia. Varying water levels, in this case, can be understood not as a substance separate from the land, but as part of the processes of that land, just as the land (the mountains, plains, river courses, deltas) is part of the processes of water. Indigenous Australians similarly developed a more integrated understanding of water, rivers, people, and landscapes (Costello 2022).

The language currently used by the media and government to describe these events can also inadvertently shape perceptions and responses. The term "displacement" suggests a passive experience, stripping agency from individuals and communities affected by natural disasters (McLeod 2023). It implies being moved apart from a place by external forces. By proactively navigating challenges, communities demonstrate the capacity for resilience and the importance of community-led recovery efforts tailored to local socio-environmental factors (Shevellar et al. 2016). To foster a sense of agency and empowerment, alternative terms like "move," "relocate," or "resituate" could be more appropriate, implying active decision-making by affected individuals and communities in responding to socio-ecological crises that we label as "natural disasters."

It is then vital to have structures and processes in place to make this more agential language a reality in terms of community-led agency in planning, mitigation, and response. Michel de Certeau distinguishes "strategies" as the "manipulation of power relationships" instigated by macro-level institutions shaping urbanism, and "tactics" as "calculated actions" responding to immediate environmental conditions (de Certeau 1984). In the interest of community resilience, governance frameworks may need to incorporate more immediately responsive tactics in a language that continually informs the strategies and policies crucial for resilience and localized agency.

With a vision where extreme landscape event preparations and responses are led by local communities possessing deep knowledge of their landscapes, the array of options would diverge significantly from prevailing "disaster" policies relying on buyouts and hard infrastructure. Given the disruption and expense associated with permanent relocation, it's prudent to explore a range of tactical responses for communities opting to stay put, prioritizing local adjustments. Relocating entire towns risks undermining community cohesion, a crucial asset during future crises. Continual localized refinement of existing systems could prove equally effective.

15.3 Methodology

Using an engaged action research methodology (Deming et al. 2011) that included desktop research, fieldwork, GIS-based analysis and design, review of community-based websites and local newspapers, discussions with government agencies, First Nations leaders, experts in hydrology, geomorphology, engineering, and urban design, and drawing on our expertise in disaster resilience, our preliminary research was designed to better understand the complex causes and impacts of the 2022 floods in Lismore. To elicit community feedback on their adaptive responses, we prepared two exhibitions for Lismore's Living Lab Northern Rivers[1] over six months (between November 2022 and July 2023) and showcased a wide range of adaptive responses to flooding around the world, together with data and visualizations recording the most recent floods. This data was then augmented by a review of the community responses documented in the 2022 NSW Independent Flood Inquiry.

This approach was designed to scope the current barriers to innovation and adaptive planning and design in that city in more depth so that we might frame future research in a more targeted and nuanced way. It is the results of this preliminary research, the subsequent framing regarding the nature and complexity of disaster displacement, and the strategies to minimize permanent displacement and inform a community place-based recovery policy, that we offer here for consideration.

15.4 The 2022 Lismore Floods

The causes of the February–March 2022 Lismore floods are scientifically complex. They involved large-scale climate drivers, the interaction of specific contingent weather events, and long-range weather system changes resulting from anthropogenic climate change. These weather and climatic factors combined with conditions on the ground—both the topography of the region produced by geological forces, as well as more recent changes to ground conditions produced by decades of

[1] The Living Lab Northern Rivers is a collaboration between Southern Cross University, based in Lismore, the authors' institution—the University of Technology Sydney, and the NSW Government, established in the aftermath of the 2021–22 floods to foster research and community engagement. The views expressed in this chapter, however, are the authors' own, and do not represent the views of the Living Lab or the organizations involved.

changes in land use and vegetation cover. In the spring and summer of 2021–2022, the La Niña climate pattern, combined with a positive phase of the Southern Annual Mode and a positive Indian Ocean Dipole, created warmer ocean temperatures around Australia, more moisture in the atmosphere, and increased levels of precipitation (Bureau of Meteorology 2022). Between 23 February and 9 March 2022, these complex interactions produced an atmospheric river of moisture-laden air from the tropical north-east over the Pacific Ocean, which collided with a trough along the east coast of Australia, blocked by a high-pressure system in the south-east over the Tasman Sea. In the Northern Rivers region, this "stuck-in-place" rain system interacted with the region's distinctive orography (mountains and high ground) to generate huge and persistent rainfalls (Weber et al. 2022). The 7,000-square-kilometre Richmond River catchment, largely formed by the Great Dividing Range and the flanks of the Tweed Caldera (Wollumbin/Mt Warning—the 20 million-year-old remains of an enormous shield volcano hundreds of kilometers wide), directed large volumes of water down multiple streams and creeks to converge on Lismore, where water levels rose well above street levels, then slowly discharged through a narrow pass in the hills into another massive floodplain, before finally entering the sea at Ballina, 150 km away (Pitman 2022). See Fig. 15.1.

The environmental historian O'Gorman (2012) notes that British colonists arrived with an understanding of the terms "flood" and "drought" based on lived experiences that bore little resemblance to the environmental realities of Australian conditions. In Britain, "droughts" lasted a few days, or maybe weeks, but not years; while "floods" were part of the life cycle of rivers with consistent and abundant flows, and well-known topographies and soil conditions. In Australia, local First Nations peoples have developed a different understanding of the environment of river catchments over millennia and were aware that Australian river flows and weather have always been highly variable and extremely unpredictable (O'Gorman 2012). The first European colonists saw only the landscape's resource potential. In the catchment of the Richmond River, the Land Act enabled extensive logging of *Toona australis*, the Australian Red Cedar (Northern Star 1938), and further clearing of the endemic rainforest (known locally as the Big Scrub), rapidly changed the landscape from dense forest to pasture (Gibson 2019). Along the river's floodplain, natural wetlands were drained and turned into farmland. These changes reduced the capacity of the soil to absorb and store rainfall and increased run-off, meaning that larger volumes of water flow into creeks and rivers in a shorter amount of time, exacerbating the effects of heavy and prolonged rainfall events (Gibson 2019). See Fig. 15.2.

While the 2022 Lismore floods owe much to global and geophysical conditions, it is ultimately at this local scale, in the very place-specific qualities of a "disaster," that these factors find expression. In and around Lismore, a set of infrastructural and organizational failures—levees that were only built to withstand the last known maximum flood level; the failure of rain gauges and centralized reporting systems; errors and lack of coordination in the early-warning systems—shaped the immediate impact of the flood on the local community (Weber et al. 2022). Despite the countless institutional and policy strategies in the aftermath of the floods, on the ground, much of the recovery was shaped by the self-organized acts of local communities.

Fig. 15.1 Map of the Richmond River on the east coast of Australia, shown here in flood in March 2022. *Source* Authors

The community has evolved from the 1850s when Lismore was a service town for loggers, saw millers, pastoralists, and transportation agents. In the 1970s, an influx of alternative lifestyle seekers and environmentalists, attracted at first by the Aquarius Festival in Nimbin upstream from Lismore, significantly altered the demographic constituency of Lismore. Today, the region is characterized by a unique blend of farmers and environmentalists, its status as an important regional, education and

Fig. 15.2 A timeline of Lismore floods and La Nina events since 1890, exhibited at the first Living Lab exhibition (2022), curated by the authors. The timeline shows that La Niña events have been regular, but flooding increased in the 1950s after land clearing in the catchment and whenever La Niña and a positive Indian ocean dipole coincided. The biggest flood occurred in 2022, reaching a height of 14 m. *Source* Authors

health services center, and an extensive arts community, attracted by the cheap rents in the flood-prone city center (Ferguson et al. 2017).

Lismore's distinctively hybrid community was, even before the floods, developing community-led initiatives to support social and environmental resilience, in fora like Resilient Lismore. After the 2022 floods, this cohesiveness became apparent: in our fieldwork, we witnessed flags with hearts that communicated solidarity. See Fig. 15.3. Elsewhere, shopfronts displayed hand-written notes saying, "I'll be back," accolades were signposted for the "tinnie army" who had used boats to save people stranded in their houses, and clean-ups were structured around community kitchens. See Fig. 15.4. In the absence of timely coordinated government response, these initiatives assumed a new significance: community as the main driver of the city's rapid disaster response and ongoing recovery in the region.

Fig. 15.3 Resilient Lismore. Lismore communities generated solidarity with 'heart' flags, displayed across the city. *Source* Authors

Fig. 15.4 One of the community kitchens operating during the post flood clean-up. *Source* Authors

15.5 Lismore's Adaptive Response

The data about displacement caused by floods is highly uncertain. According to Australian Federal Government data, over 10,000 people were temporarily displaced (Regional Development Australia 2023). However, local reporting has suggested 31,000 (Morrison 2023). Other national reports documented 15,600 people across the region seeking shelter in evacuation centers (Kurmelovs 2023). A survey of 800 flood-affected Lismore residents at the end of 2022 found that while 52 percent had been able to return to their homes; 26 percent were still living either in temporary accommodation (including caravans, sheds, or evacuation "pods") or with friends or family; 18 percent were still living in other highly insecure temporary or crisis accommodation; 4 percent had left the region entirely (King 2023).

To see what was happening on the ground, we looked at two neighborhoods in the high-risk flood zone: the first in North Lismore at the confluence of Leycester Creek and Wilsons River; and the second, a community of twenty households on the other side of the river beside one of its tributary creeks. In both locations, most timber houses had been raised off the ground and then raised again on steel stilts with government funding after the 2017 floods.

Living so close to the river and being exposed to more episodes of minor flooding seemed to encourage a degree of acceptance of the river's variability and a kind of adaptive 'mentality' where flooding was just another part of life, or perhaps a fair exchange for cheap rent and a beautiful landscape so close to the city. Some owners had already adapted by maximizing the use of flood-proof materials in their dwellings—hardwood and fiber-cement paneling allowing them to hose down the interior of their house and return home as quickly as possible after a major flood. In some cases, fences between properties swept away by previous floods had not been replaced, and the ground plane operated like a shared landscape or commons through which people and water variously moved. See Fig. 15.5. Some who remembered past floods had already minimized the accumulation of belongings so that they might move them quickly out of the way, staying at home to monitor the speed of rising flood levels, moving possessions to upper floors as necessary, making the decision to stay or evacuate based on a knowledge of the closest evacuation route to higher ground

Fig. 15.5 Open fenceless landscapes with houses on stilts in north Lismore. *Source* Elise Derwin

and accumulated knowledge of the river and its hazards. The "tinnie army," those that had boats in their garages, rescued neighbors in the night (Maxwell 2023) when it became clear that the SES were unable to come and that flood proofing preparations based on predicted flood levels would be inadequate.

These responses show that the hazards of the floodplain landscape can catalyze community resilience founded on the goal of minimal displacement of belongings and people. The two Lismore neighborhoods suggest approaches that are not for everyone: they require a heightened level of awareness of the signs that a flood is imminent and how to respond when it arrives, but their adaptations give clues that suggest why people choose to stay, what they do, and how they continue to adapt and perhaps how governments might assist. They offer possible approaches that could help minimize the trauma and disruption associated with permanent displacement. It is worth keeping these case studies and the tactics they reveal in mind when considering the full range of resilience responses that might be available to address "disasters."

Despite a 2023 survey of the local community conducted by the Northern Rivers Living Lab, which revealed community values around the urgency of maintaining a sense of community and the sense of connection to the river and its landscape, six months after the floods, both neighborhoods were incentivized to move by lucrative government "buybacks," and disincentivized to stay by insurance companies. Twelve months later, only six percent of the eligible 1,100 households identified by the Reconstruction Corporation had resolved buybacks (Herbert 2023). However, in the neighborhood we studied in North Lismore, half the residents had sold; yet across the river, only one—the other 19 have resisted for the time being. For the government, the cost is the loss of affordable housing assets, significant in a region with chronic housing shortages. For Lismore, the cost is that progressive community fragmentation has encouraged more people to leave. This has left pockets of depopulation in sections of Lismore, further fragmenting the hybrid community and its coherence with its landscape.

15.6 Precedents for Adaptive Response from Localized Tactics with Minimal Displacement to Permanent Retreat

To contextualize our Lismore research, we now move beyond Lismore to examine global precedents that suggest a spectrum of adaptive responses to extreme landscape events. Some of these responses include rural and urban community-led tactics for physical displacement, community building, and communication. Others show the value of community involvement in institutional strategies. All resonate with the discussions of the landscape and community issues arising from the Lismore floods, to which we return later.

Making a temporary retreat to high ground is an effective tactic for subsistence societies or farming communities who operate in time and space continuums as collectives. Indigenous people on the Mississippi floodplain, for example, constructed a series of earthen mounds to create high ground when water levels rose (Da Cuhna 2019). Retreat and resettlement followed the pulses of the high- and low-flow movement of water. During low flow, the river plains provided food and shelter. During high flow, the floodplain was recharged with sediment, nutrients, and new channels, and the mounds offered refuge. Temporary relocation was not an inconvenience but a way of staying in place. A similar construction of high ground was used more recently in Bangladesh on the Brahmaputra River's floodplain, where architect Kashef Chowdhury designed and built a series of plinth islands, tear-shaped to divert water flow around them, and large enough to accommodate housing and livestock and thereby resettle the community temporarily in times of high flow. Water ponds in the center of these high-ground refuges provide water security when the non-potable floodwaters rise. During low flow, when the lower-lying land is not only workable but also abundantly fertile, the high land is cared for by guardians (Graber 2019; Economist 2023). In both these examples, community-driven temporary retreat was and is a part of living in their landscape, and is informed by knowledge of landscape flux.

In many urban environments, this awareness is underdeveloped. Inspired by the prevalence of social media in Indonesia and the strong local cultural spirit of *gotong royong* (mutual aid), designers responded to the communication failures of the 2013 floods by creating an open-source platform that uses bots to gather information on floods (Turpin et al. 2014). The app tracks social media posts with keywords such as "flood," allowing individuals to add their information to a digital map in real time. Combined with information added by government sources, such as official warnings, *PetaBencana.id* allows anyone to access a free map and view the data during a flood event, helping to guide and structure decision-making and humanitarian efforts. Real-time community knowledge sharing like this gives communities agency, improving decision-making about when it is necessary to leave and when it is safe to return in a way that is more fine-grained than centralized government directives (Mahtani et al. 2018).

Building community capacity for food security is another way to give agency back to locals. The R-urban initiative in Paris, for example, developed in response to a lack of localized government action on climate, social, and economic issues, helps communities adapt to an over-reliance on global food markets and supply chains for materials by supporting local production and consumption. A pilot project of urban hubs in Colombes Paris, 500 m apart, was facilitated by residents (Atelier d'architecture autogérée 2013) (see Fig. 15.5). The hubs included spaces for urban agriculture to provide local food security, employment, and engagement; recycling stations to store and recycle locally salvaged materials, drive the circular economy, and build local expertise; and public housing and temporary residences with shared facilities to provide affordable housing and reduce homelessness (Innovation in Politics Institute 2023). Although not exclusively directed toward flooding issues, initiatives like these bind communities, enabling them to mobilize when crisis strikes.

In Kuku, Aotearoa/New Zealand, a revival of Indigenous knowledge systems is providing Māori coastal farming communities with alternative, holistic approaches to address coastal flooding exacerbated by Western systems of land management. Because wetlands have been drained, coastal lowland forests cleared, and biodiversity lost to dairy monocultures, the community needed to find adaptive solutions to address flooding associated with sea-level rise. Māori communities engaged with scientists and landscape architects to develop a series of climate indicators that could trigger the adoption of new farming practices, the regeneration of wetlands, and a shift to high ground if and when impacts occurred (see Fig. 15.6). Transformation would be intentionally slow, to provide process and time for the community to build capacity, develop ownership of ecological issues and build economic sustainability (Allan et al. 2017).

Large-scale institutional programs around the world are beginning to highlight the interdependencies between landscape and community resilience. The Netherlands' Room for the River is a government initiative based on the Dutch culture of co-operative action (Dutch Water Sector 2019) that originated in the Middle Ages, when collective organizations were established to maintain polders (areas of low-lying farmland protected by dykes). In the 1980s, the Dutch term *poldermodel* was applied to consensus-based politics where different social partners were encouraged to cooperate with the government to address social and economic conflicts. The science of Room for the River was in 2006 coupled with consensus-based consultation, fostering landscape knowledge capacity, community respect, and ecological resilience by de-emphasizing risks, hazards and disasters as institutional governance problems, and re-emphasizing that a landscape of extreme flux can be inhabitable by cooperative collectives.

In contexts where institutional approaches have come to dominate, technical and engineered infrastructural solutions and/or permanent retreat are more typical solutions. Buyouts, the primary vehicle in the United States for facilitating retreat (with the same meaning as the Australian term "buyback"), do not typically accommodate a collective approach: wealthier communities whose home values are higher and, accordingly, more expensive to buy out, tend to remain in place, supported by infrastructure upgrades, while more economically challenged communities are offered a

Fig. 15.6 Kuku Farm. Illustrated map of Māori ancestral lands co-designed to address sea-level-rise impacts over time as needed. 1. Protect coastal dunes as a first line of defence 2. Make room for water 3. Protect habitat and biodiversity 4. Diversify farming practices: Aquaculture 5. Protect the most arable land 6. Diversify farming practices: Harakeke. 7. Diversify farming practices: Manuka 8. Resettling the high ground 9. Adaptive infrastructure source: authors drawn by Charlie Curtin, Miguel Guilarte. *Source* Authors

relocation package, often at an undervalue (Yarina et al. 2019). A recent review of the 50 government-led community relocations from towns that had experienced extreme floods over the last 140 years in the US cited problems of disruption to and neglect of social dynamics, bureaucratic hurdles, significant delays, and a long-term lack of resources "that would allow a community to recover its long-term vitality" (Pinter 2021). The result was that, of the 50, only six had increased their population at all, while 22 were in decline. Clearly, there are significant social and environmental justice issues: coordinated retreats "leave vulnerable populations to carry the burdens of migration away from hazardous areas, while wealthy inhabitants are allowed to stay" (Yarina et al. 2019).

15.7 A Suite of Options

If, as Costello (2022) suggests, people, not the river, are both the problem and the solution, moving people away from the river simply displaces the problem. As both the Lismore and international cases suggest, permanent relocation solutions break up communities, intensify social justice issues, and encourage deleterious business-as-usual approaches to land management. While climate change impacts are likely to drive retreat in many vulnerable places, if the aim is to minimize harm in the short term while enhancing long-term social and environmental resilience, simply moving to high ground in a timely fashion while floods recede may be enough to obviate the need for the construction or reinforcement of defensive infrastructure, or even the relocation of communities. To do so will necessitate re-establishing a connection to the river and its floodplain, discouraging a sense of complacency about the unpredictable rhythms and flows of floods, thereby building a community's resilience while empowering communities to make landscapes more resilient in their own right.

From our case study analyses, the best resilience strategies offer a suite of options from staying in place to temporary spatial relocation and, as a last resort, permanent displacement, or retreat. We have identified tactics that locals have used to minimize the spatial and temporal extent of displacement, such as knowing the river and how it operates, learning to read its signs, staying connected, listening to the radio or monitoring social media, making informed decisions about if and when leave, knowing evacuation routes and where local places of high ground are, minimizing possessions, storing important possessions up high, using appropriate building materials to ensure one's house is flood resilient, having a boat, and even preparing for the worst (these were all tactics that emerged from our discussions with community members over the course of this research). They illustrate an attitude that is perhaps the hallmark of resilience: the capacity to make small micro-adjustments after a disturbance, using what is at hand to adapt and, in the long term, to learn from it to survive.

These options suggest that, in the aftermath of an event, individuals or communities can be resourceful under pressure and manage to be minimally displaced during those events that governments and media call "disasters." Those who are not so prepared or endowed might find inspiration from an archive of stories like these; and it might help designers and governments take a more nuanced, local approach when designing or developing tactics and strategies for vulnerable places and communities.

15.8 Policy and Governance

What, then, might governments do to give communities agency?

Firstly, they might address the language of these major events. When language invokes the idea of "natural disasters," inevitably, the response invites an abstract perspective, which leads to stratagems to design "outcomes" or "solutions" to combat them. Permanent "displacement" is one of these, as is hard infrastructure. Rejecting "natural disaster" language might encourage society to rely more on the immediacy that comes from reading the landscape and to develop a language of landscape and weather events that works with that reading in any form of response or policymaking.

Secondly, governments might employ strategies that empower communities. These might include: extensive consultation to identify close-knit communities; enhancing community awareness of the river and its unpredictable floods through an archive of stories and events that encourage the local community to celebrate and learn from the river and each other; identifying and building community capacity and cohesion; ensuring early warning systems are networked and functioning; subsidizing the retrofitting of appropriate flood-proof building materials and other resources; and bringing collaboratives of agencies and stakeholders together to discuss, even if not to agree on, community resilience strategies. At a whole-of-catchment scale, financial support for community-based environmental restoration projects, such as reforestation and riparian planting that slow and infiltrate water across the entire catchment, can mitigate the severity and frequency of floods and improve water quality. These projects that link social and environmental resilience in a mutually reinforcing way may be far more effective in the long run than displacing functioning communities living in high-risk zones.

Finally, policy might be undertaken to empower local communities to deploy tactics that enable them to work with their landscape and its flux, nurturing a suite of options that connect them to the landscape, and not just in the aftermath of an event. With this established, permanent relocation has its place but should be seen as a last resort in these options. Offering a buyout package to communities that don't have much money gives them little choice, especially as they see the rest of their community begin to fragment. Once a tipping point is reached, the community is lost, and it becomes difficult to stay and difficult to start new communities elsewhere. In this context, displacement can be socially destructive, whereas instating agency in a floodplain landscape can be empowering.

All this needs to be place-based and catchment-aware. We can no longer treat nationally significant emergencies as needing national solutions. Permanent displacement is, at its core, an unsustainable practice. It arises from language, society's increasing failure to read the landscape, and the inability to think tactically about resolving complex problems with less red tape, possibly less efficiently, but with more self-sufficiency and compassion. The evidence we have from Lismore, from other flood-affected places, and from the local community we engaged with all indicate that there may be another way, but it will need a shift in the way we talk about and plan for cities and regional landscape systems. Otherwise, we are prone to a new wave of "natural disasters," conjured by our self-extinction tendencies, that ignore the lessons of people and look to blame the river.

Acknowledgements The research described in this chapter has been supported, in part, by the Northern Rivers Living Lab, a collaboration between the Northern Rivers Reconstruction Corporation (NSW Government), Southern Cross University, and the University of Technology Sydney. Nevertheless, the views and opinions expressed in this chapter are solely those of the authors.

References

Adams V, Van Hattum T, English D (2009) Chronic disaster syndrome: displacement, disaster capitalism, and the eviction of the poor from New Orleans. Am Ethnol 36(4):615–636. https://doi.org/10.1111/j.1548-1425.2009.01199.x

Allan P, Bryant M, Smith H (2017) Climate change adaptation for coastal farms: bridging science and indigenous knowledge with art and design. Plan J 2(2017):497–518

Atelier d'architecture autogérée (2013) R-urban: resilient agencies, short ciruits, and civic practices in metropolitan suburbs. Harvard design magazine 37 urbanisms core. https://www.harvarddesignmagazine.org/issues/urbanisms-core/

Bryant M, Allan P (2013) Open space innovation in earthquake affected cities. In: Tiefenbacher J (ed) Approaches to disaster management—examining the implications of Hazards, emergencies and disasters. Intech, pp 183–204. http://www.intechopen.com/books/. ISBN 978–953–51–1093–4

Bureau of Meteorology (2022). http://www.bom.gov.au/. Accessed 19 Mar 2022

Claghorn J, Werthmann C (2015) Shifting ground: landslide risk mitigation through community-based landscape interventions. J Landsc Arch 10(1):6–15. https://doi.org/10.1080/18626033.2015.1011419

Costello O (2022) Buuban butherun (flood stories). In: 2022 New south wales independent flood inquiry volume three: appendices, State of New South Wales, Sydney, pp 150–171

Da Cunha D (2019) The invention of rivers: Alexander's eye and Ganga's descent. University of Pennsylvania Press, Philadelphia

de Certeau M (1984) The practice of everyday life (trans: Rendall SF). University of California Press, Berkeley

Deming M, Swaffield S (2011) Landscape architecture research: inquiry, strategy, design. John Wiley & Sons, Hoboken, NJ

Dutch Water Sector (2019) Room for the river explained. https://www.dutchwatersector.com/news/room-for-the-river-explained

Economist (2023) Walking on water. Econ 447(9342):70–71

Ferguson H, Kijas J, Wessell A (2017) Towards reflexive localism: exploring the diverse co-creators of alternative food across time in the Northern rivers of New South Wales, Australia. J Hist Geogr 56:14–21. https://doi.org/10.1016/j.jhg.2016.12.005

Gibson G (2019) The Richmond: a wounded river. http://www.morethanjusttalk.com.au/the-richmond-a-wounded-river/

Graber N (2019) Retrospective: Kashef Chowdhury. Arch Rev 1460:72–81

Herbert B (2023) Eighteen months on from record NSW Northern rivers flooding, just 70 eligible homes have buyback settled. https://www.abc.net.au/news/2023-08-28/18-months-on-nsw-northern-rivers-flood-only-70-buybacks-settled/102783030. Accessed at 09 Sept 2023
Innovation in Politics Institute (2023) R-urban-ecological transition movement involving citizens in civic resilience networks. https://innovationinpolitics.eu/showroom/project/r-urban-ecological-transition-movement-involving-citizens-in-civic-resilience-networks/
King S (2023) New southern cross study reveals ongoing housing and mental health challenges for flood-affected. https://www.scu.edu.au/news/2023/new-southern-cross-study-reveals-ongoing-housing-and-mental-health-challenges-for-flood-affected/. Accessed 1 Aug 2023
Kurmelovs R (2023) Lismore one year after the floods. The saturday paper. https://www.thesaturdaypaper.com.au/news/economy/2023/03/04/lismore-one-year-after-the-floods. Accessed 14 Aug 2023
Lunghino B, Santiago Tate AF, Mazereeuw M, Muhari A, Giraldo FX, Marras S, Suckale J (2020) The protective benefits of tsunami mitigation parks and ramifications for their strategic design. Proc Natl Acad Sci USA 117(20):10740–10745. https://doi.org/10.1073/pnas.1911857117. Epub 2020 May 4. PMID: 32366652; PMCID: PMC7245067
Mahtani N, Turpin E (2018) Neuroecologies of attention and intelligence in the Megacity. Perspecta 51:107–118. http://www.jstor.org/stable/45215607
Maxwell R (2023) Tinnie army saved many from Lismore flood catastrophe. In: Mandurah mail. https://www.mandurahmail.com.au/story/8104196/tinnie-army-saved-many-from-lismore-flood-catastrophe/
Mazereeuw M, Yarina E (2017) Emergency preparedness hub: designing decentralized systems for disaster resilience. J Arch Educ 71(1):65–72. https://doi.org/10.1080/10464883.2017.1260928
McBride BB, Brewer CA, Berkowitz AR, Borrie WT (2013) Environmental literacy, ecological literacy, ecoliteracy: what do we mean and how did we get here? Ecosphere 4(5):67. https://doi.org/10.1890/ES13-00075.1
McLeod C (2023) NSW floods: just 11% of applications approved for buybacks and no homes retrofitted nearly 20 months on. The guardian. https://www.theguardian.com/australia-news/2023/oct/20/nsw-floods-just-11-of-applications-approved-for-buybacks-and-no-homes-retrofitted-nearly-20-months-on?utm_source=pocket_reader. Accessed 19 Oct 2023
Morrison G (2023) As Lismore slowly repairs it's preparing for higher floods under climate change. Cosmos magazine. https://www.echo.net.au/2023/02/as-lismore-slowly-repairs-its-preparing-for-higher-floods-under-climate-change/. Accessed 14 Aug 2023
Mosneaga A (2023) Solutions discourse in disaster displacement: implications for policy and practice. Disasters 47(3):676–699. https://doi.org/10.1111/disa.12566
Northern Star (1938) Early days of cedar getting on the Richmond. https://trove.nla.gov.au/newspaper/article/94569226
Nurandini BA, Mukminan (2019) Disaster management based on geo-literacy. IOP Conf Ser Earth Environ Sci 271:012032
O'Gorman E (2012) Flood country an environmental history of the Murray-darling basin. CSIRO Publishing
Pinter N (2021) The lost history of managed retreat and community relocation in the United States. Elementa 9(1):00036. https://doi.org/10.1525/elementa.2021.00036
Pitman A (2022) The background and causes of the extreme rainfall and flooding in NSW in 2022. In: 2022 New south wales independent flood inquiry volume three: appendices. State of New South Wales, Sydney, p 2–22
Regional Development Australia (2023) Northern rivers, regional economic recovery plan 2025. https://www.rdanorthernrivers.org.au/recovery2025/wp-content/uploads/2023/04/RDA-Regional-Recovery-Plan-Northern-Rivers-2023.pdf. Accessed 1 Aug 2023
Rhodes A, Besbris M (2022) Best laid plans: how the middle class make residential decisions post-disaster. Social Problems 69(4):1137–1153. https://doi.org/10.1093/socpro/spab026
Shevellar L, Westoby P (eds) (2016) The routledge handbook of community development research. Routledge, London

Šakić Trogrlić R, Duncan M, Wright G, van den Homberg M, Adeloye A, Mwale F (2022) Why does community-based disaster risk reduction fail to learn from local knowledge? Experiences from Malawi. International Journal of Disaster Risk Reduction 83:103405. https://doi.org/10.1016/j.ijdrr.2022.103405

Turpin E, Holderness T, Quaggiotto G (2014) "combining 'big' and 'small' data to build urban resilience in Jakarta. United Nations Global Pulse Blog

Weber T, Barnes C, Flook I, Chandler F, Charlton-Henderson S, Calley L, Vaze J, Flett D (2022) Rapid project prioritisation for flood resilience in the northern rivers region. CSIRO, Australia

Yarina L, Mazereeuw M, Ovalles L (2019) A retreat critique: deliberations on design and ethics in the flood zone. J Landsc Arch 14(3):8–23. https://doi.org/10.1080/18626033.2019.1705570

Chapter 16
Displacements Due to Disaster Risk and Sense of Belonging in Andean Communities: The Case of Maca, Peru

Carlos Zeballos-Velarde, Cinthya Butrón-Revilla, Gabriela Manchego-Huaquipaco, and Sandra Carrasco

Abstract This chapter investigates the complex dynamics of post-disaster resettlement initiatives through a case study of Maca, Peru. Maca is a rural town located in the Colca Valley, prone to volcanic and seismic hazards. The increasing volcanic activity in the valley triggered the resettlement of local communities, which have developed a strong attachment to their ancestral lands, with strong socio-cultural connections and community bonds. The research is based on documentary review, on-site observation, and semi-structured questionnaires. The study evidences a profound distrust of governmental authorities due to unfulfilled promises and lack of transparent communication, hindering effective disaster management. However, residents demonstrate resilience and adaptability rooted in their deep cultural attachment to the Colca Valley, fostering community cohesion and collective identity despite perceived risks. The findings highlight the importance of community involvement and empowerment in disaster management, advocating for collaborative approaches that integrate local knowledge and resources to develop sustainable strategies that respect cultural values and enhance community resilience. This chapter offers valuable insights for policymakers and practitioners navigating the complexities of post-disaster resettlement while preserving community heritage and promoting sustainable development.

16.1 Introduction

In the aftermath of disasters, the resettlement or relocation of disaster-affected communities emerges as a strategy to mitigate risks associated with site-specific vulnerabilities (Jha et al. 2010). Resettlement, however, is a complex and multidimensional process not limited to providing housing alternatives in a new site (Gaillard

---

C. Zeballos-Velarde (✉) · C. Butrón-Revilla · G. Manchego-Huaquipaco
Faculty of Architecture and Urbanism, National University of San Agustín, Arequipa, Peru
e-mail: czeballos@unsa.edu.pe; czeballosv@gmail.com

S. Carrasco
School of Humanities and Social Sciences, Deakin University, Melbourne, Australia

C. Ochiai et al. (eds.), *Disaster and Displacement*, Disaster Risk Reduction,
https://doi.org/10.1007/978-981-96-2583-3_16

2015). Resettlement profoundly impacts the social, cultural, and economic fabric of communities, challenging their ability to rebuild social ties, livelihoods, and political and cultural structures (Oliver-Smith 1991; Carrasco et al. 2016b; Gaillard 2015). Therefore, even though in some cases resettlement is a viable option, it remains the less desirable option for post-disaster recovery as it implies creating permanently displaced communities in which people are rarely engaged in the decision-making in the top-down processes often led by governments in coordination with multiple stakeholders (Jha et al. 2010; Carrasco et al. 2016a).

Governments have a crucial role in coordinating such endeavors, utilizing their authority, resources, and responsibility for public safety. Nevertheless, the intricacies involved in resettlement after a disaster provide substantial obstacles, requiring a thorough analysis of the government's involvement in this undertaking. In the context of resettlement initiatives, governments at various levels are confronted with the challenging issue of striking a balance between urgent humanitarian demands and long-term resilience aims. It is necessary to deal with complex sociopolitical factors, ethical considerations, and environmental imperatives to accomplish this successfully. Further complicating the effort to ensure fair outcomes for affected communities is the historical legacy of displacement, which, when combined with discrepancies in resource distribution, makes the task even more difficult to accomplish (Opdyke et al. 2018; Kaushal and Piyatadsananon 2013).

Scholarship on post-disaster resettlement highlights the intricate nature of government actions and their impact on social justice and sustainable development. Blaikie et al. (2003) emphasize the complex relationship between political, economic, and environmental elements in influencing resettlement results. They stress the importance of developing sophisticated policy solutions to address this complexity. Oliver-Smith (2009) examines the social aspects of displacement, specifically exploring the imbalances in power and cultural dynamics that impact the experiences of the impacted populations.

The arguments around decision-making processes and community participation are crucial in discussing post-disaster resettlement governance. Wisner et al. (2012) criticize the use of top-down strategies in resettlement, asserting that these methods typically neglect local knowledge and priorities, which in turn worsens vulnerabilities. On the other hand, Aldrich (2011) supports using participatory models that provide communities with the capacity to make decisions, emphasizing their ability to promote social unity and adaptability.

Government-led resettlement attempts can involve significant ethical considerations. Cernea (2000) emphasizes the significance of obtaining informed permission, providing equitable compensation, and preserving cultural heritage throughout displacement processes. Cernea also cautions against the potential dangers of pressure and marginalization. Hugo (1996) examines the societal consequences of forced removal, illuminating the lasting psychological distress and societal upheaval experienced by those impacted.

Conversely, there have been very few studies that have investigated the problem of resettlement through the lens of identity, cultural attachment, and distinctive landscape features. Rural communities around tend to develop a significant emotional connection with their land (Baldwin et al. 2017), which they regard to be the place where their ancestors formerly lived (Xu et al. 2019). This connection can sometimes become more significant than their personal sense of safety when confronted with the possibility of disaster. The Peruvian Andes are one of these regions, with a strong connection between the community and the landscape and coexistence with natural hazards. In these Andes, communal bond rituals are preserved that date back centuries, reflecting pre-colonial heritage. Zeballos et al. (2021a, 2021b) investigated the heritage values, both tangible and intangible heritage values present in the Colca Valley, which is located in the Andes of Arequipa, Peru. They also investigated the vulnerabilities of the region's own fragility or inadequate resilience, as well as the influence of agricultural techniques and ancestral values on the community organization.

16.1.1 Background on Disaster Risk in Andean Communities

Peru is located in the Pacific Ring of Fire and is transversed by the Andes Mountain range. Thus, it is prone to multiple hazards, and its vulnerabilities result in various disasters such as landslides, volcanic eruptions, earthquakes, and floods. Disaster risks in Peru have historical, socioeconomic, and cultural components, which particularly threaten Andean communities due to the characteristics of their natural environments located in remote mountainous areas, with limited infrastructure, and historically marginalized and neglected by authorities.

The merging of the South American and Nazca tectonic plates generates significant seismic and tectonic activity, resulting in frequent earthquakes caused by subduction and crustal movement. These seismic events can wreak havoc on local communities, disrupting daily life, livelihoods, and infrastructure. Volcanic eruptions, such as those of the Sabancaya volcano in the Colca Valley, also inflict substantial damage on crops, livestock, and community health. Proximity to these volcanic formations puts human populations at risk from ash clouds, lava flows, and pyroclastic flows. Furthermore, the steep slopes of the Andes make landslides a constant threat, particularly during the rainy season.

Disaster risk in Andean communities has deep historical roots. Since ancient times, indigenous groups have managed these environmental hazards by creating unique risk-reduction techniques. The Spanish colonization and subsequent development of infrastructure altered these communities' vulnerability patterns, contributing to the severity of disasters in some cases. Poverty, limited access to education and healthcare, and subsistence agriculture further expose communities to hazards. Because they lack proper infrastructure and services and are frequently situated on unstable hillsides, informal settlements are more susceptible to disasters. Nonetheless, a large number of indigenous Andean people continue to practice their traditional ways and pass down their expertise in disaster resistance.

Maca is a picturesque town in southern Peru's scenic Colca Valley with roots in Andean colonial miscegenation. Maca boasts an extraordinary cultural legacy.

However, the community's position is hazardous because of its proximity to the Sabancaya volcano and location on an active geological fault. The Peruvian central government has proposed a relocation project that would move the population to a desert area in Majes, roughly 150 km from their original home, due to the ongoing threat of earthquakes and volcanic eruptions. Despite being driven by a desire to ensure public safety, this resettlement project presents a wicked problem with complex ramifications and impacts. Forcibly removing the residents from their ancestral homes in Maca would lose their rich cultural legacy, which is ingrained in many generations of local history and traditions. Conversely, resettlement would protect the community, reducing the exposure to hazards at the expense of creating uprooted communities. Therefore, protecting the community and preserving the region's history and traditions are significant concerns that call for a careful balance and a multidisciplinary approach to solving the problems facing Maca.

16.2 Literature Review

The concept of disaster risk has been defined in the literature using a variety of approaches. The United Nations Office for Disaster Risk Reduction defines disaster risk (UNISDR 2015) as "the potential loss of life, injury, property, or the environment that could occur in a specific place and over a specific period of time due to the interaction between natural or human-induced hazards and the vulnerability and exposure of an exposed population." Similarly, The United Nations Development Programme (UNDP 2004) characterizes it as "the potential for loss or harm associated with a particular hazard, which is influenced by the exposure, vulnerability, and capacity of the affected population." The World Bank offers an economic perspective, defining disaster risk as "the expected annual loss (EAL) from specific hazards" (Hallegatte 2012). Wisner et al. (2004) underline the socio-economic dimension of disaster risk as "the potential for loss and damage of life, livelihoods, health, property, infrastructure, and environmental resources" (Wisner et al. 2004).

Displacement due to disaster risk can profoundly affect individuals and communities, significantly reducing their sense of belonging. Kibreab (1997) discussed displacement as the involuntary movement of people from their homes and communities due to disasters. People frequently experience a break in their social and cultural ties—essential to their sense of belonging—when they are forced to flee their homes and communities due to disasters. The uprooting of individuals or communities due to unexpected events forces people to leave familiar surroundings, losing the social and cultural ties that sustain their sense of identity (Cernea 2000). Displacement's physical and psychological upheaval can cause emotions of loneliness, alienation, and identity loss. These disruptions in familiar environments and social networks can erode the sense of belonging, making it difficult for people to develop a sense of connection to a new location or group. This exacerbates the psychological and social effects of being uprooted.

16.3 Methodology

To understand why the Maca community's residents stay in their region despite the possibility of volcanic risks, this article will address two research questions: (a) What are the factors influencing the relationship between the community of Maca and the authorities in the context of risk management, and how do these dynamics impact disaster preparedness and response strategies? (b) What is the correlation between the perception of community risk and the sense of belonging to the place, and how does this relationship impact community resilience and disaster preparedness?

The semi-structured interview method was the foundation of this study's methodology, which had the following background: (i) A simultaneous participatory workshop carried out by the Emergency Urbanism research project in the Colca Valley was conducted in three districts: Maca, Achoma, and Cabanaconde. This allowed us to establish initial contact with the Colca Valley population and solidify the diagnosis that would be made for the population study. (ii) Strategic citizens participated in a focus group that helped to strengthen the survey questionnaire that was going to be prepared and address social issues that had come up during the participatory workshop. (iii) surveys conducted in ten Colca Valley areas, revealed a variety of information about the resettlement process, the value of ancestral activities, and the resilience of Andean rural communities, among other topics. Applying these three techniques led to the conclusion that semi-structured interviews were necessary to learn more about how the Maca district's relocation process was perceived.

The sample of this research corresponds to 14 interviewees, with a family composition of between 2 to 6 members. The sample shows agricultural activity as the main economic activity, followed by commercial and tourist activity in second and third place. See Table 16.1.

16.4 Characteristics of the Site

The district of Maca is located in the Colca Valley; it has an area of 227.48 km^2 and a population of 701 inhabitants (Zeballos-Velarde 2021a, 2021b). This district is located on the left bank of the Colca River, situated in an area of active geodynamics (INGEMMET 2022). In addition, the district of Maca is close to the Sabancaya volcano, which since November 2016 has been experiencing an eruptive process characterized by recurrent explosions with the presence of gases and ash (INGEMMET 2017). This process has been affecting crops (Roberti et al. 2018) and causing earthquakes related to volcanic eruptions (Tavera et al. 2022).

Table 16.1 Sample characterization

ID	Age group	Gender	Family composition	Activity	Relocation decision
I1	Elderly	Female	1	Commerce	No
I2	Elderly	Female	2	Commerce	No
I3	Teenager	Female	6	–	Yes
I4	Adult	Female	2	Agriculture	Yes
I5	Adult	Male	2	Agriculture	Yes
I6	Adult	Female	3	agriculture	Yes
I7	Elderly	Male	–	–	–
I8	Elderly	Female	1	Commerce	Yes
I9	Teenager	Male	3	Agriculture	No
I10	Adult	Male	4	agriculture	Yes
I11	Adult	Female	3	Housewife	Yes
I12	Adult	Female	3	Tourism	Yes
I13	Teenager	Female	5	Agriculture	Yes
I14	Adult	Female	5	Tourism	Yes

16.4.1 Historical Disasters in Maca

Since 1991, the residents of Maca have been affected by landslides (Lacroix et al. 2015). Since then, the Peruvian Geological, Mining, and Metallurgical Institute-INGEMMET, with the cooperation of the French Research Institute for Development-IRD, has carried out different geophysical and geological studies to characterize the geological hazards in this area, concluding that part of the town of Maca is located in a geological hazardous landslide zone that cannot be mitigated (Lacroix et al. 2012).

Monitoring studies carried out between 2016 and 2017 by the Regional Government of Arequipa show the latent danger of the district. Therefore, the relocation of part of the district is proposed as a response. In this regard, the municipality of Maca requests INGEMMET to carry out a geological hazard evaluation of the Majes area, Pampas Bajas, with the purpose that the residents of Maca, affected by the landslides, could be resettled to this coastal desert area (Taipe Maquerhua and Araujo Huamán 2017).

Geological hazards are defined as "geological processes or phenomena" that can potentially cause death, injury, or other health effects in addition to property damage, loss of services and means of transportation, disruption of the social and economic system, or harm to the environment. Geological hazards include earthquakes, volcanic activity, active faults, as well as landslides. Also, hydrometeorological factors contribute considerably to some of these processes. However, the report made by INGEMMET only addresses landslide risks. In order to determine whether the chosen relocation site in the Majes region, awarded by the regional government agency-AUTODEMA, which manages the development of the area, was a safe area against the aforementioned geological hazards and may be used for Maca population resettlement, this evaluation was necessary. The resettlement area's location on the Pampa Costanera, which is separated by two main ravines that run from northeast to southwest, and features large sections of gently sloping, flat-undulating terrain typically with a slope of less than 5° to 10°, has been determined by this study. Rock falls and landslides are improbable to impact the resettlement sites (Pampas Majes) because of the low slopes and restricted height of the hills in the area (Taipe Maquerhua and Araujo Huamán 2017). Figure 16.1 shows the location of the original settlements in Maca and the resettlement sites in Majes.

Positive information about the relationship between risk mitigation and spatial configuration can be found in the relocation report from the Maca district. The population's economic activities, which are based on agriculture and the methods they have evolved throughout time, are dependent on the geographic arrangement; however, this relationship has not been taken into account. It is imperative to take into consideration that the farming methods employed on land with a 5° to 10° slope in coastal strips differ from those used on land with a slope of more than 10° in Andean regions, as seen in Fig. 16.2. Based on this view, the majority of Maca's residents still reside in the Maca district and have not relocated completely.

Figure 16.2 (left) shows the difference between Maca, which is an established settlement in the mountains with a particular landscape. Conversely, Fig. 16.2

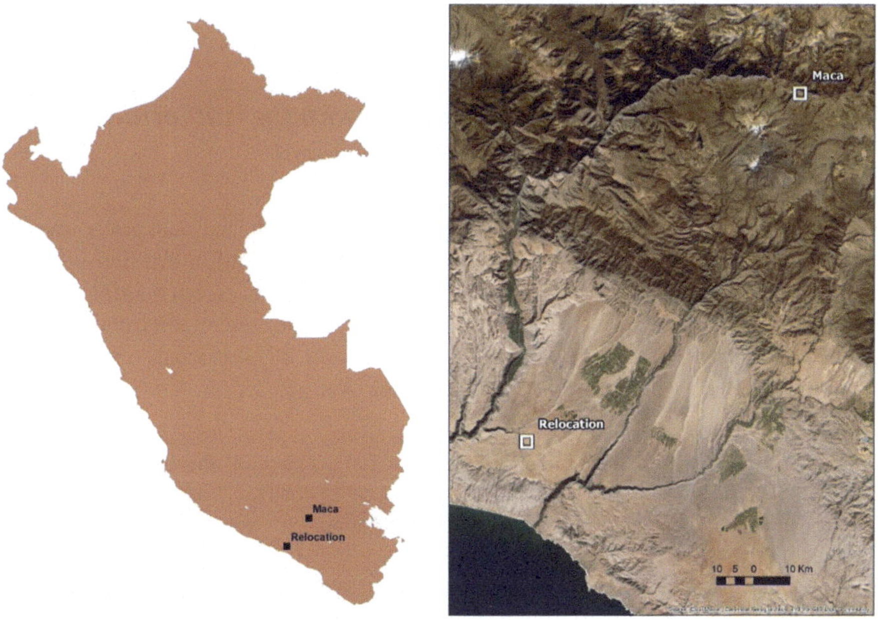

Fig. 16.1 Location of Maca and the relocation area in Majes. *Source* ESRI. Map made by authors

Fig. 16.2 Settlements in Maca (Left) and comparison with resettlement sites in Majes (Right). *Source* Carlos Zeballos

(right) shows the landscape of the resettlement site in Majes, an area with recently established settlements on the dry Peruvian coast.

16.5 Results

The analysis of the semi-structured interviews led to the discovery of two findings concerning the residents of Maca's willingness to remain in their homes despite the presence of a hazardous volcano.

16.5.1 Relationship of the Community with the Authorities Regarding Disaster Risk Management

It is crucial to highlight that the community's perception of the national government's proposed resettlement program amounts to mere rhetoric and unfulfilled pledges. Interviewed participants highlighted that they received promises, but tangible actions remain elusive, leaving affected communities in a state of uncertainty and distrust. Moreover, the interviewees expressed a notable level of skepticism towards both municipal and national authorities, as shown below:

> When they first started talking, they said they were moving to Majes. That doesn't seem to exist; instead, it seems that the authorities themselves do not mobilize, they do not demand, there is nothing, only promises, nothing more (male adult farmer).
>
> Yes, they promised us... since when have we been suffering..., it's just a promise, just words (female teenager farmer)
>
> The (national) government has explained to us how to relocate to Majes; we have even signed documents. I don' t particularly believe them, because they tell us one thing and, in the end, they come up with another. There is nothing here (female adult tourism worker).

The feeling of distrust among affected communities evidences the need for meaningful engagement and transparent communication to rebuild faith in the government's initiatives and ensure the effective implementation of any future plans.

16.5.2 Risk Perception and Sense of Community

Interviewees showed a high level of awareness regarding the vulnerabilities of their settlements in Maca and the exposure to natural hazards. Communities articulated a profound understanding of the potential impacts these events could have on their lives, properties, and belongings. From the interviewees' narratives, it became evident that they had directly experienced the disruptive impacts of such occurrences. Remarkably, during the interviews, there was unanimous consensus that the seismic event of 1991 stood out as the most impactful, leaving a lasting imprint on the town's landscape. Residents vividly recalled the widespread devastation caused by the earthquake, underscoring the need for ongoing preparedness and resilience measures within their community:

> The houses have fallen, and there were dead people, deaths (female adult tourism worker).
>
> Well, we have been left without a house; our land has been left without canals (male adult farmer).
>
> Homes and agricultural land... have been opened and collapsed... (female adult tourism housewife).

Moreover, the study demonstrates the vulnerability of the residents of Maca and the multiple hazards they face. Interviewees have expressed their concerns about potential threats. However, they have also conveyed their resilience in adapting to life amidst risk. Despite their fears, they have developed a remarkable capacity to cope with uncertainty, demonstrating a deep-seated understanding of the precarious nature of their environment. Additionally, when interviewees were questioned about their perception of risk, they provided insightful reflections on the delicate balance they maintain between the dangers surrounding them and the necessity of carrying on with their daily lives:

> We have already gotten used to it. There is no longer fear. It is normal for some to despair; this latest earthquake (2022) has knocked down houses (male adult farmer).
>
> The tremors make us scared. Not anymore, because we are already used to it... they say that the town is going to be lost, but the town is never going to be lost (female, senior adult, merchant).

Conversely, the interviewees exemplify the profound sense of community and belonging among the residents of Maca towards the Colca Valley. Their deep-rooted connection with the area's geographical and cultural milieu stands as a fundamental element of their enduring presence. Through their narratives, it becomes evident that their attachment extends beyond mere habitation; it is a bond forged over generations, steeped in reverence for the land and its rich heritage. Various remarks gathered from the interview analysis evidence this sentiment, highlighting the strong bonds between the people of Maca and the Colca Valley:

> It's my town; I need to live here (female, senior adult, merchant)
>
> … I was born here, I love my land very much, I love my Maca very much (female, senior adult, merchant)
>
> Because our ancestors gave us our lands (male adult farmer).
>
> …There is a community here (male adult farmer)

16.6 Discussion: Disaster Risk and Vulnerability in Andean Communities

The results of the semi-structured interviews conducted in Maca highlighted two crucial components of community disaster management in Andean communities.

Firstly, they revealed a complex relationship between the community and different governmental agencies at local and national levels regarding disaster risk management. Residents expressed suspicion and distrust toward municipal and national authorities, despite promises from the national government to relocate them. The government's inability to fulfill the commitments made to the community weakens disaster management efforts and undermines trust in authority. To improve disaster preparedness, there is a clear need for enhanced communication, transparency, and accountability in government actions aimed at reducing disaster risks. This distrust echoes findings by Morrow (1999), showing how past neglect or marginalization by authorities can breed deep-seated suspicion within communities, hindering their ability to engage with governmental institutions during crises. Similarly, Quarantelli (2005) suggests that perceptions of government ineptitude or corruption can further erode trust and exacerbate tensions. Miscommunication exacerbates these issues; when people feel their concerns are not addressed or their views are ignored, confidence diminishes, leading to stress, dissatisfaction, and inefficient communication pathways. Clear, effective communication is essential to avoid misconceptions and misinterpretations in disaster management.

Secondly, the juxtaposition of residents of Maca's perception of risk with their sense of community exposes a complicated relationship. Residents demonstrate resilience and adaptability, despite being aware of the high-risk nature of their location and the potential impact of disasters on their lives. They have developed coping strategies and a level of acceptance of living with danger, even though seismic events have caused them anxiety and difficulties. Residents often display a strong sense of community resilience, rooted in their shared experiences and collective identity, despite perceived threats. This aligns with Cutter et al.'s (2008) discussion of how communities leverage social capital, solidarity, and local knowledge to recover from and cope with disasters. Residents who feel a sense of belonging to their environment are more likely to assist and cooperate with one another, facilitating their ability to overcome obstacles and adapt to changing circumstances (Carrasco et al. 2023).

The profound sense of belonging and devotion to the Colca Valley underscores its cultural and historical significance to the local community, contributing significantly to their resilience. This aligns with Devine-Wright's (2013) research, which emphasizes how cultural attachment and a strong sense of place shape individuals' perceptions of risk. Emotional ties to one's community and environment influence how people perceive environmental challenges and respond to them. When individuals have a deep cultural attachment, they may be more inclined to mitigate or accept certain hazards associated with their surroundings, fostering a sense of belonging and identity.

In the context of disaster management, these findings underscore the importance of community involvement and empowerment. Effective strategies for mitigating disaster risks should not solely focus on relocating physical assets but should prioritize building trust, fostering community cohesion, and tapping into local knowledge and resources. To enhance resilience and establish sustainable disaster management practices in Andean communities like Maca, collaborative approaches involving active participation from community members, government agencies, and other stakeholders are crucial.

Furthermore, the study highlights the need for a nuanced understanding of community resilience that goes beyond infrastructure and physical preparedness. It emphasizes the importance of cultural values, traditions, and social cohesion in shaping responses to environmental challenges. By recognizing and leveraging these aspects, disaster management efforts can be more holistic and effective, leading to more resilient communities in the face of adversity.

16.7 Conclusion

The research suggests that a strong sense of belonging to the landscape can contribute positively to risk co-management efforts. This emotional connection can foster a sense of responsibility and stewardship among community members, encouraging them to actively participate in strategies to mitigate risks and manage the territory more sustainably.

Conversely, while the displacement of communities from high-risk areas is a complex process with multiple ramifications and negative impacts on affected communities, resettlement is still a common approach used by governments to minimize risks. Therefore, incorporating pre-resettlement work in disaster management and resettlement plans is crucial in communities with strong attachment to their territory, such as rural communities. By interacting with local communities, there is an opportunity to build trust and cooperation among stakeholders. This early engagement can also help identify potential risks and develop strategies to address them effectively.

Finally, the findings highlight the importance of understanding and integrating the human dimension into risk management strategies. By recognizing and valuing the emotional connection that people have with their environment, it is possible to enhance the effectiveness and sustainability of co-management efforts.

References

Aldrich DP (2011) The power of people: social capital's role in recovery from the 1995 Kobe earthquake. Nat Hazards 56(3):595–611

Baldwin C, Smith T, Jacobson C (2017) Love of the land: social-ecological connectivity of rural landholders. J Rural Stud 51:(37–52)

Blaikie P, Cannon T, Davis I, Wisner B (2003) At risk. Natural hazards, people's vulnerability and disasters. Routledge, London

Carrasco S, Ochiai C, Okazaki K (2016a) Disaster induced resettlement: multi-stakeholder interactions and decision making following tropical storm washi in Cagayan de Oro, Philippines. Procedia-Soc Behav Sci 218:(35–49)

Carrasco S, Ochiai C, Okazaki K (2016b) Influence of housing designs on resident-initiated housing modifications in resettlement sites in Cagayan de Oro, Philippines. J Soc Saf Sci 28:61–68

Carrasco S, Dangol N, Faleh M (2023) Rethinking social networks in responding to COVID-19: the case of African migrants in Melbourne's public housing. Int J Disaster Risk Reduct 98:104073

Cernea MM (2000) Risks, safeguards and reconstruction: a model for population displacement and resettlement. Econ Polit Wkly 35(41):3659–3678

Cutter SL, Barnes L, Berry M, Burton C, Evans E, Tate E, Webb J (2008) A place-based model for understanding community resilience to natural disasters. Glob Environ Change 18(4):598–606. https://doi.org/10.1016/j.gloenvcha.2008.07.013

Devine-Wright P (2013) Think global, act local? the relevance of place attachments and place identities in a climate changed world. Glob Environ Chang 23(1):61–69

Gaillard JC (2015) Post-disaster resettlement. People's response to disasters in the Philippines: vulnerability, capacities, and resilience. Palgrave Macmillan US, New York

Hallegatte S (2012) A framework to assess the economic resilience to natural disasters

Hugo G (1996) Environmental concerns and international migration. Int Migr Rev 30(1):105–131

INGEMMET (2017) Breve Evaluación de Actividad Eruptiva del Volcán Sabancaya 2016–2017 (Arequipa)

INGEMMET (2022) Monitoreo del deslizamiento de Maca, periodo 2021, distrito Maca, provincia Caylloma (Lima)

Jha AK, Barenstein JD, Phelps P, Pittet D, Sena S (2010) Safer homes, stronger communities: a handbook for reconstructing after natural disasters. World Bank Publications

Kaushal K, Piyatadsananon P (2013) Achieving success in post-disaster resettlement programmes through better coordination between spatial and socio-economic/cultural factors. Int J Disaster Resil Built Environ 4(3):352–372

Kibreab G (1997) Environmental causes and impact of refugee movements: a critique of current debate. Disasters 21(1):20–38

Lacroix P, Wathelet M, Mariño Salazar J, Taipe Maquerhua EL, Tatard L, Guillier B, Norabuena E (2012) Caracterización de los Procesos Sísmicos y Gravitacional sobre el Deslizamiento de Maca (Colca) (Lima)

Lacroix P, Berthier E, Maquerhua ET (2015) Earthquake-driven acceleration of slow-moving landslides in the Colca Valley, Peru, Detected from Pléiades Images. Remote Sens Environ 165:148–158

Morrow BH (1999) Identifying and mapping community vulnerability. Disasters 23(1):1–18

Oliver-Smith A (1991) Successes and failures in post-disaster resettlement. Disasters 15(1):12–23

Oliver-Smith A (2009) Nature, society, and population displacement. Environ Hazards 8(1):3–16

Opdyke A, Javernick-Will A, Koschmann M (2018) A comparative analysis of coordination, participation, and training in post-disaster shelter projects. Sustain 10(11):4241

Quarantelli EL (2005) A social science research agenda for the disasters of the 21st century: theoretical, methodological and empirical issues and their professional implementation. In: Perry RW, Quarantelli EL (eds) What is a Disaster?

Roberti G, Ward B, Van Wyk de Vries B, Araujo Huamán GE, Aguilar Contreras R, Manrique Llerena N, Roberts NJ (2018) Estudio Geomorfológico y Paleoambiental de las Avalanchas de Escombros de Maca y Lari, Valle de Colca

Taipe Maquerhua EL, Araujo Huamán GE (2017) Evaluación de peligros geológicos de la zona de reubicación de Maca "PB5-Majes". Región Arequipa, provincia de Caylloma, distrito de Majes, (Lima)

Tavera H, Centeno E, Mamani C (2022) El Sismo de Maca del 16 de Marzo, 2022 (M5. 5). Aspectos Sismológicos (Provincia de Caylloma-Región Arequipa). (Lima)

UNDP (2004) Reducing disaster risk: a challenge for development. A Global Report. United Nations Development Programme, Bureau for Crisis Prevention and Recovery, New York

UNISDR (2015) Sendai framework for disaster risk reduction 2015–2030. UNISDR (Sendai, Japan)

Wisner B, Gaillard JC, Kelman I (2012) Handbook of hazards and disaster risk reduction. Routledge, New York

Wisner B, Blaikie P, Cannon T, Davis I (2004) At risk: natural hazards, people's vulnerability, and disasters. Routledge

Xu G, Li Y, Hay I, Zou X, Tu X, Wang B (2019) Beyond place attachment: land attachment of resettled farmers in Jiangsu, China. Sustain 11(2):420

Zeballos-Velarde C, Butrón-Revilla C, Manchego-Huaquipaco G (2021a) Identidad territorial, globalización y patrimonio en el valle del Colca. In: Yory CM (ed) Identidad territorial, globalización y patrimonio en el valle del Colca, Arequipa. In: Perú En C, Yory M (ed) Identidad territorial: globalización y patrimonio. Universidad Católica de Colombia, B. Bogotá, pp 132–153

Zeballos-Velarde C (2021b) Identidad y Resiliencia en Paisajes Rurales en Riesgo. El Caso de Maca en el Valle del Colca, Perú. UIA 2021 Rio: 27th World Congress of Architects, 2021 Río de Janeiro. Union Internationale d'Architectes

Chapter 17
Post-Disaster Livelihood in Indigenous Communities in Taiwan

Tsai Sung Lun

Abstract There is a lack of discussion regarding post-disaster livelihood development in the indigenous context. Therefore, in this chapter, two case studies after the 2009 Typhoon Morakot are selected (relocated and non-relocated settlements) and compared regarding their post-disaster livelihood development. This chapter aims to: 1. Understand the issues and potential risks of tourism industry development after the disaster; 2. Compare the differences between relocated and non-relocated indigenous communities regarding livelihood development strategies. The chapter finds that issues related to road connections and interactions with government sections seem inevitable for both settlements. On the other hand, the relocated settlement tends to face land-ownership issues, while the non-relocated settlement can fully utilize their land and natural resources for livelihood development. That is to say, the livelihood strategies can be very distinct according to the indigenous settlement's available natural resources, community organization, and location. Therefore, there is no template to copy. However, cooperation with external stakeholders tends to be a key factor to success.

17.1 Introduction

Post-disaster reconstruction (PDR) does not end once permanent housing is completed. PDR is not merely a process of reconstructing a living environment. Instead, this process should entail socioeconomic considerations. Although recent studies have delved into post-disaster livelihoods, those in the indigenous context remain underexplored. Therefore, using the 2009 Typhoon Morakot disaster, the most disastrous typhoon in Taiwan's history, as a case study, this chapter focuses on the post-disaster livelihood development in two affected indigenous communities. Some livelihood developments in the settlements are discussed first. Subsequently, the similarities and differences are compared. Finally, suggestions for post-disaster livelihood development are provided.

T. S. Lun (✉)
Deloitte Tohmatsu Venture Support, Chuo-ku, Japan
e-mail: tsai.sung.lun@fuji.waseda.jp

C. Ochiai et al. (eds.), *Disaster and Displacement*, Disaster Risk Reduction,
https://doi.org/10.1007/978-981-96-2583-3_17

17.1.1 Post-Disaster livelihood

As mentioned previously, the importance of livelihoods in post-disaster recovery has been highlighted (Peacock et al. 1987). However, it is not a new concept in the field of development (UNDP 2015). Research has also discussed the relationship between livelihoods and post-disaster recovery, which can be categorized into three layers: individuals and households, communities, and recovery agencies (Sina et al. 2019). At the layer of individuals and households, the household size (Adams et al. 1998; Niehof 2004), early recovery of income support (Eriksen et al. 2005; Adams et al. 1998), and ability to move to other livelihoods (Marschke and Berkes 2006) are essential for disaster recovery. At the layer of communities, the importance of infrastructure and basic services (Oliver-Smith 1991), environmental resources nearby (Adams et al. 1998), and equity of resource distribution (Tellman et al. 2014) has been underscored. At the layer of recovery agencies, the governance livelihood support (Gaillard et al. 2008; Thorburn 2009), long-term livelihood support (Neef et al. 2015; Daly et al. 2017), and cultural appropriateness of livelihood support (Joakim and Wismer 2015) are deemed integral.

17.1.2 Post-Disaster Livelihood Development in the Indigenous Context

According to Joakim and Wismer (2015), culture is essential for the development of livelihoods. Therefore, discussing the local traditional knowledge and perception of local communities is inevitable in disaster risk reduction and the PDR context. In the indigenous context, traditional knowledge (indigenous knowledge) is essential for achieving development goals. Indigenous people belong to indigenous populations through self-identification as indigenous and are recognized and accepted by these populations as members (UNDESA 2004). Due to their unique indigenous social structure, some difficulties in livelihood development have been underscored. In Taiwan, Taiban et al. (2020) addressed the challenges of post-disaster livelihoods in an indigenous context. However, this study is based on short-term livelihood recovery and focuses on only one permanent housing settlement.

17.1.3 Post-Disaster Tourism

As mentioned previously, income support is considered crucial for disaster victims. Therefore, post-disaster tourism has been implemented in disaster-stricken areas. Some places already had a developed tourism industry that evolved into a post-disaster industry, whereas others developed their tourism industry after the disaster. According to Yan et al. (2016), there is an increasing trend of tourists visiting sites commemorating natural or human-made disasters. In Taiwan, relevant research has been conducted to examine the economic loss of the tourism industry after a disaster. According to Liu (2014), the Maolin National Scenic Area, a disaster-stricken attraction, lost over 700,000 visitors in the year after the disaster, resulting in a financial loss of NTD1.39 billion (USD 46.2 million[1]). From an indigenous perspective, Lin and Lin (2020) examined the Tsou indigenous settlement and suggested that indigenous culture-based tourism could serve as an essential source of livelihood, supporting long-term development and reviving post-disaster livelihoods. However, the underlying threats and issues related to the development of tourism remain unresolved.

17.2 Research Question and Methodology

According to the literature, livelihoods are indispensable when discussing PDR. However, although traditional knowledge and community culture are widely admitted to be critical in the disaster risk reduction framework, studies on livelihoods in the indigenous context remain limited, especially on how disaster-affected indigenous communities develop their tourism industry. Therefore, this chapter addresses the following aspects:

1. Understanding the issues and potential risks of livelihoods and tourism industry development after disasters.
2. Comparing the differences between relocated and non-relocated indigenous communities in terms of tourism industry development.

To explore different post-disaster livelihood strategies in the indigenous context, this study applied a comparative case study approach. The case study approach can suit the research objectives and draw critical suggestions for similar future PDR projects. Given that non-relocated reconstruction and relocation are the two major PDR patterns (Jha et al. 2010), two settlements, Rinari (relocated) and Dawu (non-relocated), were selected for this study.

Data were collected through secondary sources and semi-structured interviews. First, for secondary data, newsletters, official publications, and other documents related to the Typhoon Morakot PDR were carefully reviewed. Second, for the semi-structured interviews, a pilot field trip to Rinari and Dawu was undertaken in August 2021 to facilitate the design of the interview guide and build a social network with local residents. The primary interview survey was completed by 2022. Interviews with local residents were conducted using a purposive sampling method to diversify the standpoints and understanding of post-disaster livelihood development issues. Eventually, participants from 15 households in Rinari and seven households in Dawu were interviewed. Moreover, government officials (three people), an institutional representative, and an NGO representative were also interviewed.

[1] Based on the exchange rate of 2014/1/1.

17.3 Research Context

Typhoon Morakot hit Taiwan between August 6 and 8, 2009. During this time, precipitation broke the record in Taiwan and approached the highest rainfall record in the world (Chen 2009). Due to the typhoon's topography and structure, the stark and continuous rainfall caused severe flooding and landslides in the south-central Taiwanese mountains, which caused 699 deaths and missing people, and 1,766 households suffered from housing damage. Given that most of the rainfall was concentrated in mountainous areas, more than 73% of the disaster victims were indigenous (Hsieh et al. 2012). After the disaster, the central government swiftly launched a contingency relief countermeasure in which the local government and NGOs provided emergency shelters for affected people. Afterward, given budget concerns and the construction period, the construction of permanent housing was handed over to three major NGOs in Taiwan: Tzu Chi, the Red Cross, and World Vision. Therefore, the post-Typhoon Morakot PDR was an NGO-led project.

17.4 Case Study 1: Rinari

The Rinari settlement is a permanent housing settlement that accommodates three disaster-affected communities: 177 households from Haocha Village in Wutai Township (Rukai), 174 households from Dashe Village in Sandiman Township (Paiwan), and 132 households from Majia Village in Majia Township (Chuan 2018; Fig. 17.1). Among the three villages, Haocha Village was the total-damaged community, given that all housing in the village had been buried in mud or washed away by the flood and landslide, while the other two villages were partially affected. The Rinari settlement was completed in 2010 with the help of Taiwan World Vision (hereafter, World Vision), a major NGO in Taiwan (Fig. 17.2). World Vision participated in the construction of permanent housing in cooperation with the Architecture Office. Moreover, upon completion of the construction project, World Vision established the "livelihood restoration center" in the settlement to help the residents rebuild their post-disaster livelihood.

17.4.1 Livelihood and Tourism Development in the Rinari Settlement

In 2012, a resident (H1) of the settlement from the younger generation advocated for the *"Rukai Industry Development Association"* to be organized—an association that united some of the households in the settlement to start various tourism businesses. The association primarily focuses on developing the community's post-disaster livelihood, including the hostel industry, where households with vacant rooms on the

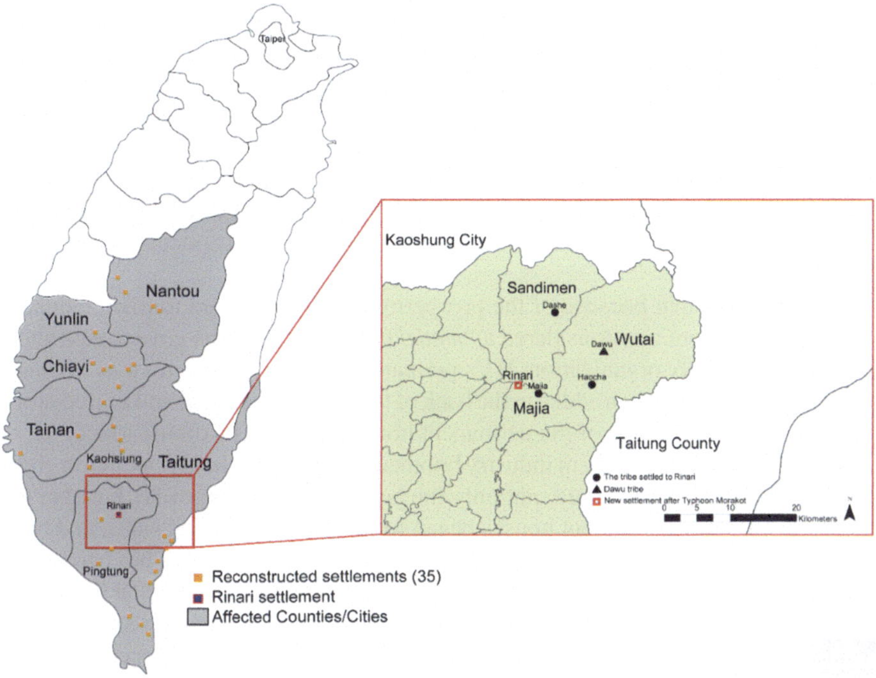

Fig. 17.1 The location of Rinari settlement and Dawu settlement (*Source* Author)

Fig. 17.2 Rinari settlement (*Source* Author)

second floor of their permanent houses can receive tourists to stay overnight. During their stay in the settlement, tourists can visit the chieftain's house and learn more about traditional Rukai culture. Moreover, to further expand the scale of the business, many restaurants and shops have opened in the settlement, including a breakfast store, a grocery store, and restaurants that sell indigenous food, one of which has been listed in the Michelin Guide. As mentioned in Chapter Three, because of successful industrial development, the settlement also has the most housing extensions compared to other post-disaster permanent housing settlements (Rinari residents H2, H3, and H4).

H1 mentioned that because of the prosperity of the hostel and tourism industry, the Rinari settlement was considered a *"model community"* by the government and a leader in community organization. H1's permanent house was also used as a restaurant, with a modern kitchen added to the back of the permanent house and a three-story extension built on the side of the permanent house to serve as a dining space.

As the community's tourism industry boomed and the household population grew, as mentioned in Chapter Three, not only the permanent housing of H1 but also the extensions of many permanent houses began to be built at the front, sides, and back of the houses. For instance, resident H2 extended the second floor of her permanent housing to accommodate more tourists and extended the space on the first floor as an indigenous traditional handicraft classroom (see Chapter Three, case study three). Resident H3 renovated his front yard to include a vending corner as a reception space for tourists staying in his permanent house (Fig. 17.3).

Fig. 17.3 Front yard with the vending corner (*Source* Author)

17.4.2 Problem of Livelihood Development and the Solution for the Rinari Settlement

Despite the relatively successful and prosperous tourism industry, according to a survey on livelihood development conducted by the author, the residents of the Rinari settlement still face several challenges: 1. Land ownership issues; 2. housing extension demolition issues; 3. concerns related to roads connecting the original and current settlements; and 4. issues with receiving industrial support in the permanent housing settlements, all of which are described below.

17.4.2.1 Land Ownership Issues

The residents stated that as stipulated by the trilateral contract[2] (a contract that specified the ownership of the permanent housing and its land, which was signed by the government, NGO, and Rinari residents), the permanent housing land belongs to the government. Therefore, although residents live in permanent housing, they only own houses without land ownership. Thus, any extension was considered an encroachment on government-owned land and was at risk of demolition by the local government at any time.

NGO representative N1 added that the lack of land ownership in the trilateral contract was due to the concern that there were too many cases in the past in which indigenous people sold their land to the Chinese at a low price. If the indigenous people would thus mortgage their properties, it may make them more vulnerable to becoming economically disadvantaged because of the mismanagement of their assets.

At the same time, Legislator L1 also believed that when considering land management issues from the local government's perspective, the county should consider the land allocation within the county based on the concept of total land control. For example, when land is provided to residents for permanent housing, it is necessary to allocate additional government-owned land to compensate for the county's quota of disaster-preparedness land.

[2] The trilateral contract regulates the rights and obligations of the permanent housing beneficiaries:

1. The construction land of the permanent house belongs to the government; the resident only has the ownership of the permanent housing.
2. The permanent housing cannot be transferred or sold to others. The permanent housing can only be inherited by the direct blood relatives.
3. The household needs to return to the permanent house if they return or reside in their original settlement.

17.4.2.2 Housing Extension Demolition Issues

As mentioned previously, the government deemed housing extensions illegal according to the building code. Thus, in early 2020, when visiting the community, Rinari resident H3 stated that the Haocha household occasionally received a "*warning*" from the local government that they would demolish the extended structures.

According to the interviews, because someone specifically reported the illegal construction by the H1 family from May to June 2020, the local government hoped that H1 would demolish the housing extension. However, the H1 family was unwilling to demolish them. Therefore, on October 15, 2020, the local government assigned hundreds of police officers to enforce the demolition of H1's housing extensions (Fig. 17.4a and b).

After the forced demolition of permanent housing extensions, Rinari residents and scholars concerned about the issue of permanent housing quickly established the "*Morakot Post-Disaster Human Rights Promotion Association.*" They had staged several protests in front of the Legislative Yuan in Taipei and advocated for revising the trilateral contract to ensure that permanent housing would truly be "*permanent.*" Although this time only the house of H1 was demolished, other residents who had extended their houses had similar worries (Rinari residents H2 and H3).

Resident H1 stated that many of the government's policies were not enforced according to the same standards. Given that almost every household had extended its permanent housing, the legitimacy of demolition was questioned. In addition, Rinari residents H1 and H3 questioned the criteria for the demolition of unauthorized buildings; the regulation did not specify the height and usage of the buildings. Therefore,

(a) (b)

Fig. 17.4 **a** Residents protesting in front of the permanent house. **b** Permanent housing extension was demolished

it is fair to say that the demolition incident of his housing extension lacked rationality and fairness.

Legislator L1[3] believed that demolition and relocation could be sensitive matters among the indigenous people because of their unique historical context and life experiences. However, he also believed that despite the past trauma of the relocation of the indigenous village, the authorities could not be lazy and lenient in enforcing the demolition of unauthorized structures.

> The indigenous used to live where no building code existed, so they did not have a building registration. The thing is different in the permanent housing settlement. There is a management system. Our government will, of course, be in a reasonable standpoint to allow, for example, the construction of rain shelters and garages. Nonetheless, using steel structures to extend buildings to three floors high and make a profit is not acceptable.

Representatives of township office T1 emphasized that the local government, village head, and H1 were good friends and that there should be some discussions before demolition. Demolishing illegal structures was not a sudden decision but an expected consequence. However, T1 hoped that the central government would understand the importance of these extended structures for residents' livelihoods.

17.4.2.3 Roads Connecting Their Original and Current Settlements

According to the trilateral contract, those living in an identified danger zone can be allocated permanent housing on the condition that the residents cannot return to their original settlements. However, the residents believe indigenous life cannot be separated from the original settlement (Rinari residents H3 and H4). Therefore, returning to their original settlements was deeply meaningful to them. Furthermore, many residents have access to agricultural land in their original settlements. Thus, even though they had a permanent house in the new settlement, they wanted to return to the mountains to farm during the non-typhoon season, as stated by Rinari resident H2:

> The government should stop lying to itself...we have TV and internet in our original house, and many people spend the night in the mountains...does the trilateral contract make sense?

In addition, the residents hoped that the government would pay attention to and allocate a budget to build and maintain roads back to their original settlement. Some residents consider repairing the road an opportunity to start ecotourism businesses, where tourists can visit their original settlements and learn about indigenous cultures (Rinari H5).

Legislator L1 believed that from the local government's standpoint, it was possible to build roads for people to return to their original settlement. However, the core issue was the difference in residents' ideas of their future living patterns and the government's consideration of the homeland plan.

---

[3] Legislator L1 is in charge of Pingtung County and is knowledgeable about the permanent housing issue.

> Because the Rukai people are used to living a sedentary lifestyle, they are gradually adopting the pattern of living in their new settlement after moving to a permanent housing settlement. Some people think they can live in permanent housing for a long time and think the government should give them the land, but some want to return (to the original settlement). Therefore, should the community reach a consensus before talking about going back to their original settlement?

17.4.2.4 Livelihood Support in the Permanent Housing Settlement

Residents complained that there was insufficient space for farming in the permanent housing settlements. Before the disaster, the local people had long practiced agricultural livelihoods. Nonetheless, minimal farmland space was allocated to settlements during the planning process because of land limitations. This situation has led some residents to use the riverbank below the Rinari settlement for farming. However, the river bank is government-owned land. Thus, there were still concerns about potential violations of the "*State Property Law.*"

In addition, concerning the community's hostel industry, Rinari resident H1 mentioned:

> The government abandoned us. Initially, the Tourism Bureau and the Indigenous People's Association do their best to support our hostel business. They generously provide us with funding for various projects. However, on the other hand, the local government's building management unit demolishes the economic lifeline (housing extensions) on which we rely for our livelihood.

Rinari resident H4 also added:

> We are not that rich. All these kinds of paperwork (registration as a legal hostel) will be time-consuming and cost a lot. We have no motivation to do it.

Legislator L1 acknowledged the government's lack of thoroughness in considering its PDR policy and inability to provide sufficient arable land for residents to cultivate. He also believes that the central government should delegate authority to local governments to establish autonomous ordinances and internal management practices according to local conditions.

Based on the interview survey results, some issues, challenges, and proposed solutions related to the stakeholders are listed in Table 17.1.

17.5 Case Study 2: Dawu

Dawu settlement is the old indigenous village established prior to Typhoon Morakot. Until 2023, the village has a population of 528, 158 households. Different from the Rinari settlement, all of the population in the Dawu settlement is Rukai. The settlement is divided into upper-Dawu (Daramaku) and lower-Dawu (Shaoshan) two communities due to the geographical condition.

Table 17.1 Important issues, difficulties, and solutions derived from the interviews (Rinari)

Issues	Challenges	Proposed solution (stakeholder in charge)
1. Land ownership issues	1. Difficulty in adjusting the trilateral contract 2. Conflicting interests of government land management 3. Security assessment of housing extension and legalization procedures	1. Coordinate the amendment of the trilateral contract (local and central governments) 2. After acquiring land ownership, try to increase the capacity building rate under the regulation's limitation (local government)
2. Housing extension demolition issues	1. Indigenous people's aversion to the concepts of demolition and relocation 2. The rationality and fairness of the law enforced are difficult to grasp	1. The government should understand the historical background of the indigenous people and the reasons for housing extension (local and central governments) 2. Establish a friendly communication mechanism between the government and residents to avoid confrontation (local government)
3. Roads connecting their original and current settlements	1. Illegal to return to the hometown due to the trilateral contract 2. Generational gap (young and old people have different views on their original communities and livelihoods)	1. Consider the livelihoods and cultures of indigenous peoples in a holistic pattern when formulating the relocation policy (central government) 2. Allowing sufficient time for dialogue and consensus-building among residents in the community (local government and settlement)

(continued)

Table 17.1 (continued)

Issues	Challenges	Proposed solution (stakeholder in charge)
4. Livelihood support in the permanent housing settlement	1. Insufficient horizontal links between local government ministries 2. Illegal hostel	1. Strengthen the cooperation and communication between local government ministries (local government) 2. The central government should reasonably authorize local governments to handle cases flexibly; for example, local governments can establish regulations for the registration and management of hostels (local and central governments)

17.5.1 Livelihood and Tourism Development in the Dawu Settlement

Similar to Haocha Village, Dawu Village is located in the Wutai Township (Fig. 17.1). It is divided into two sub-settlements: Derau and Talamakau. The settlement is approximately 500 m from the Aliou North River (Fig. 17.5). According to the settlement chief (D1), before Typhoon Morakot, there were approximately 110–120 people. However, after the typhoon struck, the population decreased to 70–80 persons. Thanks to the tourism industry, residents have increased to 500 persons (as of 2021). Regarding livelihoods, before the disaster, D1 stated that the Dawu settlement had started its tourism industry. Six large-scale guesthouses in the settlement received tourists traveling from nearby areas. Besides the tourism industry, some households practice self-sufficient agriculture. During Typhoon Morakot, three households were washed away by the flood, and access to roads to the cities was interrupted. However, the government designated the settlement as a *"safe area,"* given that most of the settlement's facilities were functioning, and the housing remained habitable. Although some residents were not satisfied with this decision, after negotiations between the villagers and the government, the people in the Dawu settlement agreed to stay, given their emotional and physical connections to the land.

After Typhoon Morakot, the settlement started promoting the "*Hayou River,*" a seasonal attraction that was only open to visitors during the dry season from January to April every year. The total number of visitors was limited to 400 per day to ensure quality. While visiting the Hayou River, visitors can experience cold springs and hot springs along the way. The most popular tourist spot for visitors was the colorful *"seven-colored rock walls."* The erosion of rivers after Typhoon Morakot revealed the colorful rock walls. Since 2016, after years of operation and promotion by visitors

Fig. 17.5 Dawu community (*Source* Author)

on SNS, Hayou River has become a famous scenic spot in Southern Taiwan. It has brought many business opportunities for the settlement (Fig. 17.6).

In terms of agriculture, Dawu residents also cooperated with the Pingtung University of Science and Technology (NPUST) to scientifically assess the land load to avoid damaging the environment. Moreover, NPUST introduced economic crops such as wild Chinese herbs "*mountain angelica,*" and mushroom cultivation. In addition, the NPUST team designed a simple chicken farming industry in the Dawu settlement for remote mountainous areas where transporting feed is challenging. The settlement also created its own brand and started selling agricultural products in metropolitan areas, such as Taipei.

Fig. 17.6 Hayou river (seven-colored rock walls) (Commercial Times 2020)

17.5.2 Problem of Livelihood Development and the Solution for the Dawu Community

According to the survey conducted by the author, the residents of Dawu settlement were facing several challenges regarding post-disaster livelihood and tourism development: 1. establishment of Natural, Cultural, and Ecological Scenic Areas; 2. independent operation of the tribal tourism industry; 3. refinement of tourism and livelihood diversification; and 4. disaster response; these issues are further described as follows.

17.5.2.1 Establishment of the Natural, Cultural, and Ecological Scenic Areas

Due to the influx of tourists into the settlement, aligned with the prosperous expansion of the Hayou River tourism business, ecological conservation and tourism quality have degraded. In response, under the cooperation and consultation of the Tourism Bureau of the Ministry of Transportation and Communications, the Pingtung County Government, and the residents, the first *"Natural, Cultural and Ecological Scenic Areas"* (NCESA) was designated in 2020 in the Dawu settlement and the Hayou River basin. The regulations stipulate that trained community residents must be tour guides for tourism activities in the area and that conservation fees can be collected as a fund for the tribe's tourism development to ensure the quality of tourism through total control and prior reservation.

Settlement chief D1 mentioned the process and reasons for the designation of the NCESA as well as the allocation of responsibilities:

> In the past, there were often some accidents when outsiders came to climb the mountain here, so we decided to control the area to protect the safety of hikers. Therefore, we introduced the NCESA, trained settlement residents as guides, and invited the instructors through the community development association and the village office. So now the guides to Hayou River are all our people, and visitors must take our shuttle bus to the settlement, part of the traffic control.

Dawu Resident D2 supported the designation of NCESA, which he believed was beneficial for the development of the community's tourism industry.

> The NCESA is a way for us to be ourselves, a tribal autonomy that gives our settlement dignity. We have a reciprocal relationship with tourists. If you want to come to our settlement, you must make reservations, meet our regulations, and hire our people to guide you through the itinerary so the revenue can stay in the settlement.

Settlement chief D1 mentioned that the success of the tourism industry revitalized the disaster-affected community.

However, Dawu resident D2 also had concerns about the designation of the NCESA because it may affect indigenous people's hunting rights and weaken traditional culture and identity:

> Although the NCESA provides legal protection, according to the law, under the viewpoint of ecological conservation, we are not allowed to hunt in the NCESA. Therefore, the establishment of the NCESA also has disadvantages… at the beginning, I did not want the tribe to establish the NCESA....

17.5.2.2 Independent Operation of the Tribal Tourism Industry

As previously mentioned, the maximum number of one-day registrations was 400. Among them, 200 were under the control of the settlement, and the other 200 were under the control of the Wutai Township office.

Dawu Resident D2 hoped that the township office would not interfere excessively. Since the Hayou River tourism was initially established by individual residents in the settlement, these households hoped to retain autonomy in their business operations.

However, the representative of the township office T1, believed that the tribal management system is not yet perfect and that leaving everything to the tribes may cause corruption and damage the quality of tourism; therefore, some government intervention is necessary.

> We found that there were many problems before; for example, if a friend of the settlement residents visited, they would skip the internet booking and take the guests up to the mountains for sightseeing without compliance, acting on their own initiative. When you let go of the whole thing, there will be problems. For example, last year, there was an accident that affected the public perception of Hayou River.[4]

17.5.2.3 Refinement of Tourism and Livelihood Diversification

A representative of the township office, T1, believed that for settlement tourism to develop sustainably, the industry must move towards a refined route and continue to improve its strengths:

> At present, the concept of eco-tourism is changing. If we try to refine the tourism industry, the residents need more training and study, and their knowledge needs to be improved. For example, in Hayou River, you need five licenses to operate the business. If the settlement does not want rivals to compete with them, then the settlement must be competent enough.

However, Dawu Resident D3 stated that the current Hayou River tourism industry can only operate from January to April. Although income was high during the four months, the settlement had no productivity during the rest of the year, and many residents had to leave the settlement for part-time jobs in the cities. Hence, the settlement needed to establish a stable, long-term source of income. In addition, the landscape of the colorful rock wall was formed by riverbed erosion. Thus, landscape resources may change after the next typhoon. Therefore, the tribe needed to diversify its livelihood to hedge against risks.

---

[4] On April 23, 2021, a total of four tourists from Taoyuan visited Hayou River. On the return trip, two cars collided in a hot spring area; their car crashed into a mountain and flipped over.

In response to Dawu resident D2, settlement chief D1 suggested using cultural assets to diversify tourism resources. For example, residents lived in Old Dawu (a settlement of stone slate houses that had not been rebuilt) before the 1940s, which is located at a higher altitude than the current Dawu settlement. Thus, the settlement is currently negotiating with the government, hoping to reclaim the land ownership of Old Dawu and expand the range of the NCESA to Old Dawu to diversify the options of the eco-tourism itinerary.

Dawu Resident D3 mentioned that the settlements' agricultural industry and tourism should complement each other. Although the agricultural industry had already established its own brand, its production cost was too high, and the output was low. D3 believed that the development of tribal agriculture depends on tourism. People can visit the landscapes and enjoy food.

17.5.2.4 Disaster Response

Settlement chief D1 mentioned that although the government approved that the Dawu settlement did not need to be relocated, there was much discussion among the residents about whether to move immediately after Typhoon Morakot:

> The land is free in Dawu. After the family in Dawu split up or expanded, they could draw up another piece of land for housing construction. The house structure (in the Rinari settlement) was planned at the standard of temporary housing, which we did not like. Since our land is safe, we should not leave (the Dawu settlement). Of course, there are also cultural considerations. Moving to flat land is not suitable for our culture.

Settlement chief D1 also stated that the key to staying in the mountains and living sustainably was the road connection with the cities and the road back to Old Dawu.

> The government needs to ensure that our roads are accessible to the cities. That is why we require the government to use concrete instead of asphalt for our roads because concrete is more durable, and some parts are still under repair. However, the capacity of our roads here is too low. For example, the width of Gurenren Bridge[5] can only accommodate one vehicle, so medium-sized buses cannot get in. Therefore, it is an obstacle to our tourism industry. Moreover, we are now suggesting to the government to pave the track back to Old Dawu. It can be a very potential tourist spot to diversify our tourism resources and be meaningful for our cultural identities.

During the typhoon season, settlement chief D1 mentioned that the community already had a disaster response mechanism.

> Every year, some of the bridge structures are washed away; thus, the settlement's communication with other settlements is interrupted. However, we do not need to evacuate during typhoons because the settlement is on higher ground, and a three-day food supply is sufficient. If the settlement is isolated for more than three days, the village office will distribute food to everyone. In fact, developing the agriculture industry in the settlement is also a way to make the settlement self-sufficient in terms of food security so that we can be more resilient in the face of disasters.

[5] Dawu Tribe's only bridge to the other settlements.

Settlement chief D1 also discussed disaster response measures for tourists with the hostels in the settlement. In cases where tourists are trapped in the mountains due to typhoons, a cooperative mechanism allows trapped tourists visiting the settlement to be housed in one of the hostels as a temporary shelter and provided with sufficient food.

According to the interview survey results, some of the issues, challenges, and proposed solutions that the related stakeholders could implement are listed in Table 17.2.

Table 17.2 Important issues, difficulties, and solutions obtained from the interviews (Dawu)

Issues	Challenges	Proposed solution (stakeholder in charge)
1. Establishment of the natural, cultural, and ecological scenic areas	1. The influence on the traditional hunting culture and lifestyle of the settlement	1. The NCESA area is only a short-term objective. In the future, the settlement will have more autonomy in managing their territory and livelihood development (local and central governments)
2. Independent operation of the tribal tourism industry	1. Residents do not want too much government intervention 2. It is difficult to coordinate the business and the distribution of revenue	1. The settlement's business system is not mature enough. The government still needs to intervene and assist (local government) 2. In the long term, set up a joint management organization and reasonable benefit distribution mechanism (settlement)
3. Refinement of tourism, livelihood diversification	1. The Hayou River tourism industry is only for short-term economic income 2. Tribal agricultural productivity is too low	1. The tourism industry should pay attention to the investment of human resources and refinement of the content (settlement) 2. Diversify the livelihood and develop different sightseeing spots. The tourism industry and the agriculture industry should be complementary (settlement)
4. Disaster response	1. Interruption of access roads affects the livelihoods of residents and the tourism industry 2. Response mechanism in the event of a disaster	1. Regularly repair and keep access roads functioning (local government) 2. Develop disaster response plans for the residents and tourists (local government and settlement)

17.6 Discussion and Conclusion

This chapter discusses long-term livelihood issues in disaster-affected indigenous communities by comparing case studies based on two settlement conditions: relocated and non-relocated. Given the lack of discussion on the post-disaster long-term livelihood development issue in the indigenous context, this chapter aims to list some of the critical livelihood development challenges faced by indigenous communities in the aftermath of a disaster.

Regarding the differences, it is evident that although the two indigenous settlements faced the same typhoon disaster, the relocation and non-relocation settlements developed distinct post-disaster livelihood strategies and tourism programs. Given that the residents had been relocated far from their original settlements, they utilized their permanent housing to attract tourism. Moreover, some households modified and extended their housing to maximize revenue for post-disaster livelihoods. However, significant challenges such as land ownership and the lack of agricultural land pose threats to future livelihood development. The issue of land ownership escalated into a conflict between residents and the government, hampering the trust of stakeholders. Meanwhile, as a non-relocation settlement, Dawu residents took advantage of natural resources such as the Hayou River to attract tourists and bring back some residents. No ownership issues arose. Nonetheless, residents also found themselves relying on single industries, which is risky. Isolation from cities and poor road connections can be challenging during disasters.

In terms of similarity, the two case studies first underscored the issue of access roads. In the Rinari settlement, despite the residents being relocated to the new settlement, their strong connection and the willingness to continue the agricultural livelihood led them to request that the government retain the road back to their original settlement and revise the trilateral contract. In Dawu, the community remained in the same location even after the disaster. Nonetheless, they hoped for a road back to Old Dawu to diversify their tourism resources and boost their cultural identity. Roads to cities are also a critical issue for residents. Second, the two case studies underlined the necessity of government intervention and supervision. In the Rinari settlement, the legalization of housing extensions and hostels required going through the administrative process. This conflict could not be resolved without the assistance of the government. The government cooperated with the Dawu community to establish the unprecedented NCESA, which protected the ecosystem and secured residents' livelihoods—a win-win situation. Moreover, government intervention ensures that the tourism industry's revenue can be fairly distributed, an essential element for post-disaster livelihoods (Tellman et al. 2014).

This chapter proposes some suggestions. In the planning stage of the PDR, the government should discuss relocation patterns with residents to envisage several possible post-disaster livelihood strategies. The government must realize the indigenous people's inextricable relationships with nature and culture. Furthermore, some NGOs (such as World Vision in the case of Rinari) can be ideal mediators in resolving conflicts between stakeholders. After the construction of permanent housing, continuous and periodic meetings among stakeholders can help residents address issues related to livelihood development. Given that livelihood development is a dynamic and ever-changing process, assistance from external stakeholders (the government and NGOs) will be beneficial.

References

Adams AM, Cekan J, Sauerborn R (1998) Towards a conceptual framework of household coping: reflections from rural West Africa. Africa 68(2):263–283

Chen YL (2009) Report on Typhoon Morakot (0908) of 2009. Weather Forecast Center Central Weather Bureau. Taipei, Taiwan

Chuan KC (2018) A study on the post disaster village relocation at the Town of Rinari in Majia township pingtung county. Degree Thesis of Department of Public Admin-istration and Policy, National Jinan International University

Commercial Times (2020) Pingtung Wutai Hayou river is open for reservation and will be full in February. Commercial Times, Taipei, Taiwan

Daly P, Halim A, Hundlani D, Ho E, Mahdi S (2017) Rehabilitating coastal agri-culture and aqua-culture after inundation events: spatial analysis of livelihood recovery in post-tsunami Aceh, Indonesia. Ocean Coast Manag 142:218–232

Eriksen SH, Brown K, Kelly PM (2005) The dynamics of vulnerability: locating coping strategies in Kenya and Tanzania. Geogr J 171(4):287–305

Gaillard J-C, Texier P, Régnier P, Neri B, Scuteri S, Miniati S (2008) From emergency relief to livelihood recovery: lessons learned from post-tsunami experiences in Indonesia and India. Disaster Prev Manag: Int J 17(3):410–430

Hsieh ZC, Fu ZC, Chen ZC, Lin WY (2012) A road farther and farther from hometown? rethinking the policy of rebuilding after typhoon Morakot. Natl Tai-Wan Univ Soc Work J 26:41–86

Jha AK (2010) Safer homes, stronger communities: a handbook for reconstructing after natural disasters. World Bank Publications

Joakim EP, Wismer SK (2015) Livelihood recovery after disaster. Dev Pract 25(3):401–418

Lin PSS, Lin WC (2020) Rebuilding relocated tribal communities better via cul-ture: livelihood and social resilience for disaster risk reduction. Sustain 12(11):4538

Liu TM (2014) Analysis of the economic impact of meteorological disasters on tour-ism: the case of typhoon Morakot's impact on the Maolin national scenic Area in Tai-wan. Tour Econ 20(1):143–156

Marschke MJ, Berkes F (2006) Exploring strategies that build livelihood resilience: a case from Cambodia. Ecol Soc 11(1)

Neef A, Panyakotkaew A, Elstner P (2015) Post-tsunami recovery and rehabilitation of small enterprises in Phang Nga Province, southern Thailand. In: Recovery from the Indian Ocean tsunami: a ten-year journey. Springer, Tokyo, pp 487−503

Niehof A (2004) The significance of diversification for rural livelihood systems. Food Policy 29(4):321–338

Oliver-Smith A (1991) Successes and failures in post-disaster resettle-ment. Disasters 15(1):12–23

Peacock WG, Killian CD, Bates FL (1987) The effects of disaster damage and housing aid on household recovery following the 1976 Guatemalan earthquake. Int J Mass Emergencies & Disasters 5(1):63–88

Sina D, Chang-Richards AY, Wilkinson S, Potangaroa R (2019) What does the future hold for relocated communities post-disaster? factors affecting livelihood resilience. Int J Disaster Risk Reduct 34:173–183

Taiban S, Lin HN, Ko CC (2020) Disaster, relocation, and resilience: recovery and adaptation of Karamemedesane in Lily Tribal Community after Typhoon Morakot, Taiwan. Environ Hazards 19(2):209–222

Tellman B, Rivera A, Alaniz R, Contreras D (2014) Violence as an obstacle to livelihood resilience in the context of climate change. UNU-EHS Working Paper

Thorburn C (2009) Livelihood recovery in the wake of the tsunami in Aceh. Bull Indones Econ Stud 45(1):85–105

UNDP (2015) Sustainable development goal

UNDSA (2004) Workshop on data collection and disaggregation for indigenous peoples: the concept of indigenous peoples. NY, US

Yan BJ, Zhang J, Zhang HL, Lu SJ, Guo YR (2016) Investigating the motivation–experience relationship in a dark tourism space: a case study of the Beichuan earthquake relics, China. Tour Manage 53:108–121

Chapter 18
Home in Transition: Exploring Integration Among Refugees from the Horn of Africa in Melbourne's Public Housing Estates

Sandra Carrasco, Majdi Faleh, and Neeraj Dangol

Abstract Public housing estates in Melbourne have emerged as central points in the housing discourse, characterized by precarious conditions and unsafe environments that perpetuate social stigma, disproportionately impacting socio-economically vulnerable groups. The demographic composition of public housing residents has evolved, notably with a growing number of residents from the Horn of Africa who were permanently displaced due to conflicts in their home countries and resettled in Australia. Public housing provides vital housing support for migrants and refugees facing economic disadvantages and plays a crucial role in their integration into Australian society. Nonetheless, these communities encounter significant challenges in adapting to a new lifestyle while preserving their cultural identity, often contending with outsider status within the local community and confronting discrimination and exclusion. This study aims to investigate the relationship between public housing residents' living conditions and their interactions with both physical and social environments, seeking to uncover barriers and opportunities for nurturing cohesive communities striving to maintain their cultural identity while exploring avenues for enhanced integration. The preliminary findings of this research offer implications for further study, policy formulation, and practical interventions for community integration and rebuilding a sense of home and social supporting structures.

S. Carrasco (✉)
School of Humanities and Social Sciences, Deakin University, Melbourne, Australia
e-mail: s.carrasco@deakin.edu.au

M. Faleh
College of Engineering, Architectural Engineering, University of Sharjah, Sharjah, United Arab Emirates
e-mail: mfaleh@sharjah.ac.ae

N. Dangol
Melbourne School of Design, The University of Melbourne, Melbourne, Australia

C. Ochiai et al. (eds.), *Disaster and Displacement*, Disaster Risk Reduction,
https://doi.org/10.1007/978-981-96-2583-3_18

18.1 Introduction

The Office of the United Nations High Commissioner for Refugees (UNHCR) recorded unprecedentedly high levels of global displacement with 110 million forcibly displaced people worldwide and more than 36 million refugees (UNHCR 2024). Factors such as conflicts, political instability, deteriorating security, and grinding poverty have contributed to the increasing numbers of refugees worldwide (Amuedo-Dorantes et al. 2018; Victoria 2011). Despite ongoing controversies and contentious political debates, Australia maintains a consistent history of migrant and refugee resettlement, having resettled some 900,000 refugees since 1947 (Shelton 2019). The number of humanitarian migrants from the Horn of Africa and Sudan, however, only increased in numbers from the 1990s due to the conflicts in these countries (Abur and Spaaij 2016). The State of Victoria currently houses 41 percent of the 38,614 migrants from the Horn of Africa and Sudan in Australia, with more than a quarter of them being resettled through humanitarian stream visas (ABS 2018).

Among the essential services that resettled refugees require, housing plays a crucial role in their journey toward integration into the host country (UNHCR 2013; Flatau et al. 2015, p 8). Access to housing, employment, and education are deemed essential for refugees' long-term stability and adaptation (UNHCR 2002). Housing not only provides shelter and security but also fosters opportunities for community building, social networking, and establishing links to employment and education (Flatau et al. 2015). The dynamics between migrants, refugees, and their residential environment significantly shape their integration trajectories. Seethaler-Wari (2018) claims that integration happens at the local level of the city, where the characteristics of the host neighborhood directly influence the refugees' and migrants' behaviors (Fonseka and McGarrigle 2012; Hinze 2013; Wari 2017).

This chapter explores the residential experiences of migrants and refugees from the Horn of Africa and Sudan, specifically those residing in challenging environments such as public housing estates in inner Melbourne (North Melbourne and Carlton). The chapter examines the concept of identity within these communities and evaluates the feasibility of reconnecting with their own social and cultural practices in a rigid housing and neighborhood environment. Moreover, public housing is often associated with the poorest and most marginalized groups of society. Therefore, the process of integration for migrants and refugees might be hindered. This process would eventually jeopardize the creation of a sense of home and community in Melbourne, which is one of the largest and most diverse Australian cities.

18.2 Between Integration and Identity

African refugees and migrants have faced significant challenges resettling in Australia over the past six decades, with settlement and integration support playing crucial roles. Media coverage and political discourse have often highlighted concerns about the "integration potential" of African refugees, contributing to controversies (Harte et al. 2009; Dhanji 2009; Farouque 2007).

Integration is seen as a process where migrants become part of the social, institutional, and cultural fabric of society (Valtonen 2004). However, in Australia, controversies often revolve around migrants' ability to adopt the values and lifestyles of their host country while preserving aspects of their identities and cultures (Al Huraibi 2014). The United Nations High Commissioner for Refugees emphasizes that integration should allow newcomers to adapt to the host society's lifestyle without losing their cultural identity (UNHCR 2002). Despite challenges and negative portrayals, African refugees strive to rebuild their lives in Australia. For example, refugees from the Horn of Africa have successfully settled in Australia and established organizations like the Horn of Africa Relief and Development Agency (HARDA) to assist others in their community (Pittaway et al. 2009).

Housing and community spaces play a crucial role in refugees' resettlement, providing shelter, security, and opportunities for social connections and education (Flatau et al. 2015). Community networks are facilitated by a strong sense of community enabling the emergence of social capital necessary to withstand crises and disasters such as the COVID-19 pandemic (Carrasco et al. 2023). Adequate living environments are essential for physical and mental health, while inadequate conditions can lead to disconnection from the host society's cultural values (Naimi-Gasser 2013; Chitrakar 2016). Moreover, precarious living conditions can compromise safety due to a lack of neighborhood cohesion and fear of crime (Brown and Perkins 1992). Examining the experiences of African migrants and refugees in Melbourne can provide insights into how space and architecture can enhance their living experiences.

18.2.1 Multiculturalism in Melbourne and Experiences of African Identities

Historically, Australia has ranked as one of the largest immigrant-receiving nations (Burnley 2001) that changed from the first British settlement in 1788 and the controversial White Australia Policy, known as the Immigration Restriction Act of 1901, to a multicultural society with a large number of ethnic minority groups (Forrest et al. 2006). Today, about one-third of Australia's population is born overseas (ABS 2023). Australia has become more ethnically diverse through immigration and the diverse backgrounds of Australian-born citizens.

Studies emphasized the experiences of multiculturalism in Australia where diverse cultural influences are coagulated and hybridized national practices take place (Edensor and Sumartojo 2018). The multicultural experience acknowledges the links between everyday experiences of nationhood and the political and institutional structures that help to recreate a diverse identity (Edensor and Sumartojo 2018). In becoming progressively embedded in an emerging Australian national identity, multiculturalism has been crucial for handling diversity (Edensor and Sumartojo

2018). Metropolitan areas such as Sydney and Melbourne are the most diverse in Australia. Melbourne is currently home to more than 5 million, where 40% were born overseas and diverse ancestries exist within this metropolitan city.

African migrants, particularly black Africans, are a relatively new ethnic group that settled in Melbourne as they have arrived since the 1990s. Although similar challenges were faced by previous refugee groups resettled in Australia in the past, African refugees face additional barriers due to their distinguished physical attributes and their different religious and cultural backgrounds (Dhanji 2009). Mapedzahama and Kwansah-Aidoo (2017) explored the experiences and burdens of black Africans in Australia, referred to as "the burden of blackness," which coexist with racism and discrimination emerging from the fear of unbalancing the "white homogeneity" in Australia.

There are almost 40,000 immigrants from the Horn of Africa and Sudan residing in Australia. More than half of the migrants from these countries reside in metropolitan Melbourne, which makes it the largest community from the Horn of Africa in the country. The majority of Ethiopian, Somalian, and Eritrean communities residing in Australia made Melbourne their home. Additionally, 98% of these migrants were relocated to Melbourne through humanitarian visas or family visas (ABS 2018). Furthermore, migrants from the Horn of Africa and Sudan are distributed within greater Melbourne, while Somalian and Eritrean communities are concentrated in the inner Melbourne suburbs, particularly in the suburb of Carlton where this research is focused.

18.3 Research Design

This study examines the experiences of African migrants and refugees in Melbourne by employing a participatory approach at the intersection of social and cultural anthropology, behavioral geography, environmental psychology, urban planning, and architecture. The data for this study has been collected between 2018 and 2019 and focused on two integrated methods:

(1) Semi-structured interviews with a group ranging from 28 households from Horn of African origins, including Australian-born participants with African backgrounds, to understand their socio-economic challenges and the transition to their new lives as residents of Australia through constructed images of the private and public environments, compared to home. Through storytelling, the participants shared their stories about the physical environment they lived in, the sense of attachment to their homes in Africa, the adjustments made when moving to Melbourne, and the interaction with their new space (i.e. the urban environment surrounding their social housing and/or their own private space). The second stage of the research utilizes a graphic (sketching) and photo-survey method to capture their living conditions, connecting their daily experiences with the storytelling of integration.

(2) This study also relies on a creative component. One art workshop was conducted on the 17th of August 2019 with 12 participants who were migrants from various African countries. This workshop aims to engage participants in a dialogue about the meaning of home through their participation in Somali handicrafts and photography. Discussions with the community will be organized based on cognitive and social maps, which will be used to record their "perceptions, feelings, sentiments, prejudices, wants, needs, and suggestions" (Sharma 2010). To understand the experience of African migrants in Australia, we first analyzed the demographics of Australian society and the role of multiculturalism in shaping its social and cultural structure.

18.3.1 Limitations of the Study

The research was conducted prior to the COVID-19 pandemic, and the issues emerging from the challenges during the health crisis that particularly impacted Melbourne's public housing and its residents (see Carrasco et al. 2020; Raynor et al. 2019) are not part of this study. Dedicated data collection and analysis of the impact of COVID-19 on people's perceptions of the sense of home and barriers to integration are required.

18.4 Public Housing Estates as Incubators for Migrants' Integration

In the 1960 and 1970s, the development of public housing estates in Australian cities progressively led to the concentration of what is often referred to as "multiple disadvantage" (Morris et al. 2012) This term describes communities that commonly face issues such as unemployment, anti-social behavior, negative stigma, poor educational outcomes, and high crime rates (Morris et al. 2012). In many Western countries, a significant policy response over the past two decades has focused on altering the socioeconomic composition of communities aiming to promote diversity (Chaskin et al. 2012). This has been achieved through mixed housing tenure policies intended to reduce poverty concentration, revitalize public housing, and foster the development of mixed-income communities in place of the most deteriorated and problematic public housing estates (Chaskin et al. 2012). These policies are grounded in the belief that the isolation of disadvantaged communities contributes to various social issues (Arthurson 2010). By integrating these disadvantaged communities with people from broader society, it is expected that social outcomes will improve, leading to better employment prospects and greater participation and engagement within society.

Research by Lelévrier (2013) has highlighted the importance of spatial layout and the physical arrangement of different housing tenure groups in influencing the social interactions within mixed communities. However, the models that shape the physical structure of these communities have often been overlooked in the literature. Many studies within this field discuss the social effects of these developments, such as integration, segregation, social cohesion (Bolt et al. 2010; Kleit and Carnegie 2011), reputation, and stigma (Arthurson 2013; Keene and Padilla 2010), social well-being (Graham et al. 2009), and residents' perceptions (Joseph and Chaskin 2010), but few have examined the physical design aspects. Thus, a gap remains in the understanding of the design of social mix and the physical formation of socially mixed communities.

18.4.1 Carlton Public Housing Estates

Carlton is a gentrified inner-city neighborhood located on the northeastern boundary of the City of Melbourne. The 2021 Census indicates more than 16,000 residents in an area of 1.8 km^2 (ABS 2021), and approximately 13% of its residents are public and social housing tenants, with the majority living in the Carlton Housing Estate (ABS 2013). The Carlton Housing Estate spans 7.5 hectares across three sites: Lygon, Keppel, and Elgin Streets. Redevelopment work in the Carlton Public Housing Estate in Elgin Street commenced in 2006 and was completed over three stages in 2017 (DHS, 2013). Some other renewal works were conducted at the Carlton Public Housing Estates in Lygon Street in recent years. However, in 2023, a large plan to redevelop public housing towers across Melbourne was announced (Butler and Kolovos 2023). The plan includes the demolition of the existing public housing estates, which includes temporary or permanent relocation of current tenants (Homes Victoria 2023) and found criticism based on fears of unfair trade-off and for targeting some of the most vulnerable groups in the community (Dexter 2024).

This study focuses on the Carlton Public Housing Estate, which is located between Lygon and Rathdowne Streets, located within two kilometers of Melbourne's CBD. Its location close to the city center makes it convenient and physically inserted into a multicultural area. Before its redevelopment in 2006, this site included four large high-rise towers. In 2014, the towers were gradually renovated, each facing a different street and together enclosing a communal outdoor garden (Fig. 18.1). The juxtaposition of these different functions together turns the site into a multi-functional environment, where accessibility to educational, residential, and recreational services exists. However, we also need to evaluate how the typologies of these residential blocks and their connections with the public domain create or not a sense of home.

18.5 Re-Creating a Sense of Home

Within the limits of the rigid design of public housing, we observed that the possibilities for customization of the homes were reduced. However, we observed how residents adapt their units to reflect their cultural and religious backgrounds. Living

Fig. 18.1 Scheme of Carlton public housing

rooms are a typical example of how they adapted the use of the spaces and the artifacts they keep in this space that reflect their culture. It is also important to note how they adopt other foreign traditions, from the Middle East as an example, which, despite the differences, creates a sense of familiarity and belonging. The element of borrowing from similar cultures that share similar values helps migrants to create a sense of belonging within their homes. Whether it is a low seating, as seen in Fig. 18.2c, or the name of Allah (God) and the prophet Muhammad on a frame (b), these tangible elements play the role of a reminder, chosen by the dwellers, to connect with a balanced material world that they are longing for and a spiritual world that they would like to keep up with. One of the respondents supported these ideas and claimed that:

> The Quran, displayed on the top of my computer shelf and also, I have my prayer mats in different colors, and then I have curtains which are replicas of African curtains. The Middle Eastern people use the same but that is the way I can feel I can be on my own still respecting and feeling at home (Fig. 18.2).

Others also considered traditional furniture and prayer mats as tangible elements that are dear to them. In fact, re-creating a home outside of the borders of their home countries enhances a connection with the new place they live in and creates a somehow comfortable environment in such a foreign land and foreign context. Low seating, in particular, is a very common feature in the traditional homes of the Middle East. One of the respondents claimed the following:

> Getting a Somali like Somali sofas and curtains and maybe prayer mats all these stuff (Fig. 18.2a and c)

Fig. 18.2 Housing cultural customization

In the context of the interviews, we noted how the idea of recreating a home in Melbourne becomes very common amongst African migrants, by using furniture, decoration, and other tangible features. Language is also primordial in creating that sense of connection, where images and media also contribute to re-enacting what has been forgotten (memories from home, news). This idea is supported by the claims of one of the respondents:

> We keep the furniture look and even the decoration or any other like even curtains similar to what we had in back home…also we have like the TV satellite we watch all the news from back home in Arabic language and African language. (Fig. 18.2)

Beyond the units and the ambiance created within their interiors, the shared spaces inside the public housing estates play important roles in community interaction. For instance, corridors have become social spaces, and similarly, floor waiting spaces and laundry areas have become spaces for interaction or exchange. In some other situations or places, such as in elevators and waiting areas, the interaction is less spontaneous. Residents also use the diverse community facilities and public spaces. Playgrounds and community markets are popular areas for neighborhood interaction.

Together with the use of the spaces, residents appropriate them and make their own neighborhood. For instance, community gatherings take place during Ramadan, especially after prayers, or during celebrations (Eid al Fitr or Eid al Idha), so public spaces become venues for greetings, interactions, and long discussions. It also became relevant to observe the activities taking place at the nearby Albanian Mosque of

Carlton. During the holy month of Ramadan, Muslims, including African residents of the public housing, gather in the mosque for *Taraweeh* (additional prayer during Ramadan) and night prayers. *Iftaar*, or breaking the fast, is also held there, and many Somalis bring home-cooked food to share, thus creating stronger community ties during a very symbolic time.

During the daily and casual meetings in public spaces, women share tea and look after their children, in a similar way as they used to do in their home countries. In Carlton, the community also actively participates in neighborhood activities such as "Harmony Day" (an annual celebration in Australia that began in 1999, coinciding with the United Nations International Day for the Elimination of Racial Discrimination) in Carlton. Residents welcome outsiders, although they are cautious as they are aware of the misleading media reports "racializing Melbourne's African gang problem" (Budarick 2018). Religious events, celebrations, social events, and religious gatherings, are opportunities for African migrants to connect with home through shared activities they grew up with. In fact, this is highly relevant for the interviewees as one of them explained this as follows:

> "We together have a big meal during Ramadan. We have festivals during Eid and others and also there's a few meetings with the community." And another person adds: "We go together in whenever someone's sick we go together and read the Quran, we learn the Quran in community center."

In fact, this phenomenon can also be seen in the study of the Somali refugees of Nashville, Tennessee, in the US, where distinct places with which the community identifies are created and the geographic proximity supports feelings of cultural familiarity and security, as well as a sense of community and more solidarity (Chaney et al. 2018).

Festivals and religion constitute an integral part of the life of these migrants, and in particular men. Women, on the other hand, also organize their own circles to strengthen their social and cultural values as well as their connections. As explained by one of the participants, female-only groups are a common way in their small circles. She says:

> I have a group called only female group, which we do every Tuesday afternoon. We connect all the females in the building, and others in the surroundings as well.

In general, as one of the interviewees mentions, *"I like living here because I'am close to a wide range of Somali people."* Cultural connectivity amongst people from the same group, in this case scenario of Somalis, strengthens the sense of identity and confirms their belonging to similar roots. In a way, members of the Somali Muslim community take pride in connecting through not only social events but also their Muslim identity at large.

Residents, in general, feel comfortable in Carlton's public housing, but they also interact beyond the boundaries of the estates. Thus, we explored the permeability and the connection with the rest of the city, mapping their points of interest and preferred places, shown in Fig. 18.4. For residents, there is a constant search for familiar places in the city or appropriate spaces that would accommodate their culture and religion.

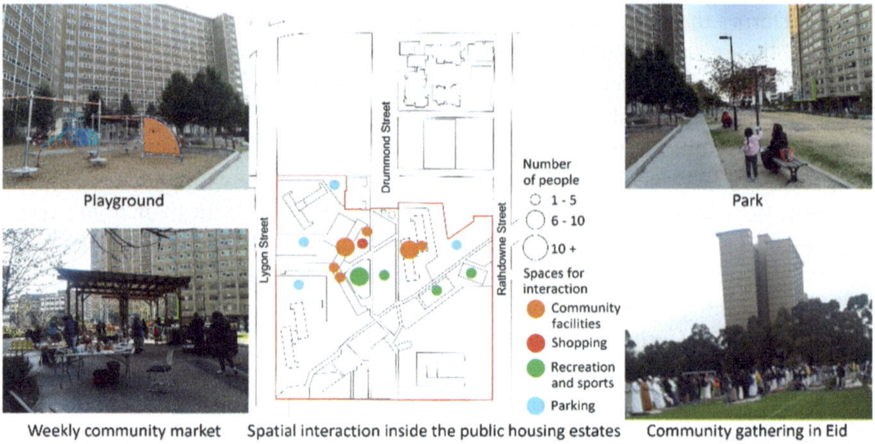

Fig. 18.3 Spatial interaction inside the public housing estates

As seen in Fig. 18.3, weekly markets and community gatherings help create a sense of community under the umbrella of social and cultural activities, which are typically open to the diverse groups in the estates. An interesting example is sports centers or pools that have women-only sessions. Similarly, neighboring and public spaces in Melbourne's CBD are frequently visited by the residents.

Community facilities such as neighbor centers and public libraries became popular among the community where the young and children with their parents (mostly mothers) regularly attend. Local cafes and restaurants that offer vegetarian or 'halal' options became centers for socialization, where it is common to see families and groups of young Africans and Australian-Africans mixed with other Australian residents from diverse ethnic groups.

Residents of the Carlton public housing estates also frequent neighboring shops. Some of the closest supermarkets have added 'halal' and African products attending to the demands of the African community. Shopping has also become a social activity for some residents who share cars to buy meat and visit larger shopping centers and city markets. Religion is an important part of the daily life of the community. For instance, the Albanian mosque, which is the closest to the Carlton public housing estates, is regularly visited by the residents every Friday. However, many residents frequently visit the mosque in North Melbourne (more than 2 km from the site) because the Imam is part of the Somali community, and the residents feel identified. Moreover, this is one of the most important places where the Somali community in Melbourne converges. The non-Muslim Ethiopian community also referred to the St. Mary Ethiopian Orthodox Church as their meeting place for their community.

An important part of the Somali diaspora strives to support their relatives living in the homeland. Moreover, as described by some Somali participants, solidarity in the community is part of the values that identify them. In case a community member requires financial assistance, in most cases to support the medical treatment of a

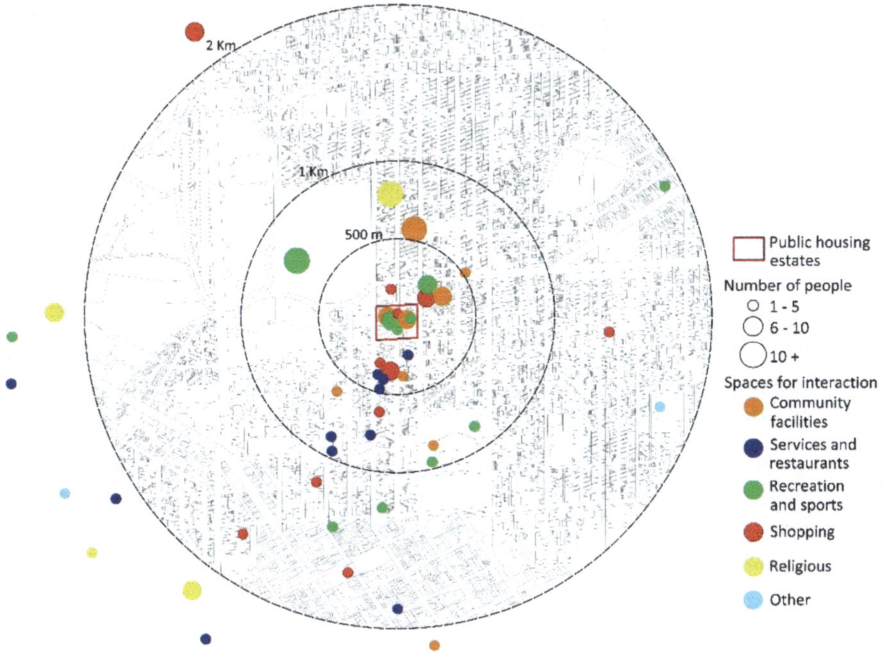

Fig. 18.4 City-housing estates spatial interaction

relative in Somalia, the community organizes a fundraising event, and the collected money is sent to Somalia. Thus, participants regularly send remittances to Somalia through businesses opened by community members.

18.6 Findings

18.6.1 'Keeping to Your Own'

Belonging, as a migrant, even when born in Australia, is often related to being part of a community of migrants from the country of origin. Particularly, the public housing estates in Carlton have become a city reference for the communities from the Horn of Africa in Melbourne, especially the Somali community. This is currently a symbol of their identity as it strengthens the sense of familiarity and belonging that they create and foster through the multiple spaces they use and the sense of place they establish through connections, interaction, and cultural events. Another big community worth mentioning is the Ethiopian community living in public housing. Due to cultural and religious affinities, this group found strong links with Somalis in their search for identity in their new home. These experiences are presented as follows:

> When I see a Somali person, nothing changed from when I lived in Somalia. When I see a Somali person that I've never seen, I call him brother, he calls me sister. We immediately feel like we have known each other our whole lives, and he asked me if, you know, he would immediately move to help, and I would immediately do the same and so I like that.

In this particular instance, migrants and refugees from Africa connect through the places they come from, or in general their continent, Africa. This sense of 'family' created outside the borders of their home strengthens the sense of attachment not only to each other but also to the environment they live in. Therefore, the sense of community could be referred to as creating a community within a larger community. This idea is also supported by the claim of one of the interviewees from the Ethiopian community. He states:

> The good thing, they make it like for us like a family. …everybody likes from Somalia and Ethiopia. The only thing I'm worried about is the other people.

Ethnicity also establishes another level of attachment throughout the years. What is noticeable here is the use of terms such as family, brother, and sister to denote strong community ties, without necessarily seeking frequent or regular connections with the outside world:

> …growing up in Carlton, that's where I would describe as my home. I think a strong memory of that would just be the black community, the strong sense of community, all the neighbors sort of becoming family. I think that's one of the greatest memories that I hold.

18.6.2 Understanding the Meaning of "Home"

Many African migrants come from extremely different experiences of 'home' (including dwellings, neighborhoods, and communities). Some of them have experienced long periods of temporary settlement in a third country. In fact, 20 of the 28 participants lived from one to eleven years in a third country. The following resident quotes present the diversity of the meaning of home prior to settling in Melbourne:

> The nomadic life was fun, but children have very... a very heavy load in terms of responsibility. You have to raise the goats... You have to walk two days to get water from the water catchment.

> Back home, it was nice. We have a flat and they have free to play around. I'm not the only person I raise my kids safe. My family can raise, my neighbors can raise, my friend she can raise. [Here] I can't trust anyone, I can't leave my kids…."

> … I think language was probably a big barrier [for my parents] but they've overcome that. Another thing is probably like cultural clothing and like the rise of Islamophobia. But then that's just more current, and not a struggle then because Australia was a very safe place before. It was much more welcoming then.

Based on these accounts, the meaning of home changes gradually to meet the requirements of a new lifestyle where the newcomers are expected to participate in this modern society, particularly through education. The feeling of security and cultural adaptability seems to be partially disrupted for migrants and refugees of African origin. As explained earlier, the sense of place is established in the public housing estates when people find a comfortable space where cultural and/or religious values are empowered. However, this notion of home, based on their accounts, is disrupted by the absence of family, weaker connections with neighbors, a lack of trust, and discrimination.

18.6.3 Isolation and the Social Design of Public Housing

While African migrants expressed some warm views of neighbors, there were still issues with how public housing is designed socially, with heterogeneous people living next to each other and little scope for forming deeper relationships.

> I wish neighbors that come to my house, ask for things, ask for something, that we can come and have tea, you know, sometimes at my house, sometimes at their house, that kind of neighbor. But we are all different cultures. So, when I had a baby, no one even knew about it. They could see I was pregnant, but they didn't know when I had it

The social disconnection between newcomers (refugees or migrants) and locals seems to be predominant in Australian society. Social bridges, or networks formed across different communities within society, are somehow limited or broken. As explained by Pittaway et al. (2009), in a study related to the integration of refugees from the Horn of Africa in Australia, the dilemma lies in the lack of acceptance from Australians and the lack of knowledge about the way of life in Australia.

However, the sense of isolation can be present within the same community, and it might cause tensions for the other African residents that do not share the same religious views. For instance, one non-Muslim participant referred to his relationships with the Somali and Ethiopian residents as restricted and strengthened the sense of exclusion:

> … the Asians are concealed to themselves. Asians are very few as well, but predominantly, people living in old buildings. Somalians are the greater number, and they're all Muslim. Somalians are Muslim, no Christianity in their country. Among Ethiopia, there are Christians and there are Muslims. They can interact with you, but Somalis are just different, even if we're African. You can't go to their house, like, you visit your neighbor.

The dynamics of social interaction for migrants are, however, a complex topic. Therefore, further research should examine the reasons behind weak social bridges between community members. Identifying these layers of social and cultural interactivity would help rethink the underlying issues of public housing, and how interactivity can be enhanced within the spaces of the estates.

18.7 Conclusion

The diversity of African migrants residing in Melbourne's public housing is not limited to their nationalities or their religions but to their previous experiences at home. These experiences might take place either in their home countries or in transitional places where they lived before coming to Australia. In this study, we found that the sense of home is closely related to the sense of belonging to a certain group. The need to feel part of a group is equally necessary for African Australians, even if they were born in Australia, as this is the basis for developing a sense of community.

Participants expressed a positive view of life in Carlton public housing. However, there are still issues with how public housing is designed and how migrant residents, especially those from African backgrounds, would deal with the isolation and stigma associated with social housing in Melbourne. Participants' perception of safety inside the estates has been repeatedly mentioned during the interviews. This sense of "exclusiveness" has created a favorable environment to develop a sense of belonging to a community, a house, and a neighborhood. However, residents also expressed their worries about the transition to other areas in the city. Thus, it is necessary to explore planning alternatives to create urban spaces for integration and security for multicultural communities.

This study explored the experiences of African residents of the Carlton public housing estate that served as a converging point for majority groups inside the estates such as the Somali residents. However, the positive findings might not be applied to other housing estates in Melbourne that are the recurrent topic of media coverage about the precariousness and integration challenges of African migrants and refugees. Moreover, we acknowledge the cultural, economic, and social differences of the various African migrant groups in Melbourne. Therefore, this study might not capture all the voices of African migrants and African Australians residing in various public housing estates in the city which would face a variety of challenges besides those mentioned in this study.

Finally, the housing issues of the African minority residing in Australia need to be studied using longitudinal research approaches to understand how and to what extent these groups are integrated into Australian society and how they acquire tools for independent and sustainable personal, family, and collective development.

References

ABS (2013) 2011 Census QuickStats: Carlton [Online]. Canberra. https://www.abs.gov.au/media-centre/media-releases/australias-overseas-born-population-grows-295-2022#:~:text=Jenny%20Dobak%2C%20ABS%20head%20of,2.1%20per%20cent%20from%202021.%E2%80%9D

ABS (2018) Migrant data matrices. Population Characteristics, Census 2016, ABS (Canberra)

ABS (2021) Census QuickStats: Carlton [Online]. Canberra. https://www.abs.gov.au/census/find-census-data/quickstats/2021/SAL20495

ABS (2023) Australia's overseas-born population grows to 29.5% in 2022 [Online]. Canberra. https://www.abs.gov.au/media-centre/media-releases/australias-overseas-born-population-grows-295-2022#:~:text=Jenny%20Dobak%2C%20ABS%20head%20of,2.1%20per%20cent%20from%202021.%E2%80%9D

Abur W, Spaaij R (2016) Settlement and employment experiences of south Sudanese people from refugee backgrounds in Melbourne, Australia. Australas Rev Afr Stud 37(2):12

Al Huraibi N (2014) Islam, gender and migrant integration: the case of Somali immigrant families. LFB Scholarly Publishing

Amuedo-Dorantes C, Bansak C, Pozo S (2018) Refugee admissions and public safety: are refugee settlement areas more prone to crime?

Arthurson K (2010) Operationalising social mix: spatial scale, lifestyle and stigma as mediating points in resident interaction. Urban Policy Res 28(1):49–63

Arthurson K (2013) Mixed tenure communities and the effects on neighbourhood reputation and stigma: Residents' experiences from within. Cities 35:432–438

Bolt G, Özüekren AS, Phillips D (2010) Linking integration and residential segregation. J Ethn Migr Stud 36(2):169–186

Brown BB, Perkins DD (1992) Disruptions in place attachment. Springer, Place attachment

Budarick J (2018) Why the media are to blame for racialising Melbourne's 'African gang' problem. 17 October 2019

Burnley IH (2001) The impact of immigration on Australia: a demographic approach. Oxford University Press, USA

Butler J, Kolovos B (2023) Labor announces $2bn for 'thousands' of new social rental homes and passes motion to make housing a human right. The Guardian

Carrasco S, Faleh M, Dangol N (2020) Our lives matter–Melbourne public housing residents talk about why COVID-19 hits them hard. The Conversation [Online]. https://theconversation.com/our-lives-matter-melbourne-public-housing-residents-talk-about-why-covid-19-hits-them-hard-142901

Carrasco S, Dangol N, Faleh M (2023) Rethinking social networks in responding to COVID-19: the case of African migrants in Melbourne's public housing. Int J Disaster Risk Reduct 98(104073)

Chaney J, Mohamed A, Williams S (2018) Somali refugee resettlement and residential patterns in Nashville, Tennessee. Southeast Geogr 58(1):80–103

Chaskin R, Khare A, Joseph M (2012) Participation, deliberation, and decision making: The dynamics of inclusion and exclusion in mixed-income developments. Urban Aff Rev 48(6):863–906

Chitrakar RM (2016) Meaning of public space and sense of community: the case of new neighbourhoods in the Kathmandu Valley. Int J Arch Res: Archnet-IJAR 10(1):213–227

Dexter R (2024) Last ditch legal action to stop public housing towers knockdowns. The Age

Dhanji S (2009) Welcome or unwelcome?: integration issues and the resettlement of former refugees from the Horn of Africa and Sudan in metropolitan Melbourne. Australas Rev Afr Stud 30(2):152

DHS (2013) Carlton housing redevelopment [Online]. Victorian Government, Melbourne. http://www.dhs.vic.gov.au/about-the-department/plans,-programs-andprojects/projects-and-initiatives/housing-and-accommodation/carltonhousing-redevelopment. Accessed 14 Jan 2017

Edensor T, Sumartojo S (2018) Geographies of everyday nationhood: experiencing multiculturalism in Melbourne. Nations Natl 24(3):553–578

Farah Farouque APDM (2007) Minister cuts African refugee intake. The Age

Flatau P, Smith J, Carson G, Miller J, Burvill A, Brand R (2015) The housing and homelessness journeys of refugees in Australia. Australian Housing and Urban Research Institute, Melbourne, Australia

Fonseka M, McGarrigle J (2012) Generating interethnic tolerance and neighborhood integration in European urban space. Centro de Estudos Geográficos, Lisbon

Forrest J, Poulsen M, Johnston R (2006) Peoples and spaces in a multicultural nation: cultural group segregation in Metropolitan Australia. Espac Popul Sociétés 2006(1):151–164

Graham E, Manley D, Hiscock R, Boyle P, Doherty J (2009) Mixing housing tenures: is it good for social well-being? Urban Stud 46(1):139–165
Harte W, Childs IRW, Hastings PA (2009) Settlement patterns of African refugee communities in southeast Queensland. Aust Geogr 40(1):51–67
Hinze AM (2013) Turkish Berlín: integration policy and urban space. U of Minnesota Press
Homes Victoria (2023) Which older-style high-rises will be retired, a list of which older-style high-rise housing will be retired. [Online]. Department of Families, Fairness and Housing, State Government of Victoria, Australia. Melbourne. https://www.homes.vic.gov.au/high-rise-victoria-which-towers
Joseph M, Chaskin R (2010) Living in a mixed-income development: resident perceptions of the benefits and disadvantages of two developments in Chicago. Urban Stud 47(11):2347–2366
Keene DE, Padilla MB (2010) Race, class and the stigma of place: moving to "opportunity" in Eastern Iowa. Health Place 16(6):1216–1223
Kleit RG, Carnegie NB (2011) Integrated or isolated? the impact of public housing redevelopment on social network homophily. Soc Netw 33(2):152–165
Lelévrier C (2013) Social mix neighbourhood policies and social interaction: the experience of newcomers in three new renewal developments in France. Cities 35:409–416
Mapedzahama V, Kwansah-Aidoo K (2017) Blackness as burden? the lived experience of black Africans in Australia. SAGE Open 7(3):2158244017720483
Morris A, Jamieson M, Patulny R (2012) Is social mixing of tenures a solution for public housing estates?. Exeley Inc
Naimi-Gasser J (2013) The remembered trees. Contractor-driven reconstruction and its consequences on communities' well-being in coastal Tamil Nadu'. In: Post-disaster reconstruction and change. Communities' Perspectives, CRS Press, Boca Raton
Pittaway E, Muli C, Shteir S (2009) "I have a voice—hear me!" Findings of an Australian study examining the resettlement and integration experience of refugees and migrants from the horn of Africa in Australia. Refug: Canada's J Refug/Refuge: Revue canadienne sur les réfugiés 26(2):133–146
Raynor K, Pert A, Townsend C (2019) Vertical cruise ships? Here's how we can remake housing towers to be safer and better places to live. The Conversation [Online]. https://theconversation.com/vertical-cruise-ships-heres-how-we-can-remake-housing-towers-to-be-safer-and-better-places-to-live-142381
Seethaler-Wari S (2018) Urban planning for the integration of refugees: the importance of local factors. Urban Plan 3(4):141–155
Sharma A (2010) Placemaking and the architecture of opportunity. In: Hamdi N (ed) The placemaker's guide to building community. Routledge
Shelton T (2019) Refugee week: the number of globally displaced people reaches record high of 70.8 million. ABC News [Online]. https://www.abc.net.au/news/2019-06-22/world-refugee-week-who-is-coming-to-australia-and-why/11223982. Accessed 9 Aug 2019
UNHCR (2013) The integration of resettled refugees essentials for establishing a resettlement programme and fundamentals for sustainable resettlement programmes, UNHCR. Switzerland, Geneva
UNHCR (2002) Refugee resettlement: an international handbook to guide reception and integration UNHCR
UNHCR (2024) Refugee data finder. United Nations High Commissioner for Refugees
Valtonen K (2004) From the margin to the mainstream: conceptualizing refugee settlement processes. J Refug Stud 17(1):70–96
Victoria G (2011) Refugee status report background information, methodology and data sources. Victoria, G. o., Melbourne, Victoria, Australia
Wari S (2017) Palestinian Berlin: perception and use of public space. LIT Verlag Münster

Chapter 19
Disasters by Design: Analyzing the Rio 2016 Olympic Games and the Eviction of Vila Autódromo from a Disaster Perspective

Augusto Cesar Oyama

Abstract Mega-events have been interpreted by urban studies as instruments of exceptional and entrepreneurial urbanism that, in the name of creating "world cities" and other developmental narratives, produce extensive urban interventions and impacts. Despite the possible positive effects, mega-events, such as the Olympic Games, can also be detrimental, especially in terms of housing rights. In Rio de Janeiro alone, over 77,000 favela residents were evicted and had their homes destroyed to make way for the 2016 Games. Although displacements are a defining feature of current mega-events, the social and non-infrastructural impacts of the Games are still considered marginal elements. A narrative is often created in which negative impacts, such as gentrification and displacements, are inevitable and, although perhaps unfortunate, they are apparently just a "natural" part of the cycle of development. Based on experiences of social movements fighting for housing in Brazil and Japan (which supported the author's master's thesis), this chapter aims to analyze the Olympic Games, particularly Rio 2016, from a disaster perspective, focusing on the impacts on those on the "receiving end" of urban renewal procedures, mainly socially excluded groups. To this end, interviews, counter-cartographies and systematizations were carried out, which supported the production of a mega-event disruption framework based on well-known disaster models and focusing on violations of housing rights. As result, it was understood that mega-events can often be considered "disasters by design" since joining the Games revealed a full and explicit commitment to specific agendas and discourses. In the case of Rio, the main strategy involved gentrification and removalist projects, based on concepts of privatization of public spaces in the name of "public interest", "sustainability" and "security". Whether they are considered a universal necessary evil, or simply a strategy that does not deliver "legacy", displacements in the mega-events lead us to reconsider the manner in which we conduct planning procedures.

A. C. Oyama (✉)
Disaster Prevention Research Institute, Kyoto University, Kyoto, Japan
e-mail: augusto.oyama@gmail.com

C. Ochiai et al. (eds.), *Disaster and Displacement*, Disaster Risk Reduction,
https://doi.org/10.1007/978-981-96-2583-3_19

19.1 Introduction

Global urban studies have widely interpreted mega-events as instruments of exceptional and entrepreneurial urbanism that, in the name of creating "world cities" and other developmental narratives, produce extensive interventions and impacts related to financial, institutional, political, infrastructural, symbolic and social dimensions (Vainer et al. 2013; Di Bella 2020). They include political events, such as IMF/World Bank conferences, cultural events, such as the World Fairs, and sporting events, such as the World Cup, the Asian Games, the Pan-American Games, among others. The Olympic Games are the largest and considered the most important among sporting mega-events.

Mega-events have the potential to foster cooperation and dialogue among nations, envisioning global solidarity and embracing the United Nations Sustainable Goals (SDG) and the practice of human rights. They are often used as economic and infrastructural development instruments, especially in communication and transport systems (Wilson 2012; Tamaki 2019). Mega-events are also considered opportunities to re-engineer the image of a city, nation, and sense of place (Bull and Lovell 2007; Sánchez 2010). Despite the possible positive effects that mega-events can have on urban territories, they can also be detrimental, especially to the fulfillment of housing rights.

These large-scale interventions often carry specific agendas, and there is much to learn from the stories of those on the "receiving end" of these urban implementation procedures. The existing population, especially the most marginalized and, therefore, vulnerable groups, and the places where these citizens actively create their everyday existence together, are invisible to the bureaucratic abstractions of providing major venues and new infrastructure (Porter et al. 2009).

This chapter aims to analyze the Olympic Games, particularly the Rio 2016, from a disaster perspective, focusing on the impacts on those on the "receiving end" of urban renewal procedures, mainly socially excluded groups. A "disaster x mega-event" framework is proposed as part of the discussion. The study is a direct reflection of the author's master's thesis entitled "Mega-events as disasters: the eviction system created at the Rio 2016 and Tokyo 2020–2021 Olympic Games" (Oyama 2023).

The focus on the Olympic Games as the main target of this analysis is justified by the fact that they concentrate on the largest records of mass displacements and destruction of communities among other mega-events. According to Oliver-Smith (2019), many studies highlight displacements themselves as a second disaster. The systematization of studies (Burbank et al. 2001; Lenskyj 2006; COHRE 2007; Sánchez et al. 2012; WCOPCRJ 2015; Boykoff 2021; Oyama 2023) shows that more than two million people were displaced at least in the past 30 years due to the Olympics, from Seoul, South Korea, in 1988 to Tokyo, Japan, in 2021. These dynamics of removals, either through iron-fisted displacement or market-driven gentrification (a system of class-based dis-placement), are a defining feature of the Olympic Games and affect mostly socially excluded groups, such as a myriad of poor, landless and unhoused/homeless people, and ethnic minorities.

The choice of the Rio 2016 Olympics, in turn, as a case study lies in these facts: Rio de Janeiro was the second host city for the Summer Olympics, after London in 2012, to be selected following the formal inclusion of the "legacy" concept in the 2002 Olympic Charter, generating a growing expectation that the mega-event would bring a positive socio-economic legacy for the city that receives it. However, even today, many elements remain to be uncovered about the articulations, negotiations, and violations of human rights, especially housing-related ones.

In analytical terms, an inflection point differentiates this research from others. The fact that many affected groups interviewed (Oyama 2023) were displaced up to 3 times by previous mega-events or urban renewal projects led to the following hypothesis: in many cases, mega-events operate in the same way as human-created disasters. Both are the result of socially produced vulnerabilities (including cycles of marginalization) and involve multiple similar ruptures (institutional, legal, urbanistic, social, symbolic, scalar, among others).

Hence, this research echoes a variety of scholarship, from disaster and displacement works to housing and gentrification studies.

19.2 Methodology

The methodology included: (i) a literature review related to the multiple impacts of mega-events in several host cities; (ii) more than 30 semi-structured and oral history interviews with key social actors, such as members of social movements and homeless organizations, professors, architects, lawyers, journalists, and writers researching both positive and negative impacts of the Olympics, and former government employees; (iii) analysis of relevant documents and databases such as census analyses, maps, photographs and newspapers (both mainstream and independent press) to examine historical tensions related to real state fronts in the period before and after the Olympics.

To examine the 2016 Olympic Games from a disaster perspective, this study adopted a vulnerability approach, given the central role it plays and especially its social and political aspects in the process of a disaster (Oliver-Smith et al. 2017). Through these lenses, a framework was proposed that provides an understanding of interactions, connections, and adaptations at three levels (physical, social, and individual systems) in human-created disaster contexts.

To explore the relationships between these different systems, as this study is focused on impacts on housing, land and socially excluded groups, layers of disruption highlighted by Vainer et al. (2012), Harding (2007), Wisner et al. (2004) were analyzed, including institutional, legal, urbanistic, social, symbolic, scalar and ruptures in real estate dynamics and the gradient of land values.

In the mapping approach, Geographic Information System (GIS) and design software were used to produce socio-spatial analyses. Also, the concept of counter-cartography was important to the discussion flow: it refers to the technique of revealing invisible or omitted issues (Oyama et al. 2019) and can take a variety of forms in different cultural and political situations. It is possible to identify at least three concepts of how they work:

(i) counter-mapping as a tactic for confronting asymmetrical power relations;
(ii) challenging the academic debate on spatial representations, being a key and alternative strategy to assess and communicate issues concerning environmental justice, public health, urban planning, and human rights;
(iii) the use of counter-cartographies as social technologies, integrating geotechnologies and participatory mapping.

19.3 Disasters Embedded in the Social Construction of Vulnerability: A Reading Key for Mega-Events

Relatively new perspectives assert that disasters do not simply happen; they are caused. Thus, understanding that vulnerability can be socially constructed, maintained, and reproduced by political and economic forces (Oliver-Smith 1999; Bolin and Stanford 1999) is an interpretative key for the unfolding of this chapter.

From this analytical perspective, one of the important factors in the production of vulnerability is cycles of marginalization. Marginalization leads to high vulnerability (Wisner and Luce 1993), with those at the margins suffering the most when faced with disasters. Homelessness and slum residents are one of the uttermost states of marginalization and reflect an advanced condition of destitution and denial of basic rights. As highlighted by Gaillard and Cadag (2009), disasters frequently lead to more marginalized people as the victims who have lost their livelihoods are often unable to recover.

Another important set of research works, such as Vainer et al. (2013), and Nogueira and Souza (2020), addresses the role that urban planning plays not in reducing risk, but in how public policies and urban interventions can create or aggravate conditions of social vulnerability, operating uninterruptedly by the general economic growth of the State.

In light of this, Harding (2007) defines human-made disasters as conditions that result from a range of policies and deliberate state actions. They produce adverse impacts on a country's economy and infrastructure and facilitate the disruption of social networks and the community. Examples include wars, actions by repressive regimes, failure to stop the spread of preventable diseases and epidemics, economic sanctions and neoliberal economic strategies.

Based on the systematization of disaster models and frameworks, the schematic representation proposed in Fig. 19.1 portrays key notions applied in mega-event contexts. This approach makes it possible to visualize, beyond a large urban project, mega-events are a human intervention with significant potential for disruptions.

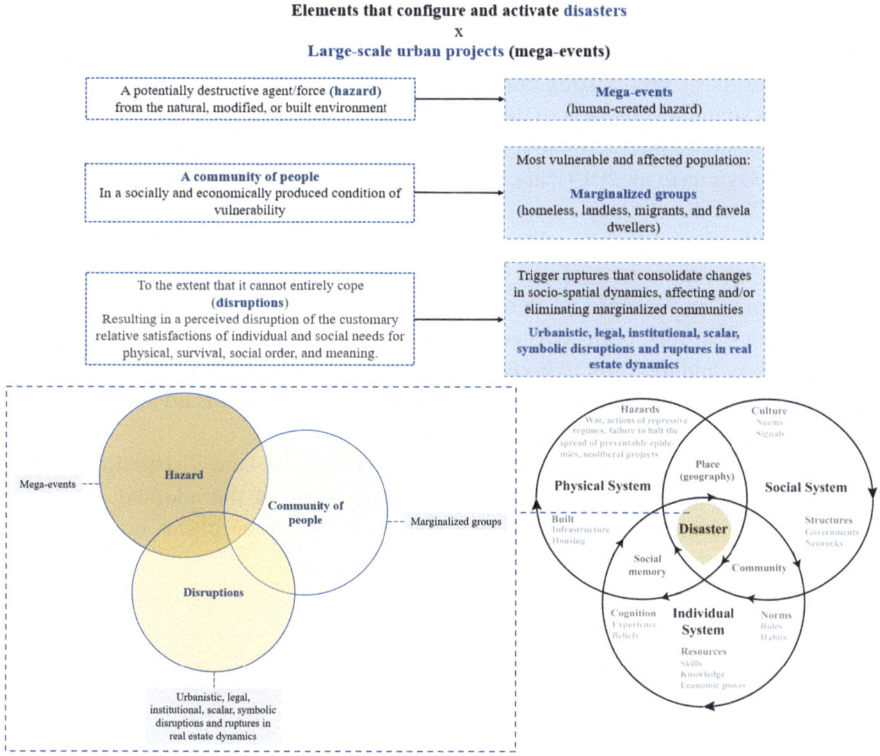

Fig. 19.1 Framework–understanding mega-events from a disaster perspective. *Source* Author

19.4 The Knot of Disasters in Brazil

The concept that disasters are embedded in sociocultural processes, or rather they are a historical product created by identifiable forces, is important to understand disasters, especially in Brazil. In this sense, an insight into the land issue and the role of the State in producing inequalities is fundamental.

As predicted by Joaquim Nabuco (Maricato 2008), an important figure in the Brazilian abolitionist movement, the weight of slavery is still present in Brazilian society long after its abolition. Socio-spatial inequality has its very firm roots in the five centuries of Brazilian society formation, especially since the privatization of land in 1850, and the legal end of slave labor in 1888. In this sense, it is essential to look at how the Brazilian structure has changed very little since then: the country still operates in a colonial logic, on the periphery of capitalism, of overexploitation of its workforce, and it is not and has never been a country of racial democracy (Fernandes 2008), myth produced by the Brazilian elite to maintain conditions of racial inequality.

In this context, as in virtually all Latin American countries, colonialism and State interventions have resulted in the unequal geographical distribution of risks and the destruction of many pre-existing cultures.

Favelas, a spatial expression of the exclusionary process of urbanization in Brazil, and where tensions, interactions, and disputes between domination and resistance materialize (Oyama et al. 2019), are currently home to around 11.4 million Brazilian people, or 6% of the country's population (IBGE 2010).

Faced with so much inequality and contradictions, Brazil, from the popular classes, has legislation to confront these issues that is very unique: the 2001 City Statute, taken as an example for the world, restricts and limits the right to property, subordinating it to the "collective good" and the "social interest". The social function of property and the right to housing are provided for in the Federal Constitution and are in line with international law principles. However, there is an abyss between rhetoric and practice (Rolnik 2019).

This complex reality is mainly reflected in the form of land distribution within the urban territory: it is imposed, by the inherent logic of the adopted economic productive process, territorial-racial segregation, which often places contradictory conditions on housing laws (Maricato 2008).

19.4.1 Rio de Janeiro and Favelas: Displacements by Color and Marginalized Condition

The city of Rio de Janeiro brings together important particularities: (i) it has generated forms of informal settlements in the face of the rigid regulations of the State; (ii) it has shaped forms of vulnerability originating from this process of segregation which is remarkable throughout the territory, and which have multiplied and become glaring especially more recently, with the emergence of mega-events; (iii) the city has been the stage for the instrumentalization of different removalist discourses, consolidating the "disaster risk" and "public safety" in the so-called mega-event phase (from the 2000s to the present time) in Rio.

In this port city, the manual labor of the slave worker during the colonial period moved all the gears of its economic and social life (Benchimol, 1985). In the gradual occupation of Rio de Janeiro, the black population was left with the worst: the unhealthy periphery and the hills (Sodré 2002). This context reflects not only Rio de Janeiro but also Brazil in general: the country is mainly a territory of housing self-construction and self-urbanization by poor classes, and it is in this historical process of urbanization that the favelas are placed.

The "favela as a problem" is an expression used by Valladares (2005) to designate the condition in which favela came to be inserted in the public and institutional debate about the city, especially in Rio de Janeiro, at the beginning of the twentieth century. According to Valladares (2005), at the center of history, the colonial heritage illuminates a linear experience: narratives of urban modernity (aesthetically and behaviorally ordered, hygienic, scientific, a place for a white people) would replace the urban image from the past (aesthetically and behaviorally chaotic, unhygienic, backward, underdeveloped, a place of favela dwellers, constituted especially by black people).

For Maricato (2008), however, the informal occupation of land neglected by the real estate market, such as hillside slopes and lowland floodplains, became a solution, rather than a problem, of housing for the marginalized groups. Especially from the 1980s onwards, favelas expressed another fundamental process: the political organization of the poor population (Tanaka 2017).

Three broad and interrelated policies were applied to Rio's favelas in the twentieth and twenty-first centuries:

(i) The first set of policies included forced evictions, mainly applied under Governor Carlos Lacerda during the military regime between 1962 and 1974, when around 140,000 people were evicted from their places.
(ii) The second set of policies included sporadic, incomprehensive, and insufficient upgrading policies, intended to provide minimal infrastructure in some favelas. The most robust program, Favela-Bairro, occurred in the 1990s.
(iii) The third set of policies criminalized and repressed the urban poor, within the scope of the Military Police's Pacifying Police Units (UPP) program, in the context of the mega-events phase from 2008 onwards.

19.4.2 Vila Autódromo: The Reflection of Removalist Discourses and Resistance Actions Throughout Rio's History

In this context of a produced socio-spatial segregation, Vila Autódromo stands out: territory that disputed this history of unequal distribution of risk and eviction narratives/projects. Located where the Olympic Park was being planned, Vila Autódromo was at the center of the conflict of hegemonic forces in the context of the Rio 2016 Olympics.

Among the more than 77,000 people displaced in preparation for the mega-event, the community became one of the main symbols of resistance against a city project that promoted the socio-ethnic cleansing (Metropolis Observatory 2016; Tanaka 2019) of large areas of Rio de Janeiro.

There were about 550 families in this location with approximately 17 hectares, mainly occupied by low-income, brown (a classification included in the black racial group) people and northeastern migrants, who had lived in the area for decades. Despite the community having been demarcated as an AEIS (Area of Special Social Interest) and most of the residents having the title of Concession of Real Right of Use, factors that safeguard the right of occupation and guarantee of tenure, Vila Autódromo faced a variety of eviction attempts by the City Hall since the 1990s.

The threats of evictions reached a new dimension in the mega-events phase, changing political discourses, but moreover, applying additional narrative layers systematized in Table 19.1, from "the favela as a problem" and hygienist dis-courses, to the instrumentalization of disaster risk narratives and public safety or "public good" narratives.

As shown in Table 19.1, the most common strategy, from the logic of modernization (second half of the nineteenth century and throughout the twentieth century) to the instrumentalization of environmental narratives (from the 90 s to the present) and mega-events, is the physical and symbolic erasure of the affected object. Vila Autódromo, from which many residents came from different past evictions, historically represents these strategies and discourses, often overlapping them.

To illustrate this scenario of instrumentalization of the Olympics to justify mass displacements, Fig. 19.2 systematizes the 119 favelas in the context of the mega-event.

From the above counter-maps, according to Faulhaber and Azevedo (2015) and Tanaka (2019), these displacements, with a marked political-ideological motivation, in general, pushed the most marginalized populations from areas with better infrastructure and more valued to peripheral and environmentally fragile regions.

In the case of Vila Autódromo, during the attempts to negotiate an alternative urbanization plan to resist evictions, the then-mayor Eduardo Paes sold the image of the Olympics using the "Inclusion Games" motto and initially stated that no one would leave the land unless they wanted to. Paes, however, completely changed the discourse as the date of the Games approached.

Table 19.1 Favelas and evictions past and present in Vila Autódromo

	"Favela as a problem": Modernization and hygienist discourses Second half of the 19th century and throughout the 20th century	**Intensification of climate events & Instrumentalization of disaster risk narratives** 1990s and 2000s	**Mega-events and the consolidation of removalist discourses** It starts in the 2000s, and consolidates itself at the present time
CONJUNCTURE	**Medical-hygienist ideology** Remodeling and sanitation of the then capital (until 1960, when the capital became Brasília) of the Republic of Brazil Black (former)slaves and favelas were an ideological obstacle to hygiene and modernization **The logic of modernization in Rio de Janeiro** Modernizing implied "Europeanizing" the urban space. The idea of evictions implied, therefore, "cleaning up", "beautifying", and building or reforming the relations of social groups with the housing space, making the presence of black people and other socially marginalized groups less noticeable	Hygienist ideology still widely used, but operated with new discursive elements **Environmental risk as one of the main threats to the city** **Favelas and the "intrinsic risk"** To face the "disaster risks", public authorities have favored, as the most common strategy, the physical elimination of the exposed element, especially if located in regions of real estate interest or urban requalification	**Business planning and the city of exception** Conjugation of narratives related to public safety, informality/illegality, disaster risk and making the city attractive for investments; the "public good", in the name of mega-events or large-scale urban interventions, becomes an instrument in the argument for removals **The security risk discourse, UPPs (Pacifying Police Units) and the militarized management of unequal territories** (i) policy of exclusion and punishment of the most socially marginalized, hidden in a security policy that controls and limits urban space; (ii) the stigma is reinforced; it portrays and guarantees only one profile, with a specific racial marker in popular territories
VILA AUTÓDROMO	**1993** **First formal eviction attempt** City hall accuses the community of causing "**damage to the natural environment**", in addition to "**aesthetic, landscape and tourist damage**"	**1996** City Hall's new argument: the community would have to be displaced for being in a **risk area**. Public Defense and Vila Autódromo residents argue in favor of permanence and win **2009** City Hall's 2009-2012 **Strategic Plan**: Goal to reduce by 3.5% the areas occupied by favelas in Rio. Vila Autódromo was included in the list of 119 favelas to be completely removed by 2012, because "**they are in places at risk of landslides or flooding, of environmental protection or intended for public places**"	**2005** Context of preparation for the 2007 **Pan American Games**: City Hall **modifies urban zoning law**, imposing that buildings could not occupy the marginal protection strip of the Jacarepaguá lagoon, with the exception of "sports equipment necessary for the Games". **2010** City Hall states that the **eviction is the IOC requirement**. In March, it says that the community would be a **threat to the security of games**

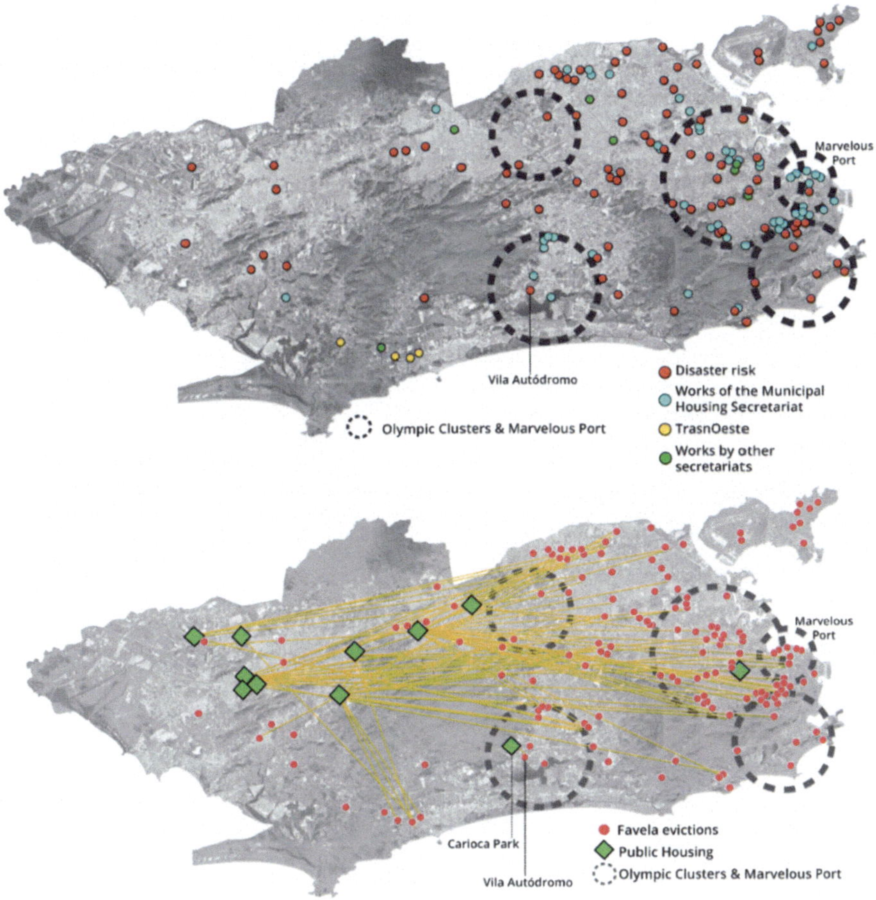

Fig. 19.2 Evictions and resettlements in preparation for the Rio 2016 olympics. *Source* Created by the author based on Faulhaber and Azevedo (2015)

Figure 19.3 shows changes in land use due to the removal and destruction of housing and social infrastructure (the spaces and places that help create and maintain social ties and trust) in Vila Autódromo over time.

The clash between Vila Autódromo and the City Hall took place in multiple dimensions and continues to be present in the struggles against ongoing evictions in other areas. The Popular Plan, the residents' alternative proposal, integrated into the process of social struggle represented a possible urbanization model around the Olympic areas: an urbanized popular neighborhood, side by side with the Olympic Park area, aligned with the Olympic principles and the perspectives of a democratic city. It demonstrated the possibility of participatory alternatives, even in a context governed by real estate forces, lack of information, and criminalization of peripheral populations.

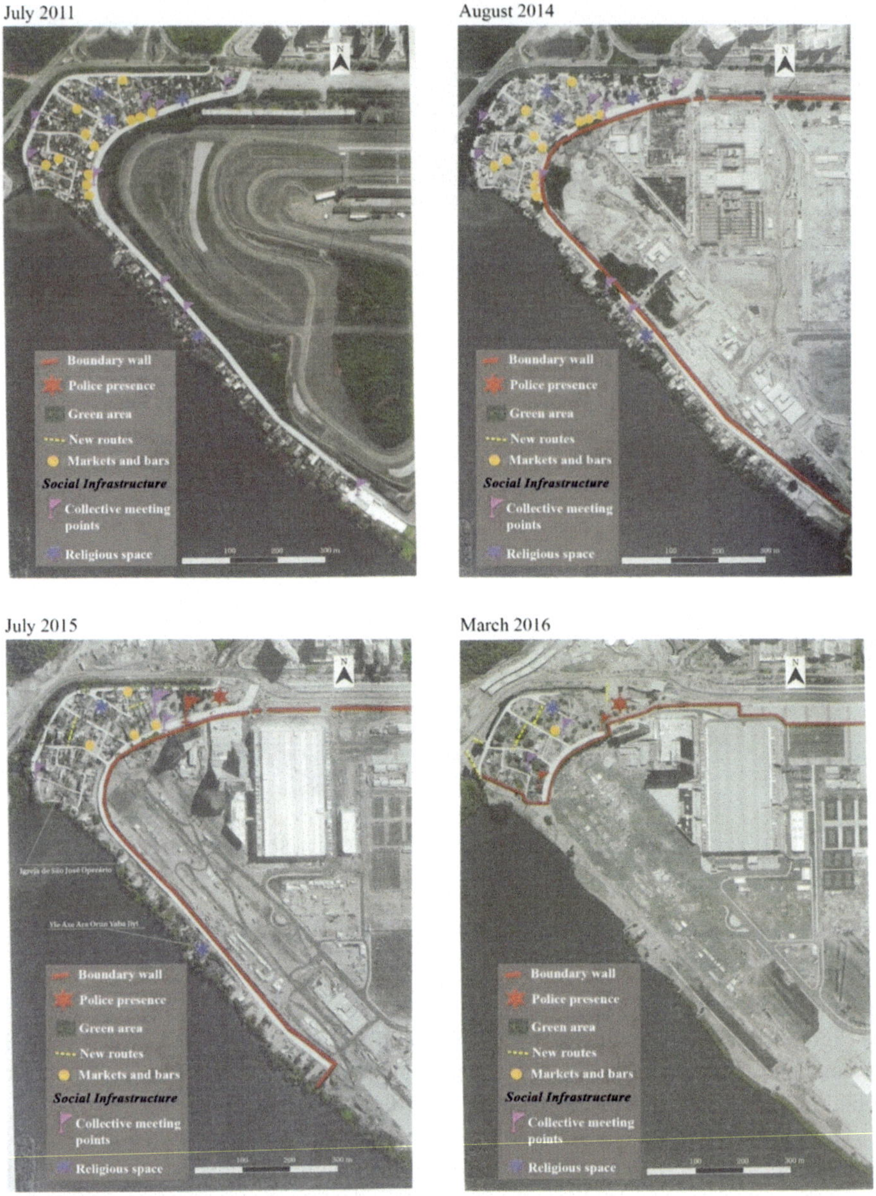

Fig. 19.3 Changes in land use and social infrastructure in Vila Autódromo in the context of the Olympic games. *Source* Created by author based on Moreira (2018)

Despite the Popular Plan, attempts at negotiations, and adjustments to the Plan, the City Hall did not incorporate any element from the residents' proposal. The Rio government's final project was the last one presented at the meetings and decided to remove at least around 70% of the residents, with worsening accessibility conditions for the remaining homes, complications for drainage solutions, and the creation of a completely new ghetto, segregated from the surroundings by a network of large roads.

Ignoring the Popular Plan and idealized by the City Hall, the "Carioca Park" housing complex, where many of the residents of Vila Autódromo were relocated, was implanted in land belonging to two companies that financially supported the electoral campaign of the then mayor Eduardo Paes. The building covers approximately 4.9 hectares and is located 3 km away from Vila Autódromo, totaling 900 housing units, of which 450 would be for residents of the community.

The situation of those who were evicted from Vila Autódromo and forced to live in the new public housing, contradicts the speech of then-Mayor Paes that "all residents are happy". A questionnaire and interviews conducted by Morgado et al. (2022) show the following reality that all those interviews said: (i) their apartments were too hot, (ii) they faced periodic flooding problems, (iii) the local traffic was intense during rush hour. Also, some interviewees complained about the lack of bus lines and excessive crowding.

As a post-mega-event collective act, a symbolic experience emerged from the ruins and what remained of Vila Autódromo: the Evictions Museum (Fig. 19.4). The open space museum aims to deal systematically with all evictions affecting Rio de Janeiro, including past and future. But also, through its presence, it focuses on expressing a potential new future where favelas are legitimate residences.

The project is managed by a collective of residents from Vila Autódromo and voluntary collaborators from the most varied backgrounds, including researchers, museologists, artists and activists. The artworks refer to the places of residence that were destroyed and celebrate those that remained.

The collection includes ruins, photographs, household items and resistance messages. While the trauma of demolition is imprinted in the territory, the collection and its representation of residents' valued bodies and souls provides a sequence of images that can become an important experience in disaster memorialization processes.

In other words, the Eviction Museum is the materialization of the Vila Autódromo's struggle that made visible gentrification, silent prejudice and the marginalization suffered by the most vulnerable in Brazil.

The description of each image is as follows:

Picture at location 1: São José do Operário Catholic Church (external área), where the community has also political meetings.

Pictures at location 2: (I) In the church, many families were able to store their belongings and take shelter during the evictions; (II) ruins of Zezinho and Inês's house.

Pictures at location 3: (I) "While housing is a right, resistance is a duty"; (II) "We resist in Common Unity. Long live the CommUnity".

Fig. 19.4 The evictions museum, a "memoryscape" that establishes a living message for many slums in Brazil. *Source* Author's elaboration based on fieldworks and interviews in Vila Autódromo, in October 2021

Picture at location 4: Open space where the former demolished community association was located.

Picture at location 5: "Welcome to Vila Autródromo". Photos are displayed regarding the situation before and after the evictions.

Picture at location 6: "Origin of the Evictions Museum".

Pictures at location 7: (I) "My house and my rights are not for sale"; (II) "Containers: where 9 families took shelter for 73 days".

Picture at location 8: "Dona Dalva's family lot. They fought and stayed".

Picture at location 9: End of Beira Rio Stree, where Iara and Gaúcho's family lived".

Picture at location 10: "Memory cannot be removed".

Picture at location 11: "Social hygienisation".

Pictures at location 12: (I) The photographic collection has varied supports, physical and digital, and is composed of different records of the history of struggle in Vila Autódromo. (II) Dona Penha's rebuilt house, one of the community's main symbols of resistance.

19.5 Creating a Foretold Disaster: The Rio 2016 Olympic Games

The mega-events phase was inaugurated in Rio de Janeiro with the 2007 Pan American Games (Sánchez et al. 2012) and extended beyond the Confederations Cup in 2013 and the World Cup in 2014, consolidating itself with the 2016 Olympic Games. These large-scale urban projects already structure a significant part of the policies in the city of Rio, whose guiding principle becomes the city of exception (Vainer et al. 2013): a model based on the concentration of large public investments in the implementation of projects that reconfigure extensive urban spaces and impact the very structure and dynamics of the city, having as a major marker the control of access to public spaces.

As previously defined, mega-events are interventions with significant potential for disruption. From this understanding, it is possible to highlight some fundamental types of ruptures that the Rio 2016 Olympic Games has produced, intentionally or not.

(1) Institutional ruptures

Just as in other cities where disasters precipitated massive population displacement, in Rio, these conditions created longstanding battles between the state's development interests and the rights of its inhabitants.

More than a simple consequence of existing conditions, the Games allow state and private actors to introduce strategies that would not be usually accepted, thriving on the social euphoria of the spectacle to privatize and commercialize public resources, increase security and policing, and militarize public spaces. In effect, the taxpayer subsidizes the commercialization and privatization of such places.

Through public–private partnerships and the concentration of contracts with few large building companies in Rio de Janeiro, the Olympics represent the transfer of public resources to the private sector, subordinating public interests to market logic (WCOPCRJ, 2015).

(2) Legal ruptures

An extensive survey by Metropolis Observatory (2016) highlighted a variety of exceptional laws produced in the context of the 2016 Olympics.

An example of the eclipse legislative achievements is the approval, just as it had happened previously in the World Cup 2014, of the General Law of the Olympics (Law nº 13.284/2016). The law restricts several rights, such as freedom of speech. There are some articles in the law limiting specific protest acts that can be held in Olympic venues-banners and signs with any "offensive messages" (Chap. 4).

The law also condemns the use of all official symbols related to competition by citizens, and the rights of informal workers exercising their activities were restricted.

There are also articles related to the closure of public areas exclusively for the circulation of previously registered persons (which affects the population's freedom of movement) and the restriction of the right to record images and sounds (which interferes with the freedom of the press), among others. In light of international legal frameworks, the main violations in the case of the Rio 2016 Olympics include: genuine consultation with those affected; adequate and reasonable notice for all affected persons sufficiently in advance with a view towards minimizing the adverse impacts of evictions; information on the proposed evictions and on the alternative purpose for which the land or housing is to be used to be made available in reasonable time to all those affected; adequate resettlement, within a reasonable distance from the original site, with access to essential services; the right to adequate compensation for the loss of any good or property.

(3) Urbanistic ruptures

Several studies have associated the holding of mega-events with the planned expulsion of the low-income population from valued areas or newly incorporated into the real estate market valuation circuit in preparation for the mega-events. In Rio de Janeiro, residents from 119 favelas were forced evicted during this period.

Santos Junior and Santos (2013) identified the following axes of urban interventions associated with the main displacement processes: (i) urban mobility, focused on the building of the Transcarioca motorway and the Transoeste highway; (ii) renovation or installation of sports equipment and facilities (e.g. the construction of a golf course in the environmentally protected area of Marapendí, in the west zone of the city, and the removals close to the Maracanã stadium); and (iii) regeneration of urban areas for touristic attractions and/or tourism promotion and infrastructure building (e.g. the 'Marvellous Port', the urbanization process in the central favelas promoted through the Federal Program Programa de Aceleração do Crescimento or the municipal program Morar Carioca).

Furthermore, it is possible to observe the growing militarization of the city, which affects mainly black populations living in favelas and the suburbs. The UPPs (Pacifying Police Units) have excluded and punishd the most socially marginalized and created visible and invisible barriers that promote the socio-spatial segregation of the city.

(4) Social ruptures

The impact of mega-events or disasters should not be measured simply in terms of physical assets destroyed and people displaced but also and perhaps primarily in terms of the extent of disruption of social relationships, networks, and assets (UN-HABITAT 2010).

In this sense, not only thousands of homes were destroyed in a city that has a housing deficit of 220,774 units (WCOPCRJ 2015), but also myriad community spaces (social infrastructures) and social relationships built over decades were disrupted during the context of the 2016 Olympics.

According to Metropolis Observatory (2016, p 21), "Since the preparation for the Pan American Games (2007), the larger process of forced evictions in Rio de Janeiro's history was initiated. These forced evictions follow a clear pattern of the expulsion of the black, poor population from areas that are undergoing massive real estate valorization".

In short, as documented by the World Cup and Olympics Popular Committee of Rio de Janeiro (WCOPCRJ 2015), many of these evictions resettled residents in new risk areas, in militia locations, and weakened social capital.

Vila Autódromo, for instance, had 97% of its area destroyed, disrupting social networks and social infrastructures built over more than 5 decades.

(5) Symbolic ruptures

A form of apathy occurs during displacements and other human rights violations caused by this particular disaster since these adverse effects on certain groups are seen as "natural" or even necessary for the "collective good" of holding the Games. Media coverage and public and political reaction to these negative impacts vary widely, from outrage to non-existence, depending on who, where and when this disaster happens.

The act of neglecting and making marginalized populations even more invisible took place, for example, in the cartographic dimension. During the 2014 World Cup and the 2016 Olympic Games, the printing and distribution of official tourist maps erased myriad of favelas from their geography. These official maps show hills that have had favelas replaced by green areas. This meant that around 1.4 million Rio residents had their presence erased from official cartographies.

Furthermore, the affected communities did not have access to official information about urban interventions, and were deprived of participating in discussions related to the projects. Effectively, in many communities, meetings between residents and representatives of the City Hall were not recorded, contrary to what is determined by the City Statute.

In the exceptional contexts produced by mega-events, chronic problems, such as housing and homelessness issues, are not being solved; instead, they are generally erased.

(6) Ruptures in real estate dynamics and the gradient of land values

The standardization of urban interventions to attract investment to the city, which hides the strategy of selling urban spaces to specific interest groups. As a result, forced displacements are the tip of a project of deep transformation of the urban land dynamics of Rio de Janeiro comprising new processes of gentrification and commercialization of the city.

In Rio, the proliferation of high-end real estate developments in the vicinity of stadiums and other emergency works provoked a double process of expulsion of the poorest population, either by the summary and violent evictions or by the expulsion resulting from the strong and unregulated consequent real estate appreciation. The mega-event phases, in this sense, intensified urban apartheid in Rio de Janeiro (Ferrerira 2014).

(7) Scalar ruptures

Through the production and control of narratives, the State creates and opens precedents of exclusion and violent agendas in different territories and scales that in other circumstances would be unfeasible.

Global expectations of civility, modernity and security, also defined in the IOC requirements, have dictated the adoption of disciplinary and social control practices. In the case of Rio de Janeiro, the security policy adopted by the city hall was followed in other territories, both at the state and municipal levels. One of the materializations of this policy is the "Pacifying Police Units" (UPP) on the boundaries and interiors of favelas, and the placement of bulkheads on the boundaries between expressways and favelas, symbolically and materially demarcating a new territorial fragmentation.

Effectively, the growing mediatization of mega-events and the alignment of agendas at national, state and municipal levels, in order to sell a specific imaginary of Brazil/Rio to the world, start to control several aspects of the urban image, with impacts on civil liberties and the right to the city.

19.6 Final Considerations

Despite the substantial evidence about the impacts of mega-events regarding gentrification and displacement, the actual practices that bring all these almost unheard stories together are rarely held up and critically examined.

In this sense, this study deals with an "underground memory", often hidden and erased, whether planned or not, which is opposed to "dominant memory", and belongs to a facet of the marginalized groups' history.

The decision to frame mega-events as disasters was an important exercise from this point of view. Disasters, understood as processes, cannot be considered accidents outside the usual functioning of society. Instead, disasters largely reflect the failure of inclusive development, where the root causes of vulnerability often merge with the origins of other development-related crises, mainly regarding land and housing chronic issues.

Especially in the Latin American case, it is fundamental to consider the structural conditions of a country with territorial and ethnic dimensions such as Brazil. The role of the racial issue and the State actions as a vulnerability-creating factor was a critical point to this analysis.

It should be a priority to listen and connect with those generally ignored at mega-events since their lives and resistance actions are a warning message to address reflections on how human rights violations are intertwined with mega-events. To date, there are no genuine practices that would be consistent with the IOC's discourse of inclusion and Olympic legacy for all citizens. An in-depth assessment of the sustainability goals and outcomes of previous Olympic Games is also essential in this regard.

Mega-events, hence, operate not only as an indicator of the host society, especially in terms of spatial inequalities and injustices. They also have the potential to reflect global situations: mega-events often exacerbate or make chronic and structural issues even more visible. According to Rocha and Xiao (2022), until now this dimension of the Olympic development has not been systematically documented throughout history, especially from the housing rights perspective. Further studies in this direction, especially regarding the future mega-events, such as Paris 2024 and LA 2028, are fundamental.

As witnessed in the case study, mega-events can often be considered "disasters by design" since joining the Games revealed a full and explicit commitment to specific agendas and discourses. In the case of Rio, the main strategy involved gentrification and removalist projects (engineered segregation, social clearance operations), based on concepts of privatization of public spaces in the name of "public interest", "sustainability" and/or "security". Whether they are considered a universal necessary evil, or simply a strategy that does not deliver "legacy", displacements in the mega-events lead us to reconsider the manner in which we conduct planning procedures.

References

Benchimol JL (1985) A modernização do Rio de Janeiro. In: Del Brenna GR (1985) Uma Cidade em questão II: O Rio de Janeiro de Pereira Passos. Rio de Janeiro: Index

Bolin R, Stanford L (1999) Constructing vulnerability in the first world: the Northridge earthquake in Southern California, 1994. In: Oliver-Smith A, Hoffman S (eds) The angry earth: disaster in anthropological perspective. Routledge, London

Boykoff J (2021) The fight against the olympics: world opposition mobilization connecting Tokyo, Paris and LA. Sakuhinsha

Bull C, Lovell J (2007) The impact of hosting major sporting events on local residents. J Sport & Tour 12:229–248

Burbank MJ, Andranovich GD, Heying CH (2001) Olympic dreams: the impact of mega-events on local politics. Lynne Rienner Publishers. https://doi.org/10.1515/9781685857615

Centre on Housing Rights and Evictions (COHRE) (2007) Fair play for housing rights: mega-events, Olympic games and housing rights. Geneva, Switzerland

di Bella A (2020) Global urbanism and mega events planning in Rio de Janeiro amid crisis and austerity. Int Plan Stud 25(1):23−37

Faulhaber L, Azevedo L (2015) SMH 2016: removals on the Olympic city. Mórula Editorial, Rio de Janeiro

Fernandes F (2008) The integration of blacks in class society [A Integração do Negro na Sociedade de Classes] [5. ed.]. Ensaio de Interpretação Sociológica, São Paulo: Globo

Ferreira JSW (2014) Um teatro milionário. In: Jennings A, et al (eds) Brasil em jogo: o que fica da Copa e das Olimpíadas? São Paulo: Boitempo Editorial, 7–15, 1st ed. https://www.boitempoeditorial.com.br/produto/brasil-em-jogo-152488

Gaillard J, Cadag JRD (2009) From marginality to further marginali-zation: experiences from the victims of the July 2000 Payatas trashslide in the Philippines. JAMBA: J Disaster Risk Stud 2(3)

Harding S (2007) Man-made disaster and development: the case of Iraq. Int Soc Work 50(3):295–306

IBGE (Brazilian Institute of Geography and Statistics) (2010) Census-11.4 million Brazilians (6.0%) live in subnormal agglomerations. https://agenciadenoticias.ibge.gov.br/agencia-sala-de-imprensa/2013-agencia-de-noticias/releases/14157-asi-censo-2010-114-milhoes-de-brasileiros-60-vivem-em-aglomerados-subnormais. Accessed 6 March 2022

Lenskyj HJ (2006) The Olympic (affordable) housing legacy and social responsibility. Int Symp Olymp Res

Maricato E (2008) Land Knot [Nó da Terra], 21st edn. Revista Piauí

Moreira PC (2018) Conflict cartography: the vila autódromo experience [Cartografia do Conflito: A experiência de Vila Autódromo]. In: Tanaka G (ed) Viva a Vila Autódromo: o plano popular e a luta contra a remoção. 1 ed. Rio de Janeiro: Letra Capital. pp 386–407
Morgado C, Bastos L, Cinelli R (2022) Comunidade Vila Autódromo: o custo social da remoção e uma ponderação sobre sua reconstituição. In: Encontro Nacional De Tecnologia Do Ambiente Construído, 19. Canela. Anais. Porto Alegre: ANTAC, pp 1–13
Nogueira FR, Souza BRM (2020) The myth of eviction as an alternative to risk elimination. In: SJacob PR, Travassos L, Santana-Chaves IM, Anjos LAP (2020) Riscos e desastres: compartilhando responsabilidades. Socio-environmental Dialogues in the macro metropolis of São Paulo, special volume, n. 08
Observatory M (2016) Human rights violations in the Olympic city. Justiça Global, Rio de Janeiro
Oliver-Smith A (1999). Peru's Five-hundred-year earthquake: vulnerability in historical context. In: Oliver-Smith A, Hoffman S (eds) The angry earth: disaster in anthropological perspective. Routledge, London
Oliver-Smith A (2019) Disasters and large-scale population dislocations: international and national responses. Oxford Research Encyclopedia, Natural Hazard Science
Oliver-Smith A, Alcántara-Ayala I, Burton I, Lavell A (2017) The social construction of disaster risk: seeking root causes. Sociology, Environmental Science. Int J Disaster Risk Reduct. https://www.sciencedirect.com/science/article/abs/pii/S2212420916305982?via%3Dihub
Oyama AC, Melo BM, Coelho ILF, Tavares JC, Pedrassoli JC, Fantin M (2019) Counter-hegemonic maps as rights consolidation tools: the case of the Banhado Community, São José dos Campos–SP. XVIII Brazilian Symposium of Applied Physical Geography (SBGFA). Fortaleza, Ceará
Oyama AC (2023) Mega-events as disasters: the eviction system created at the Rio 2016 and Tokyo 2020–2021 Olympic Games. Master's thesis: Grad-uate School of Global Environmental Studies. Kyoto University
Porter L, Jaconelli M, Cheyne J, Eby D, Wagenaar H et al (2009) Planning displacement: the real legacy of major sporting events "Just a person in a wee flat": being displaced by the Commonwealth Games in Glasgow's East End Olympian Masterplanning in London closing ceremonies: how Law, Policy and the Winter Olympics are displacing an inconveniently located low-income community in Vancouver commentary: recovering public ethos: critical analysis for policy and planning. Plan Theory Pract 10(3):395–418. https://doi.org/https://www.tandfonline.com/doi/full/10.1080/14649350903229828
Rocha CM, Xiao Z (2022) Sport mega-events and displacement of host community residents: a systematic review. Front Sport Act Living
Rolnik R (2019) War of places: the colonization of land and housing in the age of finance [Guerra dos lugares: a colonização da terra e da moradia na era das finanças]. São Paulo, Boitempo
Sánchez F, Bienenstein G, Mascarenhas G, Oliveira A (2012) Megaeventos e Metrópoles–Insumos do Pan-2007 e as perspectivas para as Olimpíadas de 2016. In: Oliveira FL et al (org.) Grandes projetos metropolitanos: Rio de Janeiro e Belo Horizonte. Rio de Janeiro, Letra Capital
Sánchez F (2010) A reinvenção das cidades para um mercado mundial Chapecó, Argos. Editora da Unochapecó
Santos Junior OA, Santos MRM (2013) The right to housing, the world cup and the Olympics: reflections on the case of Rio de Janeiro. In Territ J 64(2013):28–33
Sodré M (2002) O terreiro e a cidade: a forma social negro-brasileira. Rio de Janeiro: Imago Ed; Salvador, Fundação Cultural do Estado da Bahia, BA
Tamaki T (2019) The Olympics and Japanese national identity: multi-layered otherness in Tokyo 2016 and 2020. Contemporary Japan
Tanaka G (2017) Plan to fight and fight to plan: possibilities and limits of alternative plans. [Doctoral dissertation, IPPUR/UFRJ]. UFRJ Digital Collection
Tanaka G (2019) Conflictual planning: planejamento conflitual: O Plano Popular da Vila Autódromo. In: Klintowitz D, Moreira FF, Nisida VC (eds) Planejamento alternativo: propostas e reflexões coletivas. Instituto Pólis, São Paulo

United Nations Human Settlements Programme (UN-HABITAT) (2010) Report of the special rapporteur on adequate housing as a compo-nent of the right to an adequate standard of living, and on the right to non-discrimination in this context, Raquel Rolnik. United Nations, Geneva

Vainer C, Oliveira FL, Junior PNL (2012) Notas metodológicas sobre a análise de grandes projetos urbanos. In: Oliveira FL et al. (org.) Grandes projetos metropolitanos: Rio de Janeiro e Belo Horizonte. Rio de Janeiro: Letra Capital

Vainer C, Bienenstein R, Tanaka G et al (2013) O Plano Popular da Vila Autódromo: uma experiência de planejamento conflitual. In: XV Enanpur Anais do XV Encontro Nacional da Associação Nacional de Pós-graduação em Planejamento Urbano

Valladares LP (2005) A invenção da favela-Do mito de origem a favela.com. Rio de Janeiro: Ed. Fundação Getúlio Vargas, p 204

Wilson S (2012) Exhibiting a new Japan: the Tokyo Olympics of 1964 and Expo' 70 in Osaka. Hist Res 85(227)

Wisner B, Luce HR (1993) Disaster vulnerability: scale, power, and daily life. GeoJournal. https://link.springer.com/article/10.1007/BF00808129

Wisner B, Blaikie P, Cannon T, Davis I (2004) At risk: natural hazards. Routledge, People's Vulnerability and Disasters

World Cup and Olympics Popular Committee of Rio De Janeiro (WCOPCRJ) (2015) Mega-events and human rights violations in Rio de Janeiro Dossier. Rio de Janeiro

Chapter 20
Community Livability in Shrinking Cities: A Case Study from China

Yuqi Zhang and Chiho Ochiai

Abstract The decline of industries like coal has brought negative impacts to local communities and often force residents to relocate. In China, many shantytowns were formed near coal mines to provide housing for workers. After the decline of industries, these areas faced severe issues such as deteriorating building quality, inadequate infrastructure, and significant safety hazards. To address these housing problems, the central government implemented policies to demolish the shantytowns and relocate residents to newly built apartments in designated areas. In recent years, the government has also undertaken renovation projects for the relocated communities. This study focuses on the people left behind in cities with declining industries, evaluating the current livability of communities over ten years after relocation from the residents' perspective through questionnaires and interviews. The case study examines two relocated communities in Fuxin, a representative shrinking coal city. It also investigated residents' perceptions of shrinkage and their satisfaction with community renovation projects. Regression analysis further explores the impact of residents' perceptions of shrinkage and renovation projects on community livability assessments. The findings indicate that most residents in both communities have a strong perception of urban shrinkage, though perceptions of community decline vary significantly. Additionally, the livability of the two communities differs due to the location and surrounding facilities. While changes in community population and vacant housing currently have a minimal impact on livability, the perception of urban economic decline significantly affects community livability. Implementing renovation projects has been shown to improve community livability. However, three factors influence the effectiveness: allocation of resources across communities within the city, the contents of renovation projects, and resident involvement and post-renovation management. Future efforts should encourage multi-stakeholder participation, focus on improvements of social infrastructure, and establish resident self-management organizations.

Y. Zhang (✉) · C. Ochiai
Graduate School of Global Environmental Studies, Kyoto University, Kyoto, Japan
e-mail: zhang.yuqi.3t@kyoto-u.ac.jp

C. Ochiai
e-mail: ochiai.chiho.2x@kyoto-u.ac.jp

C. Ochiai et al. (eds.), *Disaster and Displacement*, Disaster Risk Reduction,
https://doi.org/10.1007/978-981-96-2583-3_20

20.1 Introduction

In recent decades, urban shrinkage has become a global phenomenon driven by factors such as deindustrialization, declining fertility rates, aging populations, and globalization. This trend is characterized by significant population loss, economic decline, and the deterioration of urban spaces (Richardson and Nam 2014). This trend has become particularly evident in China in recent years, especially in resource-based industrial cities. While many cities are experiencing rapid urbanization and growth, more than half of Chinese cities are facing population loss (Zhang and Ochiai 2023) (Fig. 20.1). Urban shrinkage poses significant challenges to urban development, encompassing both physical and non-physical issues. Physical issues include the vacancy and decay of buildings, land, and infrastructure, as well as the degradation of the environment and landscape. Non-physical issues include reduced investment, economic instability, exacerbated inequality and poverty, and declining urban safety (Pallagst 2014).

These problems also place unique pressures on local communities. Buildings in these areas often suffer from aging and lack of maintenance. Essential infrastructure such as heating, water supply, and sewage systems also face poor maintenance. Furthermore, the social fabric of communities suffers in shrinking cities, with fewer social interactions and weakening neighborly ties (Hollander et al. 2009). Declining mining communities, where the population is significantly affected by nearby mining operations (as defined by Veiga et al. 2001), is a typical example. Beyond the aforementioned negative impacts, mining communities, being single-industry, suffer from economic instability. Additionally, they experience a legacy of environmental degradation and negative health impacts (Martinez-Fernandez et al. 2012). Kuyek and Coumans (2003) argue that mining creates a substantial environmental footprint and impacts surface and subsurface water supplies. These communities also face the challenge of dealing with large brownfield areas (Martinez-Fernandez et al. 2012). Moreover, mining-induced land subsidence leads to structural damage to houses, resulting in dangerous living conditions for residents.

The negative impacts of declining industries often force former workers to leave their homes due to rising poverty, deteriorating living standards, declining services, and reduced employment opportunities (Siyongwana and Shabalala 2019). In China, there are many declining coal cities where shantytowns have formed near coal mines. These shantytowns, built by state-owned enterprises for their workers, consist of single-story and small houses. However, with the decline of the coal industry and the bankruptcy of enterprises, these houses have fallen into disrepair. The quality of the housing has deteriorated, with poor earthquake resistance, significant fire hazards, and inadequate basic infrastructure. Additionally, many houses located in subsidence areas have developed structural cracks. To address the housing issues faced by these residents, the Chinese central government launched a shantytown renovation policy. This initiative began in 2004 in the northeastern region, an old industrial base. The policy involves demolishing the existing shantytowns and providing either housing relocation or monetary compensation to the residents. The people living in these

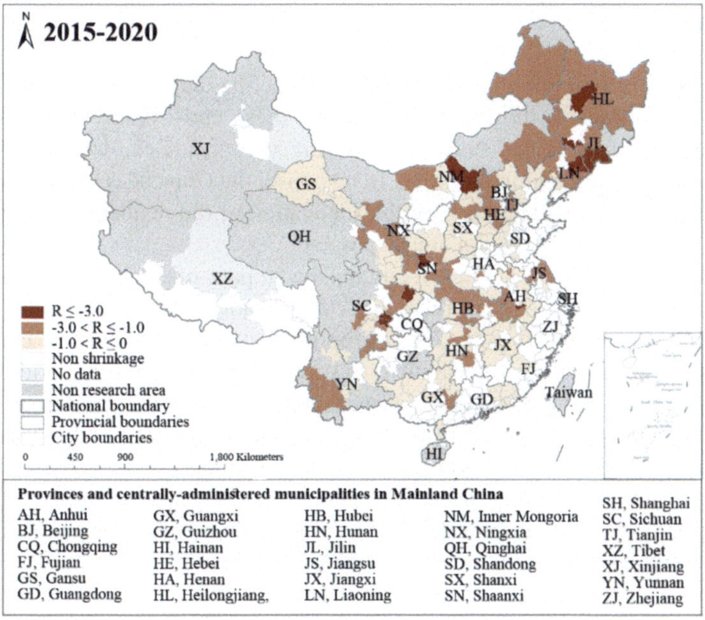

a) Cities with population loss (R is the average annual growth rate of permanent residents).

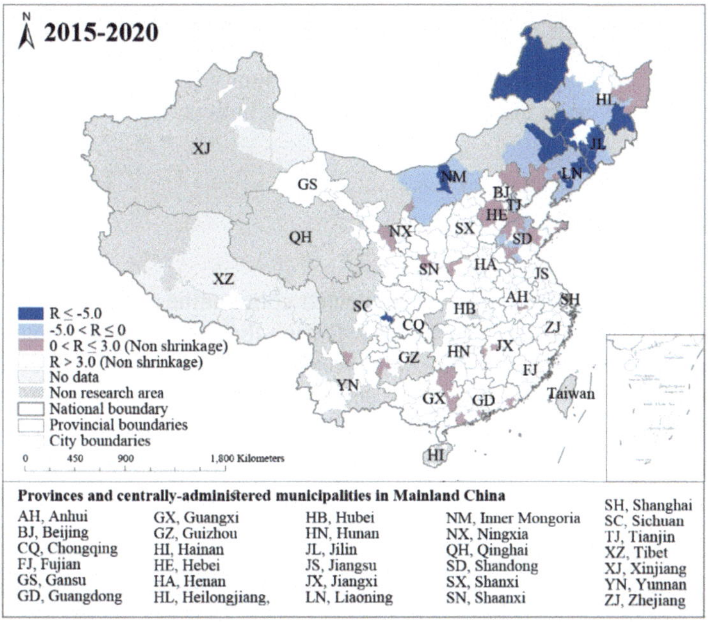

b) Cities with decreased GDP (R is the average annual growth rate of the GDP)

Fig. 20.1 Distribution of shrinking cities in China. *Source* Authors based on China City Statistical Yearbooks in 2016 and 2021

shantytowns are then relocated to newly built multi-story apartments in government-designated areas.

However, as urban populations continue to decline and economic development stagnates or declines in those shrinking cities, these early shantytown renovation residential areas face issues such as increasing vacant homes, degradation of the built environment, and aging facilities. In response, the Chinese central government has launched a nationwide policy for the renovation of old residential communities and early shantytown renovation areas.

Currently, while many studies focus on the impact of declining industries on local communities and resident relocation, few examine how to measure success in relocation. Ferris and Bower (2023) propose that success should be considered across several dimensions: for whom is it successful (relocated individuals, those who remain, etc.); comparison criteria (measured against initial conditions or other cases); timeframe (shortly after relocation or in the long term); and domains (risk reduction, livelihoods, etc.). It is crucial to include residents' perspectives and prioritize the community livability of the current community years after relocation when measuring success. Hollander (2018) also argued that evaluation metrics for the success of strategies in shrinking cities should shift from focusing mainly on population growth, income, and employment to emphasizing the subjective well-being of the people who live, work, and play in these areas. Therefore, this chapter takes two relocated shantytown communities,[1] Wenxinyuan and Mijia, in Fuxin City, a representative shrinking mining city in China, as case studies. It aims to investigate community livability after relocation in shrinking cities, understand residents' perceptions of shrinkage and its impact on community livability assessments, and explore the effectiveness of community renovation projects in enhancing livability.

20.2 Methodology

This study conducted field surveys in selected communities: Wenxinyuan (WXY) and Mijia (MJ) Community in Fuxin City. During the surveys, questionnaires were distributed, and in-depth interviews were conducted with some of the respondents. Each community has a population of approximately 5,000, and a 1% sampling rate was used. So, this study plans to collect around 50 questionnaires from each community. From each residential building, 1–2 residents were randomly selected to receive questionnaires. In total, 107 valid responses were collected. Specifically, 53 questionnaires were collected from WXY and 54 from MJ Community.

The questionnaires primarily investigated the residents' personal socio-economic attributes, residents' perceptions of urban shrinkage and community decline, their

[1] In contemporary China, the term "community (shequ)" refers to urban and rural areas defined by specific administrative boundaries. A community can be subdivided into various territory-based settlement compounds. These communities form part of the administrative and social structure managed by community committees, which provide housing, related services, and facilities for their inhabitants.

evaluations of current community livability, and their satisfaction with community renovation projects. Additionally, regression analysis was used to understand the impact of residents' perceptions of shrinkage on their evaluation of community livability and to explore whether community renovation projects have improved livability.

Additionally, during the questionnaire distribution, further interviews were randomly conducted with some respondents to understand the reasons behind their views. The interviews covered their opinions on community changes and urban shrinkage, the factors contributing to these perspectives, their satisfaction and dissatisfaction with various aspects of community livability, and their reasons for being satisfied or dissatisfied with the community renovation projects.

Regarding the evaluations of community livability, numerous studies discuss this topic. For example, Leby and Hashim (2010) evaluated livability from five dimensions: Social Relations (behavior of neighbors, community life and social contact, sense of place), Residential Environment (environment quality, maintenance of built environment, open spaces), Facilities and Services (accessibility, availability and proximity of amenities, employment opportunities), and Crime and Sense of Safety (crimes, accidents, feeling of safety). Owusu et al. (2023) constructed a framework that includes Economy (rent cost, food cost, availability of jobs), Social (crime prevalence; type of income community; opportunities the community offer), Environment (sanitation situation, flood susceptibility), Transportation and Accessibility (availability of public transport, traffic congestion, distance to work), and Culture and Recreation (availability of entertainment). Although the dimensions and indicators differ among scholars, there are seven dimensions and several indicators that are widely mentioned, as summarized in Table 20.1.

20.3 Case Study Sites

20.3.1 Overview of Fuxin City

Fuxin City, located in northeastern China, rose to prominence due to its rich coal resources. Known as the "Coal and Power City," Fuxin once boasted Asia's largest open-cast mine and the most productive thermal power station (Fuxin Development and Reform Commission Government Report 2012). The city's first coal mine was established in 1905 (Fuxin City Chronicles). Following the founding of the People's Republic of China, Fuxin saw significant development. During China's first Five-Year Plan in 1953, Fuxin was home to four of the 156 key national projects, including the Haizhou Open-Pit Coal Mine[2] and the Fuxin Power Plant, receiving over 416 million RMB (approximately US$ 58 million) in state investment (Zhang 2008). This

---

[2] The Haizhou Open-Pit Coal Mine began construction in 1951 and officially commenced production in 1953. It was the first large-scale mechanized open-pit coal mine built in the People's Republic of China and was once the largest mechanized open-pit coal mine in Asia.

Table 20.1 Framework for evaluating community livability

Evaluation dimensions	Specific evaluation indicators
Housing	• Quality of buildings • Essential utility services
Environmental quality	• Green spaces • Sanitary conditions • Sound environment
Access to services and amenities	• Primary and secondary schools • Kindergartens • Elderly care facilities • Entertainment facilities • Healthcare • Financial facilities • Shops • Restaurants • Sports facilities
Transportation	• Public transportation • Parking lots
Economic stability	• Income • Price level • Housing affordability • Employment opportunities
Social relations	• Community engagement • Neighborhood relationships • Sense of place
Safety and security	• Feeling of safety

period marked Fuxin as one of the most prosperous industrial cities (Hu and Yang 2018).

However, the market reforms beginning in 1978 led to the decline of Fuxin's centrally planned economic model. Since the 1980 and 1990s, coal resource depletion and rising extraction costs led to the decline of the coal-dominated industry, resulting in the city's economic downturn. The 1998 state-owned enterprise reform further exacerbated this, leading to numerous factory closures and worker layoffs. Fuxin suffered typical rust belt issues: firm closures, job losses, population outflow, poverty, and social unrest (Li et al. 2009). By 2000, one-third of local industrial enterprises had shut down, 23 mines were closed, and 156,000 workers (36.7% of the workforce) were laid off, with urban unemployment exceeding 7% (Fuxin City Chronicles).

In 2001, Fuxin was designated China's first pilot city for transforming resource-exhausted cities. The central government launched initiatives like the "Northeast Revitalization Strategy" (2003) and the "Transformation of Resource-exhausted Cities" (2008) to revitalize Fuxin (Hu and Yang 2018). With substantial national investment, Fuxin began its challenging transformation. Fuxin promoted wind power development. It also explored the coal chemical industry. However, this plan was not fully realized (Hu and Yang 2018). Additionally, Fuxin promoted industrial tourism by converting the abandoned Haizhou Open-Pit Coal Mine into a park in 2009. In 2019, attempts were made to use the mine pits for events like motor racing tournaments and music festivals, but these projects did not meet expectations. In summary, despite various efforts and some economic improvement, Fuxin has not achieved full revitalization. The long-term population outflow has led to numerous vacant buildings, including residential units, factories, and public facilities (Table 20.2).

20.3.2 Two Communities in Fuxin City

In Fuxin City, shantytowns were a typical form of housing during the latter half of the last century. In China, shantytowns consisted of housing areas constructed either by state-owned enterprises or by workers themselves during the rapid industrialization of the latter half of the twentieth century. These shantytowns functioned as dormitories for workers and were located near the state-owned enterprises (Gao and Ryan 2021). At that time, in Fuxin City, a large number of temporary worker housing units were built near mining areas, railways, and high-voltage power line corridors, creating a landscape known as "hundreds of miles of mines, hundreds of miles of shantytowns." These houses, with their small areas and inadequate facilities, have fallen into disrepair over the years. Additionally, coal mining subsidence caused damage to some of the houses, making living conditions extremely difficult. In 2005, Fuxin City launched a comprehensive shantytown renovation project. The shantytowns were demolished, and the original residents were relocated to newly constructed multi-story buildings in selected locations. WXY (Wenxinyuan) and MJ (Mijia) Community are representative examples of these shantytown renovation projects (Table 20.3).

WXY is located in the southwestern part of Fuxin City, adjacent to the old Fuxin Railway Station. To the southeast is the Haizhou Coal Mine, and to the south, there is a national railway. The area to the east used to house the Fuxin Mining Group's machinery manufacturing company. This area began to flourish in 1934 with the establishment of the Wulong Carbon Plant. During the 1950s, it was a typical industrial and mining living area until the coal mine closed. In recent years, as Fuxin City has gradually developed northward, with the construction of the new Fuxin High-Speed Railway Station, the old railway station area has become the fringe of the city center. The current residential buildings were constructed as part of the shantytown redevelopment project in 2006. In 2019, the community began a renovation project.

MJ is located in the northeastern part of Fuxin City, situated on the city's outskirts, approximately a 30-min drive from the city center. It is adjacent to the residential areas of the Mining Materials Plant and Coal Plant No. 12, about 5 km northwest of the Xinqiu Coal Mine. The Xinqiu Coal Mine's development in the 1950s turned it into a traditional industrial and mining living area, characterized by self-built houses. In 2011, it became a resettlement area for residents from the shantytowns of Haizhou District, Taiping District, and Xinqiu District. It was one of the last shantytown renovation areas in Fuxin City and is one of the largest residential areas

Table 20.2 Contextual information of Fuxin city

Basic information	Fuxin city
Built-up area of the city district	77 square kilometers (in 2021)
Population	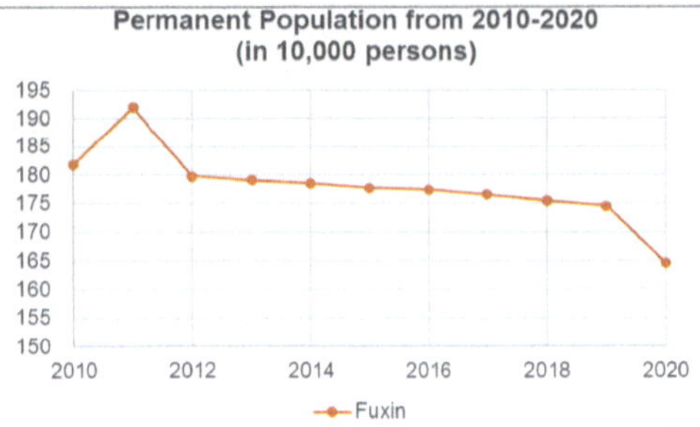
Economy	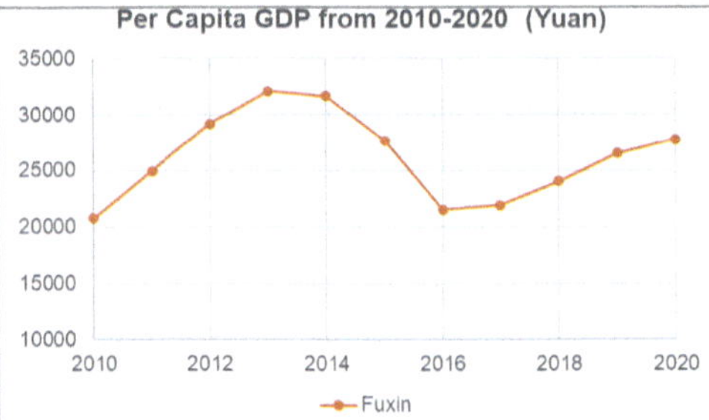
Development of leading industries	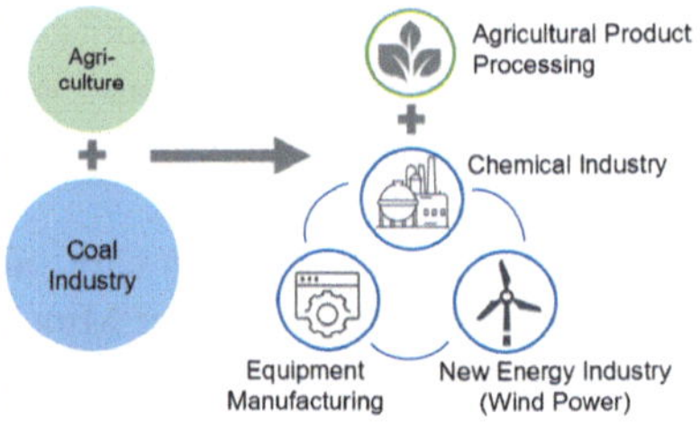
General public budget revenue of the city district	3.19 billion yuan (in 2021)
General public budget expenditure of the city district	7.50 billion yuan (in 2021)
Types of shrinkage	Both shrinkage in population and economy; long-term
Influence factors of shrinkage	Depletion of coal resources and the forced industrial transformation

(*Data source* City statistical yearbooks over the years; National population census data for 2000, 2010 and 2020)

Table 20.3 Contextual information about two communities

Community	WXY community	MJ community
Location	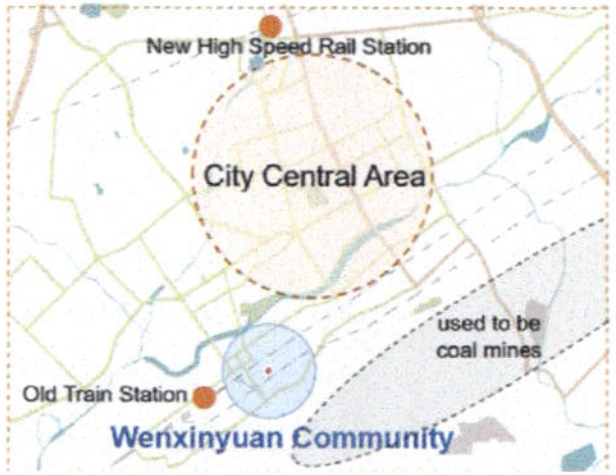	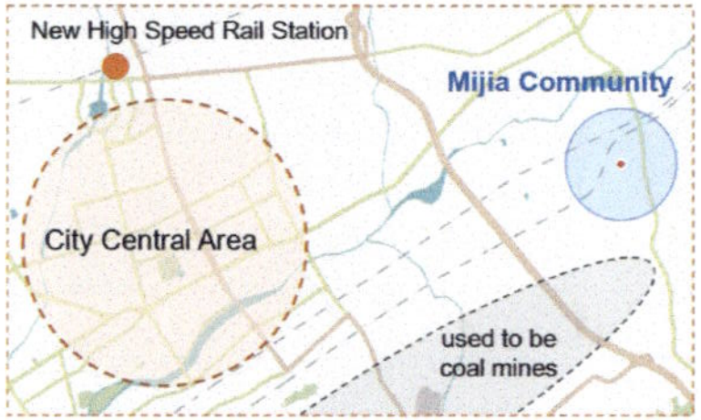
	Peripheral area of the old town	Outskirts of the city
Construction year	2006	2011
Area	0.25km^2	0.39 km^2
Residential building layout	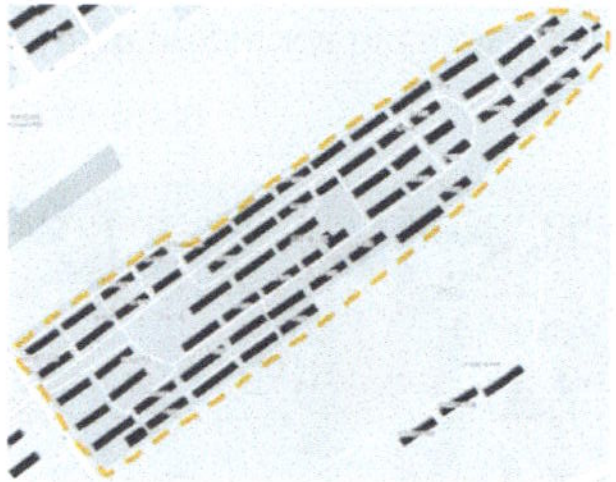	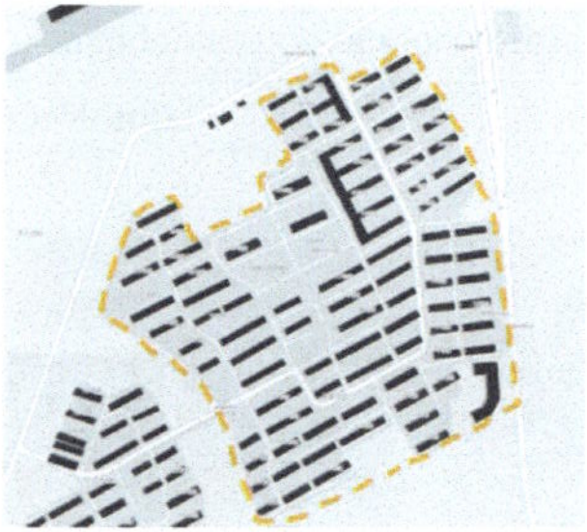

(continued)

Table 20.3 (continued)

Community	WXY community	MJ community
Residential building form		
Population	Over 5000	Over 5000
Population structure	A high proportion of elderly residents	A high proportion of elderly residents and low-income people
Residents' occupations	Mostly retired workers from coal enterprises	Mostly retired workers from coal enterprises
Year the community renovation project started	2019	2022
House price	RMB 2,000 yuan/m^2	RMB 1,400 yuan/m^2

(*Source of figures and photos in the table* Authors, 2023)

in the city. The relocation project began in 2011, with residents moved back in 2013. Due to its large size and population, this study focuses on the eastern section of MJ. Compared to WXY, MJ has limited surrounding amenities. In recent years, the low housing prices have attracted many rural residents from surrounding areas. In 2022, the community began its renovation project.

20.4 Results

20.4.1 Residents' Perceptions of Shrinkage

This study investigated residents' perceptions of shrinkage at two levels: community and city. At the community level, it includes perceptions of changes in the community's population and the number of vacant houses. At the city level, it encompasses perceptions of changes in the urban population and urban economy.

20.4.1.1 Residents' Perception of Community Decline

Regarding perceptions of changes in the community's population (Fig. 20.2a), over half of the respondents in Fuxin believed their community's population had decreased. Significant differences exist between the two communities. In WXY, a majority (71.7%, 38 out of 53) believed the population was decreasing. Respondents explained: "People with better economic conditions have moved to the north of the city, which is an area of new urban development" (WXY-2[3]). Another resident added, "Young people have left the community; after going to college or getting married, they moved elsewhere, leaving mainly the elderly behind" (WXY-3). In contrast, in MJ, only 35.19% (19 out of 54) perceived a population decline. 44.44% felt the population was increasing, compared to just 13.21% in WXY, and another 20.37% perceived it as stable. One respondent explained: "Housing prices here have been very low, and shopping in this area is also cheap, attracting many rural residents from surrounding areas. Although many elderly people have passed away, a significant number of new residents have moved in" (MJ−6). Perceptions of changes in the number of vacant houses (Fig. 20.2b) generally correspond with views on population changes.

The study further explored residents' perceptions of the impacts of a decreasing population and increasing vacancies in their communities. The majority perceived no significant impacts, though some noted effects on housing prices and heating effectiveness. In WXY, 13.21% believed increasing vacant homes contributed to lower property values, while only 3.7% held this view in MJ. Conversely, a larger proportion in MJ felt vacant houses adversely affected heating.

[3] WXY-2 represents the respondent of the second valid questionnaire from the WXY community.

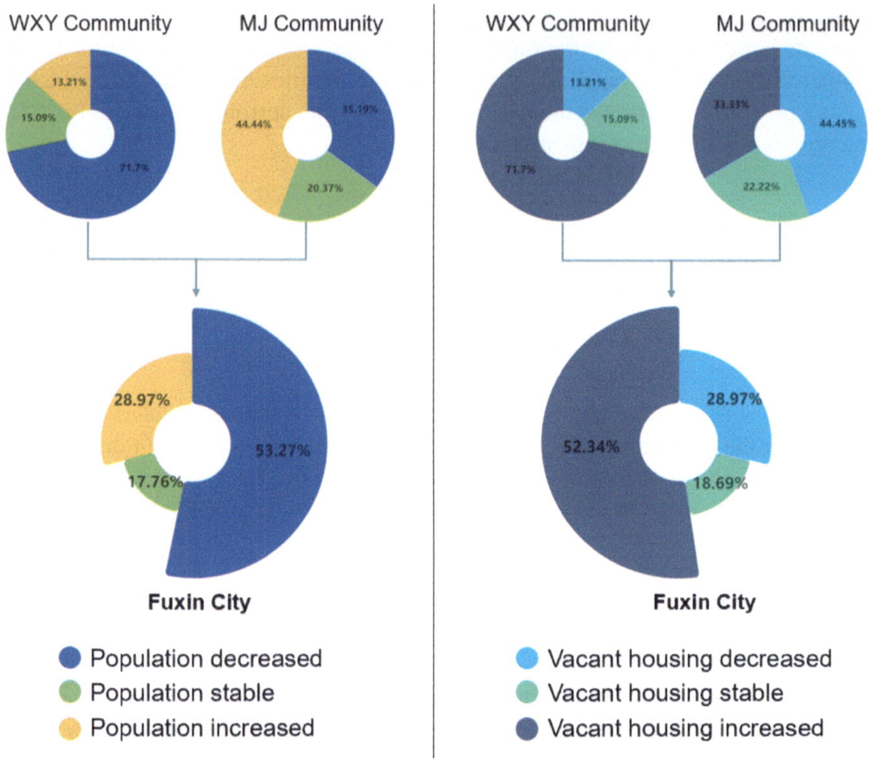

a) Changes in the community population b) Changes in the vacant houses

Fig. 20.2 Residents' perception of community decline. *Source* Authors

The analysis indicates that within shrinking cities, community development varies, with location being a significant influencing factor. Economically capable residents tend to move towards newly developed urban areas. Conversely, suburban communities attract rural residents from nearby areas. However, it is foreseeable that these suburban communities will eventually experience population decline once their attraction to rural populations diminishes. Currently, the decline of communities has not had a significant impact on residents. One reason is that most residential buildings in these communities are multi-story apartment complexes. Unlike single-family homes, vacancies in multi-story apartments may not lead to obvious environmental degradation. Additionally, the overall population in these communities remains relatively high, so issues such as diminished neighborhood interactions are not yet apparent. However, although the current impact may be minimal, early signs of its effects are shown in property values and infrastructure, particularly heating.

20.4.1.2 Residents' Perception of Urban Shrinkage

Regarding residents' perceptions of urban population changes (Fig. 20.3a), the highest proportion of respondents in both communities believed the urban population had decreased. This proportion was higher in WXY, with 64.15% compared to 53.71% in MJ. A respondent from WXY noted, "In recent years, very few college graduates have returned to work in the city. It's hard for young people to find a job here. Fuxin now has a large elderly population, with very few young people remaining" (WXY-3). Similarly, a respondent from MJ mentioned, "Many young people have moved away, although some people from rural areas have relocated to the city" (MJ-5).

Regarding perceptions of urban economic changes (Fig. 20.3b), respondents in both communities had similar views. The highest proportion believed the economy was declining, with 52.83% in WXY and 51.85% in MJ. Residents in both communities mentioned the economic decline since the bankruptcy of coal enterprises. A WXY resident said, "Since the coal mines closed down, many factories have gone bankrupt, other industries haven't developed, there are no big companies, jobs are hard to find, and salaries are low" (WXY-17). An MJ respondent added, "After the decline of the coal and power generation industries, Fuxin has faced difficulties in economic transformation and ranks last in the province's economy, with the province also ranking last nationally" (MJ-12). However, some residents who thought the economy was stable explained, "In recent years, the city has been expanding outward and has seen a lot of construction, especially around the northern high-speed rail station area" (MJ-31).

While there are significant differences in perceptions of community decline between WXY and MJ, their views on urban shrinkage are relatively similar. The primary factor contributing to the perceived population decline is the outflow of young people. After the coal industry's decline, replacement industries did not develop well, leading to fewer job opportunities. The lack of good job prospects

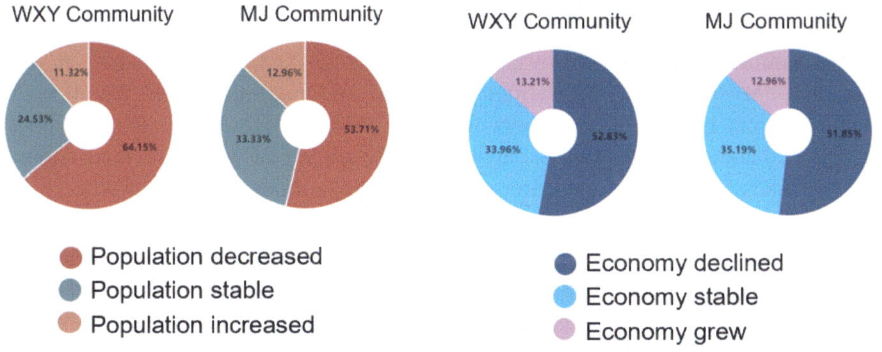

a) Changes in the urban population b) Changes in the urban economy

Fig. 20.3 Residents' perception of urban shrinkage. *Source* Authors

and low expectations for the city's future have driven capable young individuals to seek better opportunities elsewhere. Residents of MJ perceive population shrinkage less strongly, possibly due to the relative stability of their community's population, which makes them believe the overall city population is stable. More than half of the respondents in both communities believe the economy is in decline, indicating the deep impact of the coal industry's decline on residents' livelihoods. Most residents previously worked in coal-related jobs, and the industry's decline has directly affected their income and employment opportunities, fostering a sense of pessimism.

However, Fuxin is experiencing a paradox of population decrease while undergoing urban expansion. The construction of new infrastructure and projects in newly developed areas offers some hope to residents. Nevertheless, this expansion may lead to further wastage of land and resources.

20.4.2 Residents' Assessments of Community Livability

20.4.2.1 Assessments of Community Livability

This section presents residents' assessments of community livability, covering seven aspects and 24 indicators, using a five-point Likert scale. Overall, the livability rating in WXY (3.59) is higher than that of MJ (3.20), particularly in housing, environmental quality, and safety (Fig. 20.4a).

Figure 20.4b shows the average scores for each livability indicator. In WXY, the highest ratings are for Neighborhood Relationships (4.34), Safety (4.32), Shops (4.25), Restaurants (4.23), and Sanitary Conditions (4.21). One resident noted, "Our neighborhood is very clean, better than surrounding communities. We are a model community for sanitation, and the building quality is quite good for a relocated shantytown" (WXY-13). However, WXY scores lower in Income (2.91), Elderly Care (2.7), Job Opportunities (2.42), and Community Engagement (2.13).

In MJ, Neighborhood Relationships (4.17), Shops (4.11), and Housing Affordability (4.06) score well, but Building Quality (2.17), Community Engagement (2.09), Sanitary Conditions (1.69), and Green Spaces (1.67) score poorly. A resident said, "Most residents are former workers from Dongliang Mine. We all know each other and have been neighbors for years" (MJ-2). However, many residents complained about poor sanitation, saying, "There are lots of weeds, no landscaping, and the roads are in poor condition. On windy days, there's dust everywhere, and on rainy days, the roads turn to mud" (MJ-9). Building quality is also concerning. One interviewee reported, "Sometimes the roof leaks, and rainwater flows into the stairwell" (MJ-22). Regarding community engagement, residents expressed a low willingness to participate. One interviewee said, "We are very disappointed with the community committee and property management. They seldom seek residents' opinions. We've reported many issues, but nothing has been resolved, so we don't feel motivated to get involved in community affairs anymore" (MJ-37).

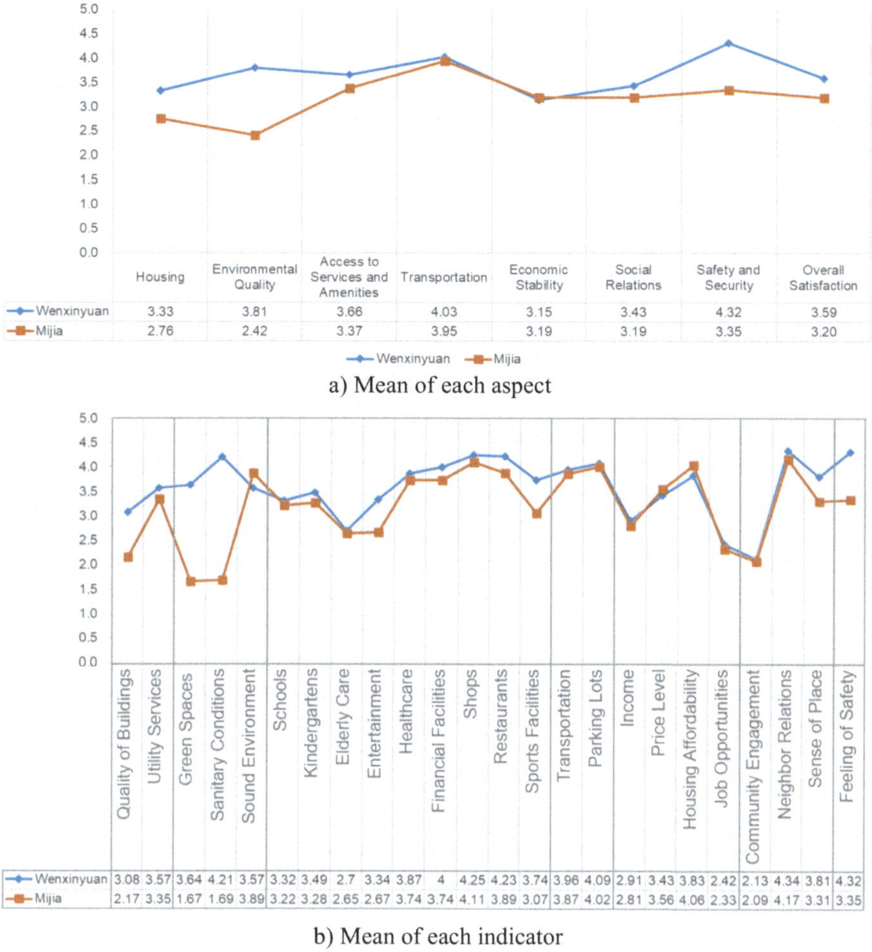

Fig. 20.4 Mean of residents' assessment in community livability. *Source* Authors

Comparing the two communities, both score highly in neighborhood relationships. Most residents were relocated from shantytowns around the coal mines, making them old neighbors or former colleagues with strong emotional bonds. Both communities score low in community engagement, with few residents participating in community activities. Many feel too old to participate or think the community committee does not seek their opinions, and their proactive feedback often goes unaddressed.

The differences between the two communities are notable. WXY performs significantly better than MJ in green spaces, landscape maintenance, supporting facilities, and a sense of security. This may be due to differing levels of government resources allocated to each community. As a model sanitation community, WXY likely receives more funding for community operations and has better-organized committees and property management. In contrast, MJ faces resource scarcity and dissatisfaction with property management. Many residents have stopped paying property management fees, leading to operational difficulties and further deterioration of community management. Additionally, WXY's proximity to the old town enhances its accessibility to various facilities.

20.4.2.2 The Influence of Shrinkage Perception on Community Livability Assessment

This study used regression analyses to examine the impact of shrinkage perceptions on community livability assessments. A regression analysis was conducted with all respondents, using perceptions of urban population change and economic change as independent variables and livability assessment as the dependent variable. In the linear regression model, perceptions of urban population decrease were coded as 1, stable population as 2, and population increase as 3. Similarly, perceptions of urban economic decline were coded as 1, stable economy as 2, and economic growth as 3.

As shown in Table 20.4, residents' perceptions of urban economic changes significantly impact community livability. When residents perceive the city's economy as growing rather than declining, they rate the community's livability higher. However, perceptions of urban population decrease do not significantly affect community livability. This may be because economic decline directly impacts residents' quality of life, leading to lower income, fewer job opportunities, reduced community investment, and deteriorating services and infrastructure. In contrast, population decline does not immediately affect daily life as visibly as economic changes do.

Second, regression analyses were conducted to investigate whether perceptions of community decline influence livability assessments. Due to different community contexts, separate analyses were performed for each community. Changes in community population and the number of vacant houses were used as independent variables, with community livability as the dependent variable. In the linear regression model, community population decrease was coded as 1, stable population as 2, and population increase as 3. A decrease in the number of vacant houses was coded as 1, stable number as 2, and increase in number as 3.

Table 20.4 The impact of urban shrinkage perceptions on livability assessments

City	Variable	Regression coefficient	95% CI
Fuxin city	Constant	3.242** (30.258)	3.032 ~ 3.452
	Perception of urban population changes	0.098 (1.545)	−0.026 ~ 0.223
	R^2	0.022	
	Constant	3.176** (28.990)	2.962 ~ 3.391
	Perception of urban economic changes	0.134* (2.155)	0.012 ~ 0.257
	R^2	0.042	

* $p < 0.05$ ** $p < 0.01$ (The values in parentheses are t-values)

Table 20.5 The impact of community decline perceptions on livability assessments

Community	Variable	Regression coefficient	95% CI
WXY	Constant	3.270** (21.736)	2.975 ~ 3.565
	Perception of population changes in the community	0.228* (2.401)	0.042 ~ 0.414
	R^2	0.102	
	Constant	4.110** (15.935)	3.604 ~ 4.615
	Perception of changes in vacant housing in the community	−0.200* (−2.080)	−0.389 ~ −0.012
	R^2	0.078	
MJ	Constant	3.027** (29.214)	2.824 ~ 3.230
	Perception of population changes in the community	0.081 (1.772)	−0.009 ~ 0.170
	R^2	0.057	
	Constant	3.331** (34.351)	3.141 ~ 3.521
	Perception of changes in vacant housing in the community	−0.071 (−1.532)	−0.163 ~ 0.020
	R^2	0.043	

* $p < 0.05$ ** $p < 0.01$ (The values in parentheses are t-values)

As shown in Table 20.5, perceptions of changes in community population and vacant houses significantly impact livability assessments in WXY but not in MJ. In WXY, residents who perceive a population decrease and an increase in vacant houses tend to rate the livability lower. This may be because the proportion of residents perceiving community decline in WXY is much higher than in MJ. A slight decline in population and an increase in vacant houses do not significantly impact community livability. However, when these changes reach a certain threshold, they affect utility services, heating, the landscape, and neighborhood interactions, reducing overall livability.

20.4.3 Community Renovation Projects and Their Effectiveness in Enhancing Community Livability

20.4.3.1 Overview of Community Renovation Projects

(1) **Overall Situation in Fuxin City**

Fuxin City has implemented several renovation policies for older residential areas and shantytown redevelopment communities in disrepair. In 2018, the Fuxin Municipal Government issued the "Maintenance Implementation Plan for Early Shantytown Renovation Communities," targeting shantytown resettlement communities approved before the end of 2009. In 2019, the "2020 Urban Old Residential Community Renovation Implementation Plan" was released. These plans focus on roof waterproofing, exterior wall insulation, pipeline replacements, and safety hazard elimination. They also include improvements to roads, drainage systems, lighting, and the environment within the communities (Fig. 20.5a).

Funding for these projects comes from three sources: (a) provincial subsidies and central budgetary investments, (b) professional operating units responsible for upgrading utilities such as water, drainage, electricity, heating, and gas, and (c) resident contributions based on the principle of "those who benefit should contribute." The renovation process is divided into four stages (Fig. 20.5b): project initiation, construction, inspection and handover, and post-management.

(2) **The Renovation in WXY Community**

WXY initiated its renovation project in 2019, making annual improvements to enhance living conditions by addressing infrastructure and environmental needs (Fig. 20.6). At the end of 2018, the Haizhou District Shantytown Renovation Office began the bidding process for construction contractors. Projects included roof waterproofing, exterior wall insulation, repairing downspouts and water funnels, and repainting stairwells. In 2020, a new perimeter wall was constructed, and an intelligent recognition barrier gate was installed at the main entrance. Landscaping was improved to create organized green spaces, and small community squares were built. Roads and public areas were paved with durable materials, and unit entrance doors were replaced for better security and appearance. Since 2021, leveraging central policies for maintaining and renovating state-owned enterprise infrastructure, the community has overhauled heating, water, and drainage systems. Wiring for electric vehicle charging stations has also been installed.

(3) **The Renovation in MJ Community**

The renovation of MJ began later than WXY. In 2022, the Taiping District Government identified MJ's renovation as a key project. The Taiping District Housing and Urban–Rural Development Bureau conducted multiple on-site surveys, held special meetings to address issues like aging heating pipelines and roof leaks, established a dedicated project team, and formulated a construction plan with project supervisors. The renovation focused on basic infrastructure projects (Fig. 20.7), including: (a) Building renovations: roof waterproofing, stairwell heating, and exterior wall painting; (b) Outdoor facilities: heating pipelines, road sewage manhole covers, and water pipelines. In 2023, the community began constructing a canteen to support its predominantly elderly population, many of whom face challenges with cooking independently. An unused building was repurposed and renovated, with a professional catering company selected to operate the canteen. Additionally, in 2023, the government planned to use small unused plots on the east and south sides of MJ to construct pocket parks, though construction has not yet started.

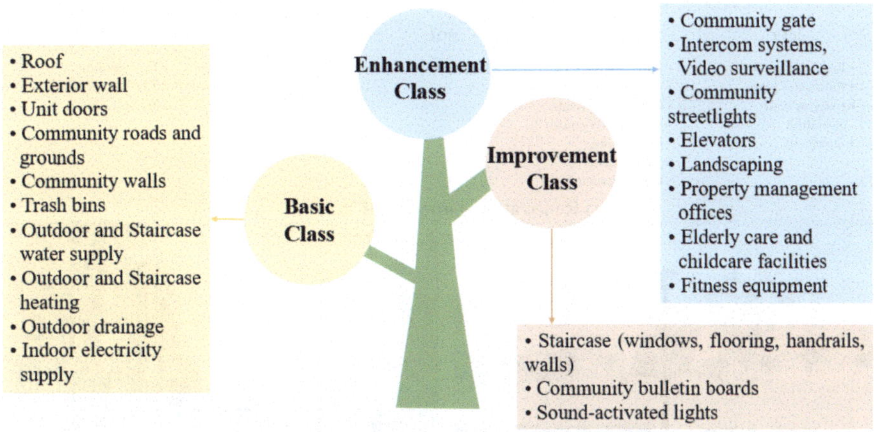

a) Categorization of specific renovation projects

Project Phase	Project Initiation	Project Construction	Project Inspection and Handover	Post-Management
Contents	• Publicize the renovation policies to residents and seek their opinions • Develop renovation plans and schedules • Financing • Obtain necessary approvals • Conduct bidding to select construction contractors	• Construction by construction contractors	• Inspection by government • Hand over the facilities to the management units • The Sub-district Offices ("Jiedao") organizes the residents' self-governance body to take over community facilities • Conduct satisfaction survey	• Establish a homeowners' association and introduce commercial property management • For communities unable to implement commercial property management, establish self-management and self-governance
Main Actors				
Municipal and district governments	Integrate promotion of implementation and process supervision			
Municipal Housing and Urban-Rural Development Bureau	Formulate plans; approve relevant procedures	Construction arrangement, management progress	Acceptance review	
Municipal Development and Reform Commission	Relevant policy development		Acceptance review	
Municipal Finance Bureau	Coordinate funding			
Water, electricity, gas and other specialized business company		Responsible for the updating, laying, and maintenance of respective pipeline facilities		
Construction company	Participate in bidding	Responsible for construction		
Sub-district Offices	Cooperate with the government to publicize policies to residents and organize residents		Organize the residents' self-governance body to take over community facilities	Community management
Community Committees	Organize residents to participate	Coordinate with residents, government and construction company	Take over community facilities	Community management
Residents	Give suggestions	Give suggestions	Satisfaction evaluation	Establishment of homeowners associations

b) Process of community renovation project

Fig. 20.5 Renovation projects in Fuxin city. *Source* Authors based on documents on the Fuxin municipal government's official website regarding the community renovation project

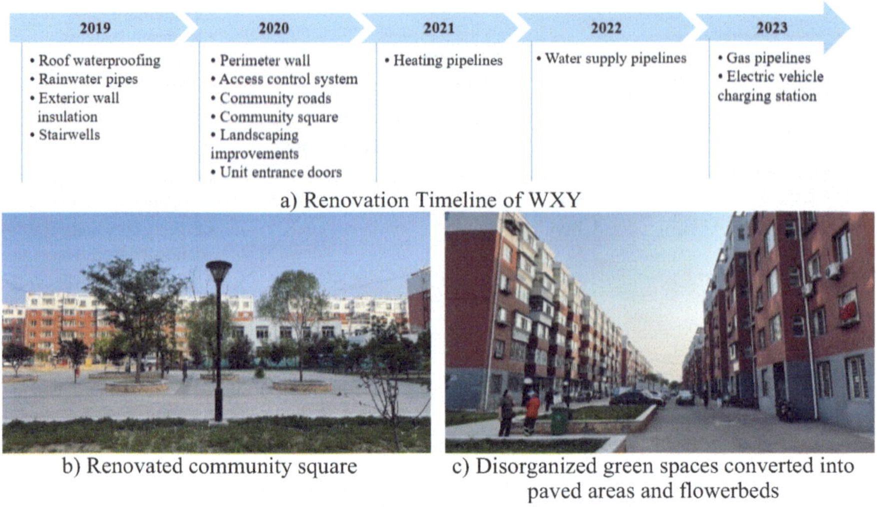

Fig. 20.6 The renovation in WXY. *Source of the figure and photos* Authors, 2023

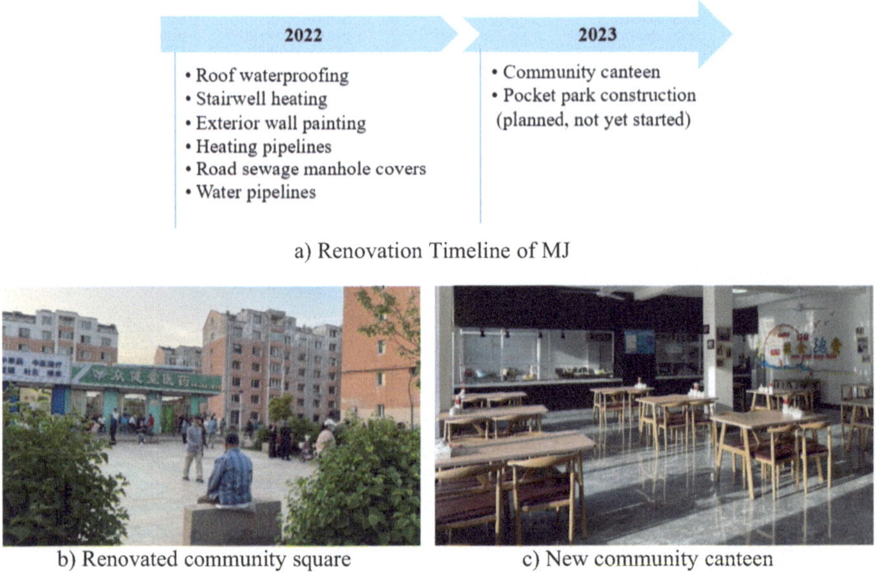

Fig. 20.7 The renovation in MJ. *Source of the figure and photos* Authors, 2023

20.4.3.2 Residents' Satisfaction with Community Renovation Projects

In the survey, respondents rated their satisfaction with the renovation projects using a five-point Likert scale, from 1 (very dissatisfied) to 5 (very satisfied). Overall satisfaction was higher in WXY (3.2) compared to MJ (2.5).

Residents were also asked to identify the aspects they were most and least satisfied with (up to three aspects). WXY residents expressed the most satisfaction with improvements in heating, sanitary conditions, greening, and roads. In contrast, MJ residents were least satisfied, with most only mentioning heating positively (Fig. 20.8a). A WXY resident commented, "The roads are now smooth, upgraded from cement to asphalt. There's no more mud, and it doesn't get waterlogged on rainy days. The heating has also improved; previously, some homes couldn't even reach 15 °C" (WXY-6). Another said, "The greenery is well-maintained now, and the sanitary conditions are much better" (WXY-21). Some WXY residents appreciated the renovated public spaces: "I really like the small square after the renovation. The environment is clean, there are pavilions and fitness equipment, and neighbors come out every day to chat" (WXY-13). In MJ, residents were primarily satisfied with the heating improvements. One noted, "After the renovation, the indoor temperature during winter has increased to above 16 °C" (MJ-2).

Regarding unsatisfactory aspects (Fig. 20.8b), besides concerns about utility services, environmental quality, and roads, many were dissatisfied with the construction process. MJ residents had more complaints. In WXY, residents were most dissatisfied with the quality and completeness of the construction work. Other concerns included long construction periods, noise and inconvenience caused by these activities, as well as a lack of benches. One resident commented, "Construction happens almost every year. They dig up the roads to replace pipes, fill them in, and then dig them up again the next year. This has severely damaged the road surfaces" (WXY-5). In MJ, residents were dissatisfied with the condition of the roads, completeness and quality of construction work, sanitation, and roofing. One resident stated, "When the pipes were replaced, the roads were left a mess and never restored properly. On rainy days, it turns into mud pits" (MJ-2). Another reported roofing issues: "After the waterproofing was redone, some buildings still leak. The construction was rushed, leading to poor quality" (MJ-18).

20.4.3.3 The Impact of Community Renovation Projects on Community Livability

This study uses regression analysis to explore the impact of renovation projects on community livability, with satisfaction with the renovation projects as the independent variable and community livability evaluation as the dependent variable. As shown in Table 20.6, the results indicate that in both communities, satisfaction with renovation projects significantly positively impacts community livability evaluation. When residents are satisfied with the renovation outcomes, they tend to rate livability higher. While this does not directly prove that renovation projects cause improved

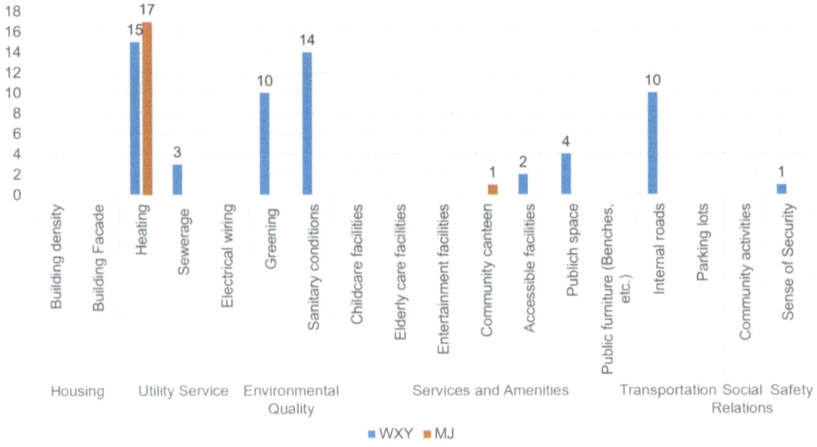

a) Satisfactory aspects

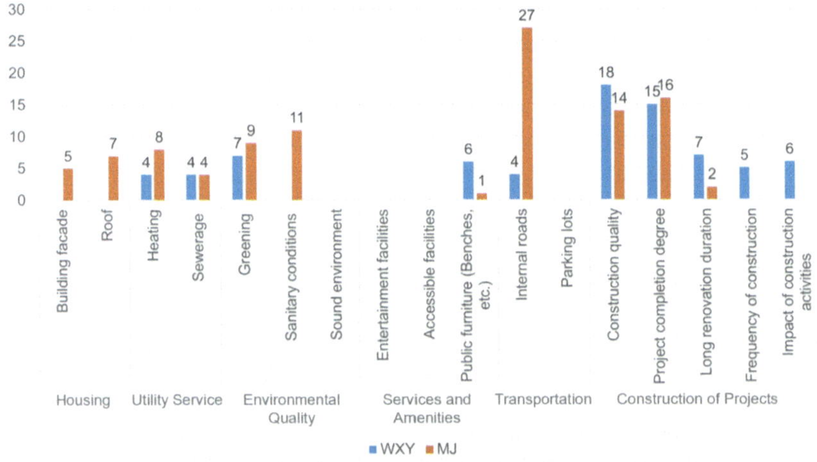

b) Unsatisfactory aspects

c) Road construction impacted residents' travel in WXY

d) Some road surfaces have not been fully restored in MJ

Fig. 20.8 Satisfactory and unsatisfactory aspects of the renovation projects. *Source of figures and photos* Authors, 2023

Table 20.6 Regression analysis results on the impact of renovation projects on livability assessment

		Regression coefficient	95% CI	VIF	Tolerance	R^2	Adjusted R^2	F-value
WXY	Constant	2.763** (17.129)	2.446 ~ 3.079	–	–	0.372	0.360	F (1,51) = 30.212, p = 0.000
	Satisfaction with renovation projects	0.257** (5.497)	0.166 ~ 0.349	1.000	1.000			
MJ	Constant	2.728** (24.552)	2.510 ~ 2.946	–	–	0.275	0.261	F (1,52) = 19.765, p = 0.000
	Satisfaction with renovation projects	0.185** (4.446)	0.103 ~ 0.266	1.000	1.000			

* $p < 0.05$ ** $p < 0.01$ (t-values in parentheses)

livability, it suggests that these projects can enhance community livability. The renovations have greatly improved community infrastructure and overall environmental quality. The reshaping of public spaces has fostered more interaction among neighbors, strengthening the sense of community and safety. Despite some negative impacts, particularly in MJ, the overall positive effects outweigh the negatives.

20.4.3.4 Factors Influencing the Implementation Effectiveness

(1) Allocation of Resources Across Communities Within the City

The primary source of renovation funding comes from government subsidies. Financial pressure on local governments in shrinking cities limits available funds, affecting resource distribution and renovation project effectiveness. As a model renovation project, WXY may receive more attention and resources compared to MJ, where progress has been slower.

(2) Contents of Renovation Projects

Comparing specific projects, WXY's renovations include building performance improvements (roofs, exterior walls, stairwells), infrastructure upgrades (water and sewage pipelines, heating lines), road enhancements, landscaping, and reshaping public activity spaces. In contrast, MJ focuses primarily on building performance and municipal infrastructure, with the addition of a community canteen but lacks landscaping and public activity space improvements. Communities with enhanced social infrastructure (such as public activity spaces) tend to have higher resident satisfaction, as these spaces promote social interactions, strengthen neighborhood ties, and foster a sense of community.

(3) Resident Involvement and Post-Management

Residents rated their involvement in the renovation process on a scale from 1 (no involvement) to 5 (active involvement). WXY's average rating (2.13) was slightly higher than MJ's (2.09), suggesting that higher resident involvement correlates with greater satisfaction.

Post-management and maintenance of renovated facilities also significantly impact satisfaction. In both communities, a community committee manages affairs, dividing residential buildings into sections called "grids," each overseen by a grid manager responsible for coordinating residents' requests. The committee also selects a property management company to provide services. Despite having the same management model, property management effectiveness varies significantly. In WXY, residents generally view property management positively. In contrast, feedback in MJ is mostly negative. One resident stated, "We've reported issues many times, like broken hallway lights, piled-up garbage, and blocked sewers, but nothing gets resolved. Eventually, people stopped paying the property fees" (MJ-2). This lack of payment has led to financial difficulties for the property management company and a decline in service quality, creating a vicious cycle.

20.5 Discussion and Conclusion

This study focuses on the people left behind after the decline of coal industries, evaluating the current livability of communities over ten years after the relocation of shantytowns near former coal mines from the residents' perspective. It also examines the impact of residents' perceptions of urban shrinkage and renovation projects on community livability. The findings can inform recommendations to improve community livability. Government-led shantytown demolition and relocation projects, as a means of providing social welfare, have significantly improved the living conditions of relocated residents compared to before. Moreover, because these relocations were conducted as whole-community moves, the original neighborhood relationships have largely been maintained. However, continuous urban shrinkage in the years following relocation has posed challenges for these relocated communities.

The research shows that residents of both communities have a strong perception of urban shrinkage, reflecting the coal industry's decline's profound impact. However, perceptions of community shrinkage differ. MJ, located on the city's outskirts, has attracted nearby rural populations, maintaining a relatively stable population compared to WXY. Unlike studies by Lorance Rall and Haase (2011) and Schilling and Logan (2008), which found vacant properties negatively influence neighborhoods, this study found that population decrease and increased vacant houses have not significantly impacted residents' lives yet. However, initial effects are evident in heating efficiency and property values, which need attention.

Community livability evaluations in shrinking cities vary, influenced by geographic location and nearby facilities. Centrally located communities with well-developed facilities typically receive higher livability ratings. This aligns with Hollander's 2011 study, which found high heterogeneity in perceptions of neighborhood quality in shrinking U.S. cities. Some cities experienced both housing and population loss while improving overall perceptions of neighborhood quality. Economic recession has a more significant impact on community livability than population decline, affecting employment opportunities, income, government investment in sanitation, green spaces, and infrastructure maintenance.

The effectiveness of renovation projects varies based on resource allocation, project content, and resident involvement and post-renovation management. Communities receiving more government attention secure more resources. Projects focusing on enhancing social infrastructure tend to yield higher resident satisfaction. Klinenberg (2018) emphasizes that robust social infrastructure fosters mutual support and collaboration among friends and neighbors. Additionally, resident involvement in the renovation process and effective post-renovation management contribute to higher satisfaction levels. Despite the varying effectiveness of renovation projects, they significantly improve community livability by enhancing infrastructure, the environment, roads, and public activity spaces. However, implementation in shrinking cities faces challenges. First, funding is a significant issue due to fiscal pressure. Second, residents' willingness to participate in community management is low, with current management overly reliant on top-down government control.

Future recommendations include promoting multi-stakeholder participation in renovation projects. Although the government should continue to play a leading role, given limited funding, it is essential to encourage the involvement of the private sector, non-governmental organizations, and community-based organizations. Additionally, prioritizing the improvements of social infrastructure, such as public activity spaces and community centers, is vital. Lastly, establishing homeowners' associations and encouraging resident self-management through training and resources can help residents effectively govern their communities and reduce reliance on the government.

References

Ferris E, Bower E (2023) Planned relocations: what we know, don't know, and need to learn, internal displacement research programme (IDRP). London, UK. https://researchinginternaldisplacement.org/short_pieces/planned-relocations-what-we-know-dont-know-and-need-to-learn/

Gao S, Ryan BD (2021) Implementation challenges of state-led redevelopment in shrinking cities: case study of shantytown redevelopment in Yichun, Northeast China. J Urban Plan Dev 147(1):05021001

Hollander JB (2011) Can a city successfully shrink? evidence from survey data on neighborhood quality. Urban Aff Rev 47(1):129–141

Hollander JB (2018) A research agenda for shrinking cities. Edward Elgar Publishing Linited. https://doi.org/10.4337/9781785366338

Hollander JB, Pallagst K, Schwarz T, Popper FJ (2009) Planning shrinking cities. Prog Plan 72(4):223–232

Hu X, Yang C (2018) Building a role model for rust belt cities? Fuxin's economic revitalization in question. Cities 72:245–251

Klinenberg E (2018) Palaces for the people: how social infrastructure can help fight inequality, polarization, and the decline of civic life. Crown

Kuyek J, Coumans C (2003) No rock unturned: revitalizing the economies of mining dependent communities. Mining Watch Canada, Ottawa

Leby JL, Hashim AH (2010) Liveability dimensions and attributes: their relative importance in the eyes of neighborhood residents. J Constr Dev Ctries 15(1):67–91

Li H, Zhang P, Cheng Y (2009) Economic vulnerability of mining city: a case study of Fuxin city, Liaoning Province, China. Chin Geogra Sci 19:211–218

Lorance Rall ED, Haase D (2011) Creative intervention in a dynamic city: a sustainability assessment of an interim use strategy for brownfields in Leipzig, Germany. Landsc Urban Plan 100:189–201

Martinez-Fernandez C, Wu CT, Schatz LK, Taira N, Vargas-Hernández JG (2012) The shrinking mining city: urban dynamics and contested territory. Int J Urban Reg Res 36(2):245–260

Owusu AB, Mensah CA, Fynn IEM, Kwang C, Arthur IK, Adu-Boahen K (2023) Indicator-based assessment of the livability of communities in the Accra Metropolitan area, Ghana: a transdisciplinary approach. Soc Sci & Hum Open 8(1):100702

Pallagst K (2014) The planning research agenda: shrinking cities—a challenge for planning cultures. Town Plan Rev 85(2):213–236

Richardson HW, Nam CW (2014) Shrinking cities. In: Shrinking cities. Routledge, pp 1–7

Schilling J, Logan J (2008) Greening the rust belt: a green infrastructure model for right sizing America's shrinking cities. J Am Plann Assoc 74(4):451–466

Siyongwana PQ, Shabalala A (2019) The socio-economic impacts of mine closure on local communities: evidence from Mpumalanga province in South Africa. GeoJ 84:367–380

Veiga MM, Scoble M, McAllister ML (2001) Mining with communities. Natural Resources Forum 25(3):191–202. Blackwell Publishing Ltd., Oxford, UK

Zhang PY (2008) Revitalizing old industrial base of Northeast China: process, policy and challenge. Chin Geogra Sci 18:109–118

Zhang Y, Ochiai C (2023) A study on the long-term spatio-temporal changes of shrinking cities in China. Urban Reg Plan Rev 10:158–178

Chapter 21
Conclusion: Displacement, Resettlement, and the Way Forward

Chiho Ochiai, Sandra Carrasco, and Tsai Sung Lun

Abstract The increasing displacement resulting from climate change, conflicts, and development projects requires multiple approaches to address the impacts and alternatives for resettlement and recovery of affected people. This book analyzes resilience and adaptation through various international case studies, highlighting emotional, psychological, and social impacts. Key lessons include the need for long-term analysis in housing changes, integrating a multi-stakeholder approach in planning, and fostering a community-centered perspective in rebuilding a sense of home. The various studies included in this book highlight cultural sensitivity in housing, stakeholder collaboration, and the role of communities in recovery. This book also defines pathways for future research, including addressing mental health, gender equity, technology in displacement, and ethical considerations. Advocating participatory governance and human rights, this book provides insights for policymakers, practitioners, and researchers focused on building resilient, inclusive communities.

21.1 Introduction

Climate change is predicted to intensify the effects of disasters on urban areas and housing worldwide, potentially causing temporary or permanent displacement of people. Furthermore, the rise in conflicts forcibly displacing people internally and internationally, together with the increase in development projects that require moving people from their original land, are calling for alternative approaches for addressing displacement and resettlement. Key issues such as validating people's roles in co-creating their recovery and adaptation remain unexplored and dismissed

C. Ochiai
Graduate School of Global Environmental Studies, Kyoto University, Kyoto, Japan

S. Carrasco (✉)
School of Humanities and Social Sciences, Deakin University, Melbourne, Australia
e-mail: s.carrasco@deakin.edu.au

T. S. Lun
Shimizu Corporation, Tokyo, Japan

C. Ochiai et al. (eds.), *Disaster and Displacement*, Disaster Risk Reduction,
https://doi.org/10.1007/978-981-96-2583-3_21

in resettlement policy and practice. In this book, the case studies observed serve the remarkable resilience and adaptability exhibited by those uprooted from their homes. The book captures the tangible and intangible impacts of displacement, such as the emotional and psychological challenges of leaving behind familiar environments and navigating the uncertainties of new, unfamiliar territories. Despite these challenges, there emerges a powerful story of hope and renewal, as individuals and communities create new bonds and build new identities during the process of resettlement. This book also transcends the mere physical act of relocating; it encompasses creating new homes and identities in the face of adversity. This acknowledges the need for empathy, solidarity, and supportive policies that facilitate successful adaptation and integration.

21.2 Lessons Learned

The book chapters cover international case studies using qualitative, quantitative, and mixed-methods research, showing diverse recovery pathways and enriched with studies addressing displacement. As mentioned in Chap. 2, given the lack of cross-country analysis, this book provides a comparison from different resettlement-related case studies to show some important lessons learned throughout the book. Moreover, each chapter has shown its uniqueness given that the relocations and resettlements are multidimensional, including deferment environmental, displacement, development, livelihood, planning, governance, human rights, human security, law, and policy considerations. Throughout this book, three critical lessons are learned, including integrating a long-term perspective to facilitate housing modifications and extensions, applying a multi-stakeholder perspective in planning and management, and adopting a community-centered perspective in rebuilding the sense of home after displacement. In the following content, these three lessons are further explained by showing the similarities and comparing the differences of these chapters.

21.2.1 A Long-Term Perspective to Facilitate Housing Modifications and Extensions

Using building structures as units of analysis, chapters in Part I reveal that modifications and extensions made by residents are commonly observed years after housing reconstruction. This indicates a gap in socio-cultural understanding during the initial reconstruction efforts of existing projects and urges a long-term perspective in the post-settlement context.

Chapters 3, 4, and 5 depict physical changes in post-disaster housing in relocated communities across different localities. Chapter 3 highlights the importance of integrating cultural components of indigenous people to housing layout and facility functions. Both government and NGOs may lack sufficient communication with local communities, leading to a mismatch between post-disaster housing and residents' livelihoods, income development, and space needs. This often prompts modifications and extensions. To examine the behavior of housing extensions and land transformation, Chapters 3 and 5 propose categories summarizing patterns of changes and recognizing a gradual process of spatial transformation. Different methods of investigating residents' perceptions are also utilized. While Chap. 4 investigates residents' perceptions by distinguishing housing extension materials, Chap. 6 examines the thermal comfort of houses.

To address the issues caused by spontaneous modifications and extensions, it is recommended to adopt a long-term perspective for post-disaster housing projects, such as the incremental design approach, to facilitate future modifications and extensions in advance. Since housing extension can lead to regulation violations and create conflicts between the community and authorities, as shown in Chap. 4, modification activities should be further complemented by technical support and legal consultation offered to the community.

21.2.2 A Multi-Stakeholder Perspective in Planning and Management

With a focus on the planning and management of resettlement and reconstruction, Part II covers a wide range of topics and issues. It includes discussions on temporary and permanent housing projects, as well as the relocation of religious facilities. These chapters examine the diverse roles of stakeholders involved, including authorities, communities, universities, NGOs, residents' associations, and religious sectors.

Addressing concerns like high costs and low recyclability of temporary housing, Chap. 7 provides an innovative viewpoint of integrating vernacular construction in housing recovery. Sustained research efforts made by civil society and policy support from the government are required for the success of this approach. In contrast, with the advancement of information technology (IT) and computational tools, Chap. 8 proposes a novel idea of using IT tools to rapidly generate spatial designs of settlements. By considering several crucial elements of temporary housing, this approach provides implementers with a guideline to follow.

While Chap. 9 proposes the idea of "regulated flexibility," emphasizing significant flexibility for community autonomy with appropriate government influence, Chaps. 10, 11, and 12 highlight the roles and characteristics of three different stakeholders in the post-disaster context—the university, the resident association, and the religious sector, respectively. Chapter 10 states that universities can provide robust social networks, multidisciplinary knowledge, and long-term commitment to disaster-impacted communities, providing a new viewpoint of understanding the role of universities in the post-disaster scenario. Chapter 11 emphasizes that the early involvement of residents and the formation of residents' associations in housing reconstruction planning benefit the establishment of community relationships and the improvement of living environments. It is evident that in the long-term recovery process, such an association can be crucial to incentivize community activities and boost the engagement of residents in the post-disaster decision-making process. From a historical point of view, Chap. 12 portrays the reconstruction process of a religious facility, highlighting the multi-stakeholder collaboration that enabled speedy recovery at the early industrial era in Japan. It is a relatively rare approach among disaster-related studies, given that most of the research focuses on the post-disaster recovery of the communities. By using historical documents and records, Chap. 12 highlights the importance of understanding the disaster countermeasures in the past. Starting from a general point of view and centering on the concept of "Build Back Better," Chap. 13 deconstructs the framework into several indicators and suggests suitable metrics for practitioners to use during post-disaster reconstruction.

Without simplifying into a dichotomy of "top-down" or "bottom-up" approaches, Part II underscores the importance of collaborative spirit among stakeholders, crossing both public and private sectors. Each stakeholder brings unique strengths and roles to the recovery process, highlighting the significance of their collective efforts. Similarly, both traditional and modern technologies can contribute to the decision-making process by offering guidelines, frameworks, or suggestions for housing recovery tailored to local contexts. Holistic viewpoints during the planning and management of reconstruction benefit from multi-stakeholder collaboration and the use of diverse technologies to aid decision-making.

21.2.3 A Community-Centered Perspective in Rebuilding the Sense of Home After Displacement

In Part III, the discussions place special emphasis on the community as a crucial agent in rebuilding the sense of home after resettlement or displacement. The definition of disaster is expanded to encompass not only natural hazards but also conflicts, mega-events such as the Olympic Games, and urban planning decisions that result in forced displacement of populations. Each chapter details specific challenges communities confront after displacement and their efforts in rebuilding cohesion and connection.

Chapter 14 specifically examines the feasibility of long-distance collective relocations and resettlements in Japan, an approach less commonly used in current post-disaster management strategies. Despite the remote distance from original settlements, some relocated and original communities still maintain an intimate relationship, largely influenced by the implementation of migration policies. By proposing the concept of community landscape knowledge, Chap. 15 provides a paradigm shift in redefining the concept of disaster, highlighting the resilience of communities and their knowledge of the biophysical context. The authors propose that "staying in place and temporary spatial relocation" can also be viable options given communities' familiarity with the local landscape. Chapters 16 and 17 explore the dynamics between community and government. Chapter 16 highlights communication and transparency issues between the two groups, while Chap. 17 examines the development of the tourism industry and how it varies in relocated and non-relocated communities after disasters, illustrating the distinct challenges each community faces and the resulting conflict with the government sector.

Chapters 18, 19, and 20 particularly focus on marginalized populations in society and deal with forced displacement caused by conflicts, mega-events, and urban planning decisions, respectively. Chapter 18 studies African refugees in Australia who have been displaced by conflicts in their home countries and their challenges in adapting to the new environment while maintaining their cultural identities. It represents a relatively novel topic exploring the issue of cross-continental displacement. Chapter 19 frames mega-events as disasters by examining the impacts of gentrification and removalist projects caused by the Olympic Games on socially excluded groups. It urges connection to those often ignored at mega-events. Chapter 20 explored the issues associated with displaced coal mine workers after the decline of the industry. It examines residents' perceptions of their living environment and explores the distinct issues in the context of shrinking cities in China.

The chapters in Part III highlight both the vulnerabilities and capacities of communities in the recovery process. Communities with strong cultural ties show adequate resilience after displacement, whereas marginalized populations may struggle with integration and adaptation to new environments. Thus, marginalized groups may require more attention from society, while self-sufficient communities benefit from greater autonomy. A community-centered perspective calls for tailored support to rebuild the sense of home after displacement aligned with communities' vulnerabilities and capacities, acknowledging that there is no universal solution.

21.3 Way Forward

Research in displacement and resettlement encompasses a diverse array of topics, each presenting unique challenges and opportunities for research. In this book, the long-term impacts of displacement have been explored, however, more studies are necessary emphasizing the effects and opportunities for resilience on individuals and communities. Additionally, besides the physical impacts and socio-economic, and cultural impacts, the effects on mental health need to be studied in displacement, resettlement, and adaptation considering pre-displacement and post-displacement factors affecting disaster-induced displaced people and refugees, and internally displaced persons.

Climate-change-induced displacement and migration are increasingly studied, however, it is important to create a cross-sectoral analysis considering the lessons learned from other drivers for displacement. Another key issue that requires more study is related to gender equality in displacement, which requires including non-binary communities and other minority groups. In this book, Indigenous perspectives were addressed, discussing the cultural appropriateness of housing in resettlement sites. However, analysis of diverse contexts is necessary to localize Indigenous perspectives and intersect with global frameworks and policies.

Legal and policy frameworks play a crucial role in addressing displacement, however, the challenges of localization of frameworks are crucial for significant impact on local communities. Additionally, the role of technology in supporting displaced populations is also gaining attention, from the analysis of social media impact in rebuilding social capital to tracking and supporting migrating communities seeking safety.

Finally, ethical considerations in displacement and resettlement research and practice are paramount, influencing methodologies and engagement with affected people. This involves the role of researchers, governments, not-for-profit organizations, and other stakeholders and redefining the priorities in analyzing the impacts of resettlement with a people's rights approach and participatory and inclusive governance rather than solely focusing on financial viability and political benefits.

Acknowledgements The Editors of the book express their gratitude to all the authors for sharing their knowledge on diverse aspects of post-disaster replacement issues and for their contributions to this book. Without their input and the inclusion of case studies in an international context, this book would not have been finalized. Furthermore, the Editors acknowledge the assistance of Yuqi Zhang, a doctoral student at Kyoto University, in editing and adjusting the format for publishing, and Jingying Wang, a doctoral student at the Chinese University of Hong Kong, in helping edit the chapters.

The manufacturer's authorised representative in the EU is Springer Nature Customer Service Centre GmbH, Europaplatz 3, 69115 Heidelberg, Germany. If you have any concerns regarding our products, please contact ProductSafety@springernature.com

Printed and bound by CPI Group (UK) Ltd, Croydon, CR0 4YY
01/10/2026
02225850-0018